Novel Strategies for Biodegradation and Detoxification of Mycotoxins

Novel Strategies for Biodegradation and Detoxification of Mycotoxins

Editors

Qiugang Ma
Desheng Qi
Lihong Zhao

MDPI • Basel • Beijing • Wuhan • Barcelona • Belgrade • Manchester • Tokyo • Cluj • Tianjin

Editors
Qiugang Ma
China Agricultural University
Beijing, China

Desheng Qi
Huazhong Agricultural University
Wuhan, China

Lihong Zhao
China Agricultural University
Beijing, China

Editorial Office
MDPI
St. Alban-Anlage 66
4052 Basel, Switzerland

This is a reprint of articles from the Special Issue published online in the open access journal *Toxins* (ISSN 2072-6651) (available at: https://www.mdpi.com/journal/toxins/special_issues/biodegradation_mycotoxins).

For citation purposes, cite each article independently as indicated on the article page online and as indicated below:

LastName, A.A.; LastName, B.B.; LastName, C.C. Article Title. *Journal Name* **Year**, *Volume Number*, Page Range.

ISBN 978-3-0365-8414-0 (Hbk)
ISBN 978-3-0365-8415-7 (PDF)

Cover image courtesy of Qiugang Ma and Desheng Qi

Contents

About the Editors

Qiugang Ma

Ph.D. holder, Professor, and Doctoral Supervisor of China Agricultural University(CAU), Position Researcher on Nutrient Requirement of Egg Layer in Modern Agro-industry Technology Research System of China, and being selected in the New Century Talents of Education Ministry of China. He obtained his Ph.D. in Animal Nutrition and Feed Science from one cooperative program between CAU and Hohenheim University (Germany), then worked as a long-term fellow at CAU. He serves as Director of the Society of Animal Nutrition in the Chinese Association of Animal and Veterinary Medicine, Standing Director of Deputy Secretary for the Beijing Association of Animal and Veterinary Medicine, and Subcommittee Head of the Donkey Industry Branch in the China Animal Agriculture Association. He also serves as the editorial board member of Toxins, Journal of Animal Physiology and Animal Nutrition, and the International Journal of Molecular Sciences. His group's major research interest includes the release, absorption, metabolism, and residues of mycotoxins in animals, developing mycotoxin-degrading bacteria, and studying their degrading mechanisms to develop safe feed additives. He earned one national prize, three ministerial prizes, and five provincial prizes for his academic achievements in relevant fields. He obtained 28 authorized invention patents, published two monographs as the Chief Editor and three as the Deputy Chief Editor, and published 62 peer-review papers in international journals and more than 100 peer-review papers in national journals as the first or corresponding author.

Desheng Qi

Ph.D. holder, Professor, and Doctoral Supervisor of Huazhong Agricultural University, and head of Hubei Feed Quality Supervision and Inspection Station. He also serves as a member of the Feed Industry, and is the Deputy Director of the Feed Professional Committee of the Chinese Society of Toxicology, Deputy President of the Animal Environmental Hygiene Branch of the Chinese Association of Animal Science and Veterinary Medicine, Managing Director of the Animal Nutritional Branch of Chinese Association of Animal Science and Veterinary Medicine, Managing Director of China Feed Industry Association, Supervisor of Hubei Feed Industry Association, and Director of the Feed Branch of the Chinese Cereals and Oils Association. He adheres to the principle of mutual benefit between teaching and learning, and the combination of "teaching, scientific research, and production," while actively participating in disciplinary construction and striving to serve the feed industry and society. Effective efforts have been made in efficiently utilizing feed raw materials, preventing and controlling anti-nutritional factors in feed, and using fungal toxins' toxicology and prevention and control techniques. He developed fifteen national and agricultural industry standards and three authorized invention patents, received a second prize for scientific and technological progress in Hubei Province, and is the Editor-In-Chief and Co-Editor of eight works.

Lihong Zhao

Associate Professor, Ph.D. supervisor at China Agricultural University, and Postdoc that graduated from the Department of Poultry Science, North Carolina State University. Her research interests are mining mycotoxin-degrading enzymes, feed and food safety, biotoxins, and enzyme engineering. She has presided over various programs, including three National Natural Science Foundation programs and one Beijing Natural Science Foundation program. She has also obtained 23 authorized invention patents. As the first or corresponding author, she has published 30 SCI papers in top journals, such as Food Chemistry, Food Control, Food Research International, etc. She earned one ministerial prize and two provincial prizes as a main researcher in relevant fields.

Preface

Mycotoxins contamination is a continuing global problem that severely affects animal health and performance and further threatens food safety. The industry's common aspiration is to eliminate feed mycotoxins contamination and control their hazards.

In recent years, animal mycotoxicosis has occurred frequently all over the world. This has brought huge economic losses to the agriculture industry. The potential damage caused by mycotoxins-induced decreased disease resistance in animals and food safety problems in human health is incalculable. There is a long way to go to prevent mycotoxins hazards.

Biodegradation is a promising strategy to eliminate mycotoxins as it can transform mycotoxins into nontoxic or less toxic metabolites under mild conditions, retaining the sensory quality and nutritional value of agricultural commodities.

On the other hand, animals themselves have a certain ability to detoxify mycotoxins, and some bioactive substances, such as lipoic acid, sporoderm-broken spores of Ganoderma lucidum, and quercetin, can improve the detoxification ability of animals to reduce the toxic effects of mycotoxins.

This reprint aims to gather contributions of original research or reviews related to novel strategies for biodegradation and detoxification of mycotoxins. Topics of interest will especially include novel mycotoxin-degrading microorganisms and enzymes, fermentation technology to reduce the mycotoxin content in cereal products, studies on alleviating the mycotoxicosis of livestock by the addition of bioactive substances or mycotoxin biodegradation agents, and any preliminary research that promotes progress in this field.

Qiugang Ma, Desheng Qi, and Lihong Zhao
Editors

toxins MDPI

Editorial

Novel Strategies for the Biodegradation and Detoxification of Mycotoxins in Post-Harvest Grain

Lihong Zhao [1], Desheng Qi [2] and Qiugang Ma [1,*]

[1] State Key Laboratory of Animal Nutrition, College of Animal Science and Technology, China Agricultural University, Beijing 100193, China; zhaolihongcau@cau.edu.cn
[2] Department of Animal Nutrition and Feed Science, College of Animal Science and Technology, Huazhong Agricultural University, Wuhan 430070, China; qds@mail.hzau.edu.cn
* Correspondence: maqiugang@cau.edu.cn

Mycotoxins are toxic secondary metabolites produced by filamentous fungi belonging, in particular, to the *Aspergillus*, *Fusarium*, and *Penicillium* genera. Aflatoxins, zearalenone (ZEN), deoxynivalenol (DON), T-2 toxin, ochratoxin A (OTA), fumonisins, patulin, and ergot alkaloids are the most contaminating mycotoxins found in food and feed, posing potentially carcinogenic, mutagenic, teratogenic, cytotoxic, neurotoxic, nephrotoxic, estrogenic, and immunosuppressant effects on both humans and animals [1]. Moreover, the broad spectrum of contamination caused by mycotoxins affects not only the economy but also poses a threat to public health due to the wide range of effects caused by contamination, which attracts the attention of researchers to explore novel approaches to detoxify mycotoxin-contaminated food and feed. This Special Issue therefore summarizes various strategies for the detoxification of mycotoxins through post-harvest detoxification methods, which are divided into physical, chemical, biological, and other developing innovative strategies. The physical strategies used to mitigate mycotoxins include rapid and proper drying, post-harvest insect control, good storage conditions, inorganic and organic adsorbents, such as montmorillonite and yeast cell walls, and advanced oxidation technology, such as irradiation and cold plasma, which allow for the rapid degradation of mycotoxins [2]. Chemical approaches involve the use of ozone, electrolyzed oxidizing water, organic acids, and natural plant extracts, which are widely accepted as safe food additives in many countries [3]. Biological approaches are defined as the microbial and enzymatic degradation of mycotoxins into non-toxic or less toxic metabolites. The biodegradation of mycotoxins is an emerging and frequently studied research topic. Many microorganisms and enzymes have been reported to degrade various mycotoxins in recent times [4].

This Special Issue aims to gather contributions of original research or reviews related to novel strategies for the biodegradation and detoxification of mycotoxins. Topics of interest will include, in particular, novel mycotoxin-degrading microorganisms and enzymes, fermentation technology to reduce the mycotoxin content in cereal products, and studies on alleviating the mycotoxicosis of livestock through the addition of bioactive substances or mycotoxin biodegradation agents.

The first study included in this Special Issue developed sustainable strategies to counteract mycotoxin contamination and cowpea weevil in chickpea seeds during the post-harvest period. The results showed that O_3 significantly decreased the incidence of *Penicillium* spp. (by an average of −50%, independent of the time of exposure) and reduced the contents of patulin and aflatoxins (−85 and −100% after 30 min of exposure, respectively). High N_2 concentrations significantly reduced mycotoxin contamination (by an average of −94%) and induced pest mortality (at 100% after 5 days of exposure). These results confirm the promising potential of O_3 and N_2 in post-harvest conservation strategies for eliminating mycotoxin contamination [5]. Five essential oils (thymol, carvacrol, cinnamaldehyde, eugenol, and citral) were tested for their inhibition effects against

Citation: Zhao, L.; Qi, D.; Ma, Q. Novel Strategies for the Biodegradation and Detoxification of Mycotoxins in Post-Harvest Grain. *Toxins* **2023**, *15*, 445. https://doi.org/10.3390/toxins15070445

Received: 6 May 2023
Revised: 12 June 2023
Accepted: 4 July 2023
Published: 5 July 2023

Aspergillus flavus and aflatoxin B1 production in broth and feed. The results showed that cinnamaldehyde and citral have a positive synergistic effect and that both of them could inhibit at least 90% of the fungal growth and aflatoxin B1 production in broth and poultry feed; thus, they could be an alternative to control aflatoxin contamination in food and feed in future [6]. Li et al. revealed that the disruption of redox genes is involved in the mechanism of coumalic acid and geraniol against *Aspergillus flavus* spore germination, and essential oils have a significant inhibitory effect on germination rates and redox gene expression [7]. Another study from the same group explored the fungistatic effect and mechanism of thymol on *Fusarium graminearum*, with the results showing that thymol can effectively inhibit the growth and toxin production of *F. graminearum* and cause an extensive transcriptome response, and the gluconeogenesis/glycolysis pathway may be a potential and important way for thymol to affect the growth of *F. graminearum* hyphae and the production of DON simultaneously [8].

Second, this Special Issue places a specific emphasis on the microbial and enzymatic transformation of mycotoxins in post-harvest detoxification strategies. A new *Alcaligenes faecalis* ANSA176 with a strong OTA-detoxifying ability was isolated from donkey intestinal chyme, which could degrade 97.43% of 1 mg/mL OTA into OTα within 12 h at 37 °C. The study of laying hens fed an OTA-contaminated diet showed that ANSA176 supplementation in their diet inhibited or attenuated the immune injury and inflammation induced by OTA through efficiently degrading OTA in the animals' intestinal tract [9]. A study was conducted to compare the potential ameliorative effects between probiotic *Bacillus subtilis* (ANSB010) and biodegradable *B. subtilis* (ANSB01G) on ZEN toxicosis in gilts. The results showed that the ZEN-contaminated diet had a harmful impact on the growth performance, plasma immune function, and hormone secretion of gilts. Although probiotic and biodegradable *B. subtilis* have similar antimicrobial capacities, only biodegradable *B. subtilis* could eliminate these negative effects through its degradation of ZEN in the intestinal tract of gilts [10]. A laccase-degrading aflatoxin B1 from *B. amyloliquefaciens* B10 was isolated, purified, and characterized by Xiong et al. Their results showed that purified laccase could degrade 79.3% of AFB1 within 36 h, and the mutation of the three key metal combined sites (H-87, C-132, and H-149) of B10 laccase resulted in the loss of AFB1-degrading activity, indicating that these three metal combined sites of B10 laccase play an essential role in the catalytic degradation of AFB1 [11]. Ery4 laccase, acetosyringone, ascorbic acid, and dehydroascorbic acid were applied to artificially contaminated corn for AFB1 reduction, which showed that AFB1 (0.1 μg/mL) was completely removed in vitro and reduced by 26% in corn [12]. Wang et al. discovered that a manganese peroxidase (MrMnP) from *Moniliophthora roreri* can efficiently degrade patulin. The recombinant MrMnP was able to completely remove 5 mg/L of pure patulin within 5 h. Moreover, up to 95% of the toxin was eliminated in simulated patulin-contaminated apple juice after 24 h. The study concluded that MrMnP can be used as an intriguing candidate useful in the enzymatic detoxification of patulin in food and beverages [13]. A novel bacterium *ketogulonicigenium vulgare* D3_3 isolated from the feces of tenebrio molitor larvae was able to efficiently degrade 50 mg/L of DON under a broad range of conditions. Furthermore, four PQQ-dependent alcohol dehydrogenases responsible for the oxidative detoxification of DON were identified from the genome of isolate D3_3. These findings suggest that bio-detoxification is a potential strategy to remit the toxicity of DON in animals [14].

Finally, some studies have focused on the toxicity of mycotoxins in cells and animals and the alleviating effects of some nutritional and enzymatic additives on some mycotoxins. Chao et al. demonstrated that AFB1 exposure impaired the proliferation of porcine alveolar macrophage 3D4/2 via the non-coding RNA-mediated pathway by whole-transcriptome analysis [15]. Wang et al. also revealed that AFB1 exposure caused pathological damage to ducklings' livers, decreased enzymatic activity and glutathione content in the liver, and increased the serum enzyme activities of alanine aminotransferase, aspartate aminotransferase, alkaline phosphatase, and γ-glutamyl transpeptidase. Moreover, the study found that dietary epigallocatechin gallate and glutathione attenuate AFB1-induced acute liver injury in ducklings via mitochondria-mediated apoptosis and the Nrf2 signaling pathway [16]. In addition, glutamine can alleviate ZEN-induced apoptosis in IPEC-J2 cells via the PI3K/Akt signaling pathway [17]. Meanwhile, dietary catalase supplementation alleviates DON-induced oxidative stress and intestinal damage in broilers, which can be associated with its ability to improve the gut microbiota, aside from its direct oxygen radical-scavenging activity [18].

Author Contributions: Writing—original draft preparation, L.Z.; writing—review and editing, D.Q. and Q.M.; supervision, Q.M. All authors have read and agreed to the published version of the manuscript.

Acknowledgments: The editors are grateful to all the authors who submitted and contributed their work to this Special Issue. Special thanks are given to the expert peer reviewers for their rigorous evaluation of all submitted manuscripts.

Conflicts of Interest: The authors declare no conflict of interest.

References

1. Kabak, B.; Dobson, A.D.W.; Var, I. Strategies to prevent mycotoxin contamination of food and animal feed: A review. *Crit. Rev. Food Sci.* **2006**, *46*, 593–619. [CrossRef] [PubMed]
2. Neme, K.; Mohammed, A. Mycotoxin occurrence in grains and the role of postharvest management as a mitigation strategies: A review. *Food Control* **2017**, *78*, 412–425. [CrossRef]
3. Guo, Y.; Zhao, L.; Ma, Q.; Ji, C. Novel strategies for degradation of aflatoxins in food and feed: A review. *Food Res. Int.* **2021**, *140*, 109878. [CrossRef] [PubMed]
4. Vanhoutte, I.; Audenaert, K.; De Gelder, L. Biodegradation of Mycotoxins: Tales from Known and Unexplored Worlds. *Front. Microbiol.* **2016**, *7*, 561. [CrossRef] [PubMed]
5. Pisuttu, C.; Risoli, S.; Moncini, L.; Nali, C.; Pellegrini, E.; Sarrocco, S. Sustainable Strategies to Counteract Mycotoxins Contamination and Cowpea Weevil in Chickpea Seeds during Post-Harvest. *Toxins* **2023**, *15*, 61. [CrossRef]
6. Han, B.; Fu, G.W.; Wang, J.Q. Inhibition of Essential Oils on Growth of Aspergillus flavus and Aflatoxin B1 Production in Broth and Poultry Feed. *Toxins* **2022**, *14*, 655. [CrossRef]
7. Li, C.; Jia, S.; Rajput, S.A.; Qi, D.; Wang, S. Transcriptional Stages of Conidia Germination and Associated Genes in Aspergillus flavus: An Essential Role for Redox Genes. *Toxins* **2022**, *14*, 560. [CrossRef] [PubMed]
8. Wang, L.Q.; Wu, K.T.; Yang, P.; Hou, F.; Rajput, S.A.; Qi, D.S.; Wang, S. Transcriptomics Reveals the Effect of Thymol on the Growth and Toxin Production of Fusarium graminearum. *Toxins* **2022**, *14*, 142. [CrossRef] [PubMed]
9. Zheng, R.; Qing, H.; Ma, Q.; Huo, X.; Huang, S.; Zhao, L.; Zhang, J.; Ji, C. A Newly Isolated Alcaligenes faecalis ANSA176 with the Capability of Alleviating Immune Injury and Inflammation through Efficiently Degrading Ochratoxin A. *Toxins* **2022**, *14*, 569. [CrossRef] [PubMed]
10. Shen, W.; Liu, Y.; Zhang, X.; Zhang, X.; Rong, X.; Zhao, L.; Ji, C.; Lei, Y.; Li, F.; Chen, J.; et al. Comparison of Ameliorative Effects between Probiotic and Biodegradable Bacillus subtilis on Zearalenone Toxicosis in Gilts. *Toxins* **2021**, *13*, 882. [CrossRef] [PubMed]
11. Xiong, D.; Wen, J.; Lu, G.; Li, T.; Long, M. Isolation, Purification, and Characterization of a Laccase-Degrading Aflatoxin B1 from Bacillus amyloliquefaciens B10. *Toxins* **2022**, *14*, 250. [CrossRef] [PubMed]
12. Loi, M.; De Leonardis, S.; Ciasca, B.; Paciolla, C.; Mulè, G.; Haidukowski, M. Aflatoxin B1 Degradation by Ery4 Laccase: From In Vitro to Contaminated Corn. *Toxins* **2023**, *15*, 310. [CrossRef] [PubMed]
13. Wang, S.; Wang, X.; Penttinen, L.; Luo, H.; Zhang, Y.; Liu, B.; Yao, B.; Hakulinen, N.; Zhang, W.; Su, X. Patulin Detoxification by Recombinant Manganese Peroxidase from Moniliophthora roreri Expressed by Pichia pastoris. *Toxins* **2022**, *14*, 440. [CrossRef] [PubMed]
14. Wang, Y.; Zhao, D.; Zhang, W.; Wang, S.; Wu, Y.; Wang, S.; Yang, Y.; Guo, B. Four PQQ-Dependent Alcohol Dehydrogenases Responsible for the Oxidative Detoxification of Deoxynivalenol in a Novel Bacterium Ketogulonicigenium vulgare D3_3 Originated from the Feces of Tenebrio molitor Larvae. *Toxins* **2023**, *15*, 367. [CrossRef] [PubMed]
15. Chao, H.; Ma, H.; Sun, J.; Yuan, S.; Dong, P.; Zhao, A.; Li, L.; Shen, W.; Zhang, X. Whole-Transcriptome Analysis of Non-Coding RNA Alteration in Porcine Alveolar Macrophage Exposed to Aflatoxin B1. *Toxins* **2022**, *14*, 373. [CrossRef] [PubMed]
16. Wang, Y.; Wu, J.; Wang, L.; Yang, P.; Liu, Z.; Rajput, S.A.; Hassan, M.; Qi, D. Epigallocatechin Gallate and Glutathione Attenuate Aflatoxin B_1-Induced Acute Liver Injury in Ducklings via Mitochondria-Mediated Apoptosis and the Nrf2 Signalling Pathway. *Toxins* **2022**, *14*, 876. [CrossRef] [PubMed]
17. Wang, T.; Wang, J.; Zhang, T.; Gu, A.; Li, J.; Shan, A. The Antagonistic Effect of Glutamine on Zearalenone-Induced Apoptosis via PI3K/Akt Signaling Pathway in IPEC-J2 Cells. *Toxins* **2021**, *13*, 891. [CrossRef] [PubMed]

18. Wang, W.; Zhu, J.; Cao, Q.; Zhang, C.; Dong, Z.; Feng, D.; Ye, H.; Zuo, J. Dietary Catalase Supplementation Alleviates Deoxynivalenol-Induced Oxidative Stress and Gut Microbiota Dysbiosis in Broiler Chickens. *Toxins* **2022**, *14*, 830. [CrossRef] [PubMed]

toxins

Article

Transcriptional Stages of Conidia Germination and Associated Genes in *Aspergillus flavus*: An Essential Role for Redox Genes

Chong Li [1], Sifan Jia [1], Shahid Ali Rajput [2], Desheng Qi [1,*] and Shuai Wang [1,*]

1 Department of Animal Nutrition and Feed Science, Huazhong Agricultural University, Wuhan 430070, China
2 Department of Animal Feed and Production, Faculty of Veterinary and Animal Sciences, Muhammad Nawaz Shareef University of Agriculture, Multan 66000, Pakistan
* Correspondence: qds@mail.hzau.edu.cn (D.Q.); wangshuai@mail.hzau.edu.cn (S.W.)

Abstract: Aflatoxin is a threatening mycotoxin primarily present in the agricultural environment, especially in food and feedstuff, and poses significant global health risks. Aflatoxins are produced mainly by *Aspergillus flavus*. Conidia germination is the first step for *A. flavus* development. In this study, the transcriptome of *A. flavus* conidia was analyzed at three different stages of conidia germination, which were characterized by two different microscopes. Dormant conidia grew isotropically with the cell size increasing up to 5 h of after being inoculated in a liquid medium. Conidia changed towards polarized growth from 5 to 10 h of germination, during which germ tubes formed. Moreover, transcriptome analyses revealed that a larger number of genes changed in the isotropic growth stages compared to polarized growth, with 1910 differentially expressed genes (DEGs) up-regulated and 969 DEGs down-regulated in isotropic growth. GO and KEGG pathway analyses and pathway enrichment demonstrated that, in the isotropic growth stage, the top three pathways were translation, amino acid and carbohydrate metabolism. The ribosome was a key pathway in translation, as *RPS28e*, *RPL53* and *RPL36e* were the top three DEGs. For polarized growth stage, lipid metabolism, amino acid metabolism and carbohydrate metabolism were the top three most active pathways. *POX1* from alpha-linolenic acid metabolism was a DEG in lipid metabolism as well. Genes related to the antioxidant system were crucial for conidia germination. Furthermore, RT-PCR results showed the same trends as the transcriptome for redox genes, and essential oils have a significant inhibitory effect on germination rate and redox gene expression. Therefore, redox genes play an important role during germination, and the disruption of redox genes is involved in the mechanism of action of coumalic acid and geraniol against *A. flavus* spore germination.

Keywords: *Aspergillus flavus*; conidia; germination; transcriptome; redox genes

Key Contribution: We have demonstrated the morphological changes, transcriptome changes, the key pathways and genes during two stages of Aspergillus. Flavus conidia germination. This study highlights that the redox genes could be a potential target to inhibit conidia germination.

Citation: Li, C.; Jia, S.; Rajput, S.A.; Qi, D.; Wang, S. Transcriptional Stages of Conidia Germination and Associated Genes in *Aspergillus flavus*: An Essential Role for Redox Genes. *Toxins* **2022**, *14*, 560. https://doi.org/10.3390/toxins14080560

Received: 5 July 2022
Accepted: 12 August 2022
Published: 18 August 2022

1. Introduction

The mycotoxin contamination of food and agricultural products is a significant threat towards human and animal health and causes enormous economic losses [1]. In particular, aflatoxins are common and one of the most toxic substances in the world [2]. Aflatoxins, including AFB_1, AFB_2, AFG_1 and AFG_2, are characterized as class A carcinogens by the International Agency for Research on Cancer [3]. *Aspergillus flavus* is one of the primary fungi that produces aflatoxins [4].

Mature *Aspergillus.* spp. produce billions of single-celled dormant conidia that are found all over the world, including in the desert, polar regions or other severe environmental conditions that are not suitable for living [5]. *A. flavus* is not only associated with food and feed spoilage but also acts as an opportunistic pathogen in plants and animals [6–8].

Conidia are the main vehicles of distribution for *A. flavus* and are reproductive structures; they are characterized by a dormant state that is essential for survival in hostile conditions [9,10]. Air-dispersed conidia are highly resistant to extreme environments and can remain viable for several years and begin to germinate as soon as they are in hospitable environmental conditions and in the presence of nutrients such as fermentable sugars, inorganic salts and a nitrogen source [11,12]. Therefore, understanding the process of *A. flavus* conidia germination is important for food and feed safety.

The germination of a fungal spore is also an important way for target organisms to be infected during the spoilage of food and feed. Dormant conidia have irregular spherical shapes. Upon the activation of germination, water uptake leads to an increase in intracellular osmotic pressure [13]. For *Aspergillus niger*, during this stage, the first morphological change in conidia germination involves swelling, with the diameter of the spore increasing two-fold or more. The swelling phase of conidia is also called isotropic growth [14]. Swelling is concomitant with many metabolic activities such as respiratory metabolism, amino acid biosynthesis, protein biosynthesis, and so on [15]. Swollen conidia are followed by polarized growth that leads to germ-tube formation. During this phase, the formation of a germ tube is also called polarized growth. A large number of metabolism activities are the same as those found in isotropic growth, and only some special metabolic activities, such as cytoskeleton formation, the vesicle trafficking system and landmark protein, are different [16]. Next, conidia complete germination when the length of the germ tube is equal to the half of the diameter of the spore. At later stages of development, the germ tube grows faster and faster and branching leads to agglomeration, mutually resulting in fungus hypha accumulation. During this phase, the secondary metabolite aflatoxins are major secreted from hypha [17].

RNA-Seq technology has been widely used in microbiology research for investigating the dynamic changes in RNA expression, including conidia germination, mycotoxin biosynthesis, environmental stress response, nutrient metabolism and so on [5,18–20]. In this study, the germination rate of *A. flavus* conidia at different times was analyzed to determine the various stages of conidia germination. Subsequently, we used different microscopes to study the morphological changes of *A. flavus* conidia during germination on Czapek–Dox (CZ) culture medium. Then, RNA-Seq technology was used to identify transcriptomic changes in developing conidia involved in various *A. flavus*, and the molecular functions of differentially expressed genes (DEGs) and their metabolic pathways were analyzed using bioinformatic methods. Most changes in the transcriptome occurred during the early phase of germination. The data showed that the transcriptome of the dormant spore is very different from that of conidia during all germination phases. Our study focused on the different changes on translation, carbohydrate and lipid metabolism.

2. Results

2.1. Conidia Germination Rate

Conidia germination of *A. flavus* has a maximal rate between 28 and 30 °C. In this study, *A. flavus* spores were inoculated in CZ culture medium at 30 °C, with approximately 50% conidia germination after 10 h (Figure 1a). Isotropic growth (swelling phase) was observed before 5 h after inoculation, and polarized growth (germ tube forming) occurred at 5 h and 10 h (Figure 1b). The cell size of dormant conidia was about 3×4 μm, but in the isotropic growth stage, the cell size was much larger than dormant conidia. The morphology of the swelling conidia was different with the dormant conidia, with wrinkle recoveries and flat cell walls. In the polarized growth stage, conidia completed germination when the length of the germ tube was equal to the conidia' radius, and some conidia had more than one germ tube.

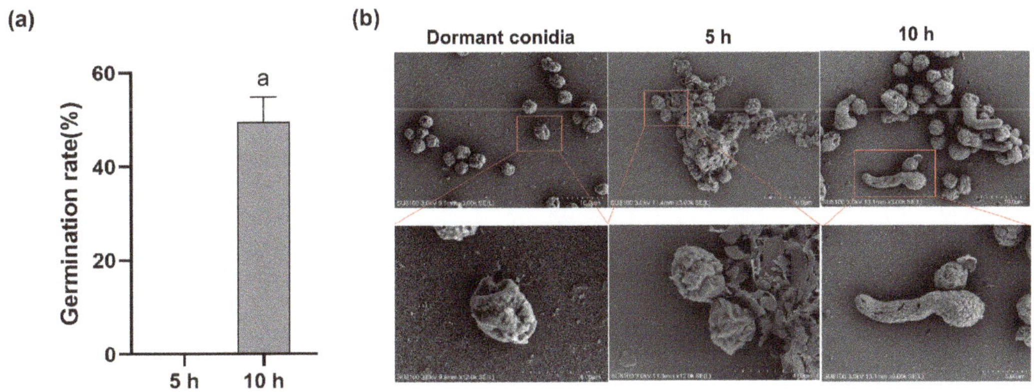

Figure 1. (**a**) Conidia germination rate of *A. flavus* in CZ culture medium. (**b**) Germination of *A. flavus* conidia as observed by SEM. Microscope images are shown for dormant conidia (0 h) and germinating conidia at 5 h and 10 h, respectively. Bar represents 10 µm, 4 µm and 5 µm (10 h). [a] Columns with different lowercase letters indicated significant differences between the compared groups ($p < 0.05$).

2.2. Flow Cytometry

Conidial samples were prepared and analyzed by flow cytometry over a 10 h period to measure the increase in the size of dormant conidia harvested with PDA. The Flowjo software provided numerical values for the FSCs of the conidial and generated a graph (Figure 2). The counts of conidia demonstrated that evident isotropic growth expansion occurs over the first few hours of germination, and polarity formation and germ tube emergence were also apparent between 5 and 10 h.

2.3. Transcriptional Profiling

In this study, nine samples of *A. flavus* NRRL 3357 were sequenced using RNA-Seq technology, averaging 24,136,399 raw sequencing reads and 24,125,185 clean reads after filtering out low quality reads. Table 1 briefly summarizes the information of sequencing data for each sample.

2.4. Gene Expression

Gene expression levels were quantified by a software package called RSEM. The number of identified expressed genes was counted and calculated in proportion to the total gene number in the database for each sample in Figure 3a. Dormant conidia averaged 10,966 transcripts. The number of expressed genes increased to 11,656 5 h after inoculation and then gradually increased to 11,702 10 h after germination. According to principal component analysis (PCA) analysis (Figure 3b), the dormant conidia sample differed from all other time points in that it contributes to the majority of the first principal component while the variation in the other time points was predominantly confined to the second principal component. The correlation of expressions (Figure 3c) showed that the RNA profile of dormant conidia was the most different when compared to other samples.

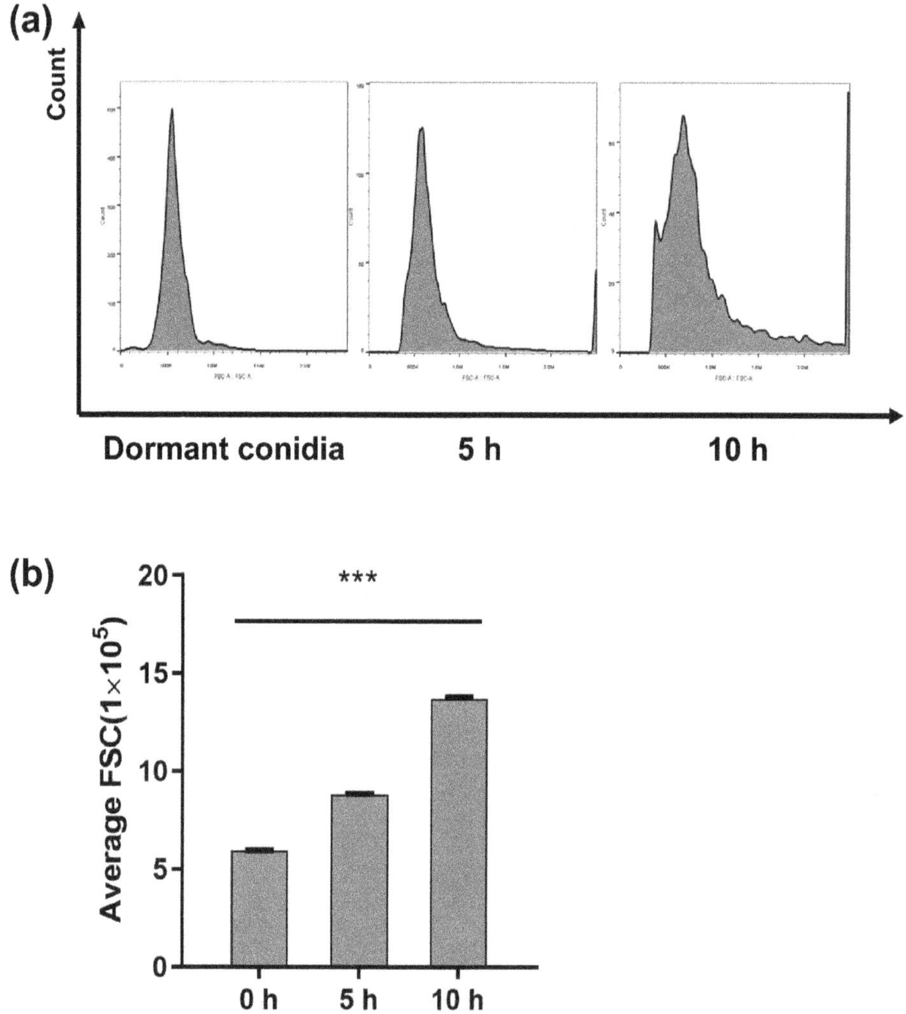

Figure 2. The differences in size of conidia germination under three different stages. The x-axis indicates forward scatter (FCS), and the y-axis indicates counts of profiles of 10,000 conidia at 0 h, 5 h and 10 h (**a**). Average size of 10,000 conidia measured as the FSC parameter (**b**). The means and standard errors of duplicate samples have been plotted (n = 3). The"*"on column diagram indicate a statistically difference of treatment at "***"means $p < 0.001$.

Table 1. Statistical summary of the different conidia RNA-Seq datasets.

Sample	Raw Data Size (bp)	Raw Reads Number	Clean Data Size (bp)	Clean Reads Number	Clean Data Rate [1] (%)
0 h	1,206,821,150	24,136,423	1,206,488,600	24,129,772	99.97
0 h	1,206,804,050	24,136,081	1,206,255,800	24,125,116	99.95
0 h	1,206,810,000	24,136,200	1,206,384,900	24,127,698	99.96
5 h	1,206,793,700	24,135,874	1,206,071,500	24,121,430	99.94
5 h	1,206,826,500	24,136,530	1,205,452,250	24,109,045	99.88
5 h	1,206,836,250	24,136,725	1,206,384,850	24,127,697	99.96
10 h	1,206,827,750	24,136,555	1,206,450,400	24,129,008	99.96
10 h	1,206,830,400	24,136,608	1,206,429,400	24,128,588	99.96
10 h	1,206,829,750	24,136,595	1,206,415,650	24,128,313	99.96

[1] Clean Data Rate (%) = Clean Reads Number/Raw Reads Number.

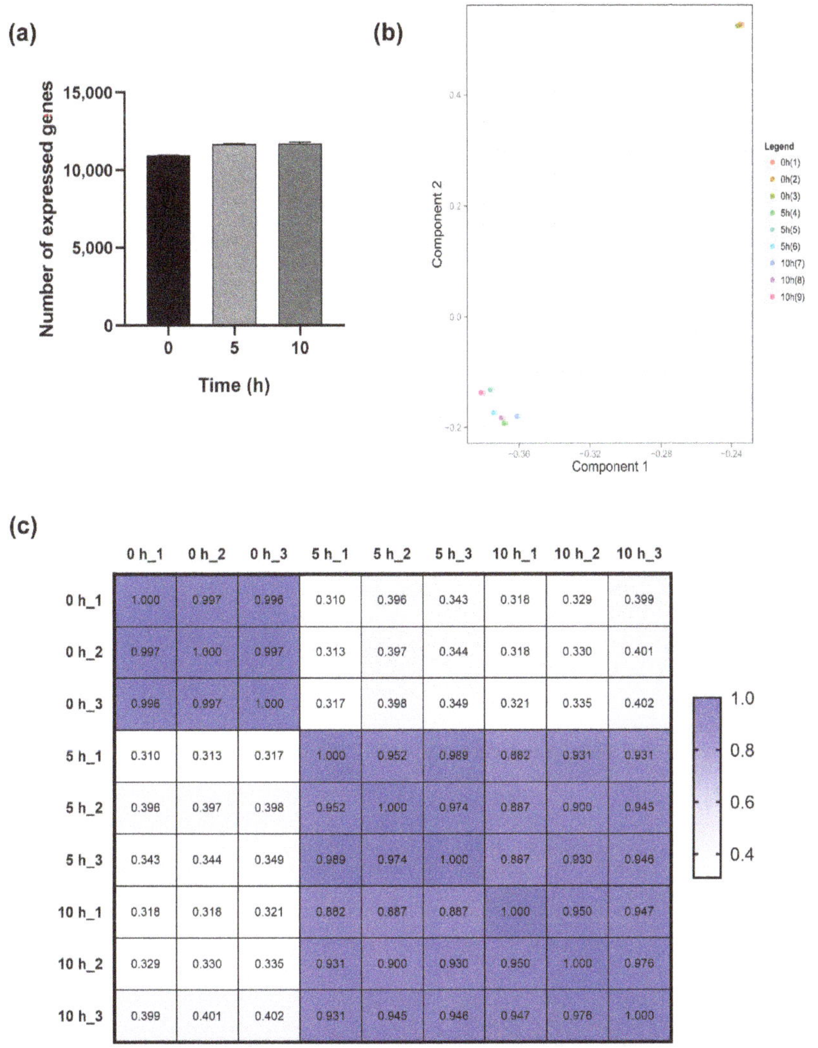

Figure 3. The number of expressed genes (**a**) during germination of *A. flavus* and the similarity of the RNA profiles of the different stages of germination represented by principal component analysis (**b**) and correlation coefficients (**c**).

2.5. Differential Gene Expression and Functional Analysis

The results of the differential gene expression analysis revealed that were many germination responsive genes existing in the spore (Figure 4). Compared to dormant conidia, 1910 genes were up-expressed and 969 genes were down-expressed with a two-fold change or greater ($p < 0.05$). Meanwhile, a number of differentially expressed genes were much lower between the 5 h and 10 h time points. Genes numbering 321 were up-regulated between 5 h and 10 h, whereas 80 genes were down-regulated.

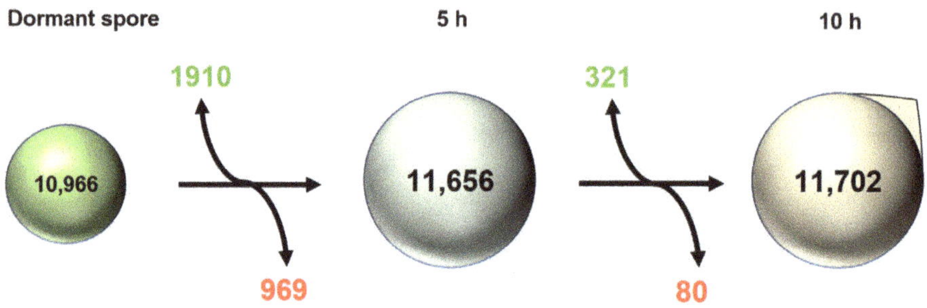

Figure 4. Overview of the global changes in the transcriptome of conidia during germination. Inside the spore, the number of expressed transcripts is provided. Green and red represent numbers of genes with fold change ≥2 up-regulated and down-regulated between two stages, respectively.

Moreover, compared to dormant and 5 h conidia, GO analysis results indicated that 726 DEGs, accounting for 27.13% of all significant DEGs, were associated with cellular compounds; 1297 DEGs, accounting for 22.32% of all significant DEGs, were annotated with molecular functions; 1207 DEGs, accounting for 22.96%, were classified with biological processes. The groups in the three main categories are shown in Figure 5a. Within the biological process category, the most highly represented groups were metabolic processes, cellular processes, single-organism processes and localization. In the cellular component, cells, cell parts, organelles, membranes and macromolecular complexes were the most abundant groups. Meanwhile, binding, catalytic, structural molecular activity, transport activity and nucleic-acid-binding transcription factor activity were the largest terms with respect to molecular functions. Likewise, the results of GO analysis revealed that metabolic processes, single-organism processes and cellular processes are the most abundant terms between the 5 h and 10 h time point (Figure 5b). Cells, cell parts, membranes, membrane parts and organelles in the cellular component and binding and catalytic activity in molecular functions were the most highly represented terms.

Genes usually interact with each other to play roles in certain biological functions. Pathway enrichment analysis of DEGs based on the KEGG database was performed. After comparisons of dormant with 5 h time point conidia, 1849 genes were annotated for 121 known metabolic and signal pathways. During the late stages of germination (5 h vs. 10 h), 238 genes were classified into 92 pathways. However, the pathway distributions of these changes in genes in both isotropic growth and polarized growth were in accordance with each other, and more genes displayed at least a 2-fold change in isotropic growth. Carbohydrate metabolism, amino acid metabolism, translation, lipid metabolism and metabolism of cofactors and vitamins involved in metabolism and genetic information processing were the most abundant groups.

In addition, the top 20 KEGG enrichment results (shown in Figure 6a) were generated. In the isotropic growth stage (Figure 6b), ribosome- and oxidative phosphorylation-related DEGs were the most significant, which indicated that the initiation of energy metabolism and translation constitute key processes in the initial stages of germination (Figure 7a). Furthermore, the map of ribosome and the most changeable genes are represented in Figure S1; RPS28e was the most up-regulated gene and RPl36e was the most down-regulated gene. During polarized growth stages (Figure 6b), organic acid metabolism and lipid metabolism were the most abundant pathways. These data showed that lipid metabolism was an important process for germ-tube growth (Figure 7b). Furthermore, alpha-linolenic acid metabolism from lipid metabolism was the most important from pathway enrichment analysis. All five DEGs in this pathway were down-regulated and they are shown in Table 2 and Figure S2.

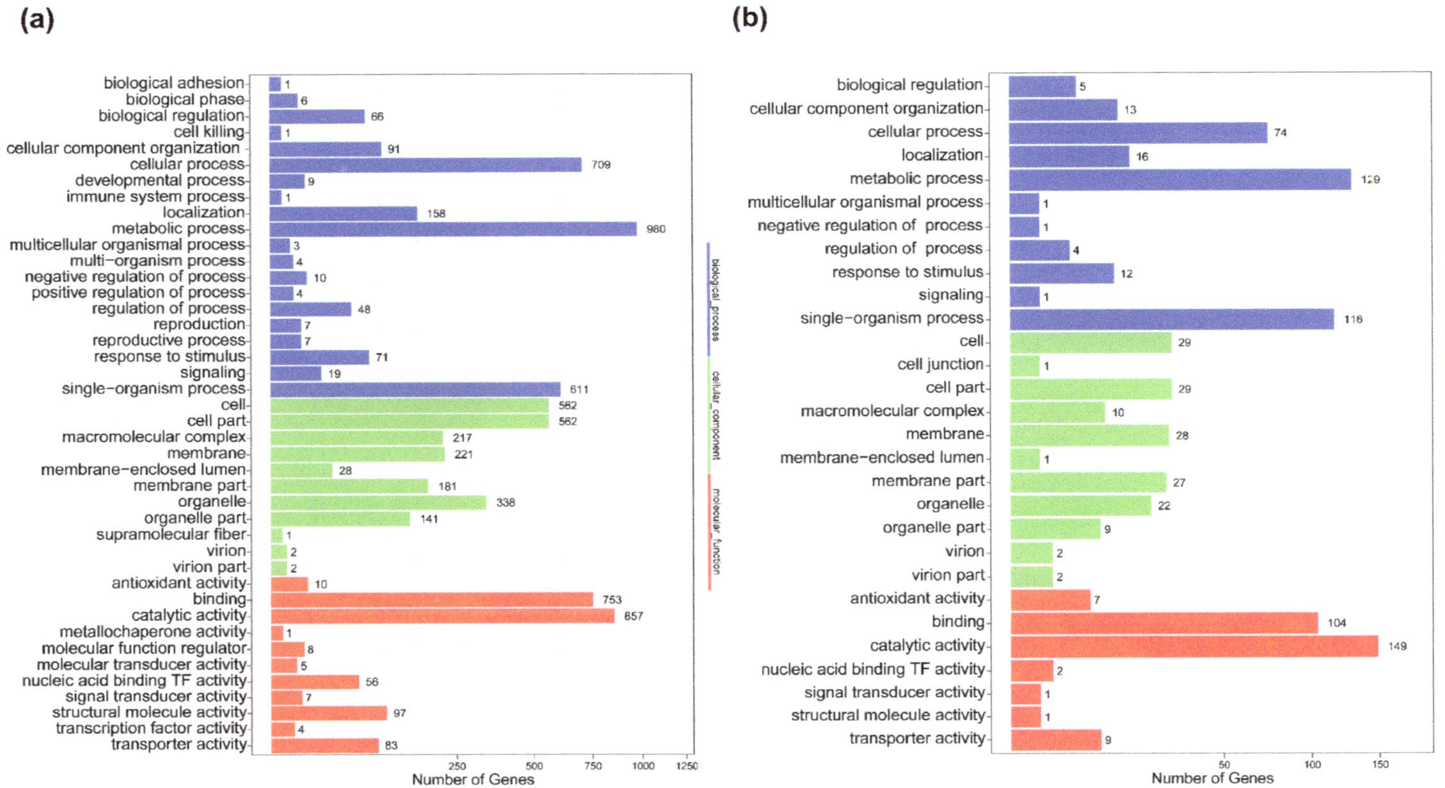

Figure 5. GO functional enrichment analysis of DEGs in 0 h vs. 5 h (**a**) and 5 h vs. 10 h (**b**) during conidia germination. All GO terms are grouped into three ontologies: blue is for biological processes, green is for cellular components and red is for molecular function. The y-axis indicates the subcategories, and the x-axis indicates the number of genes in the same category.

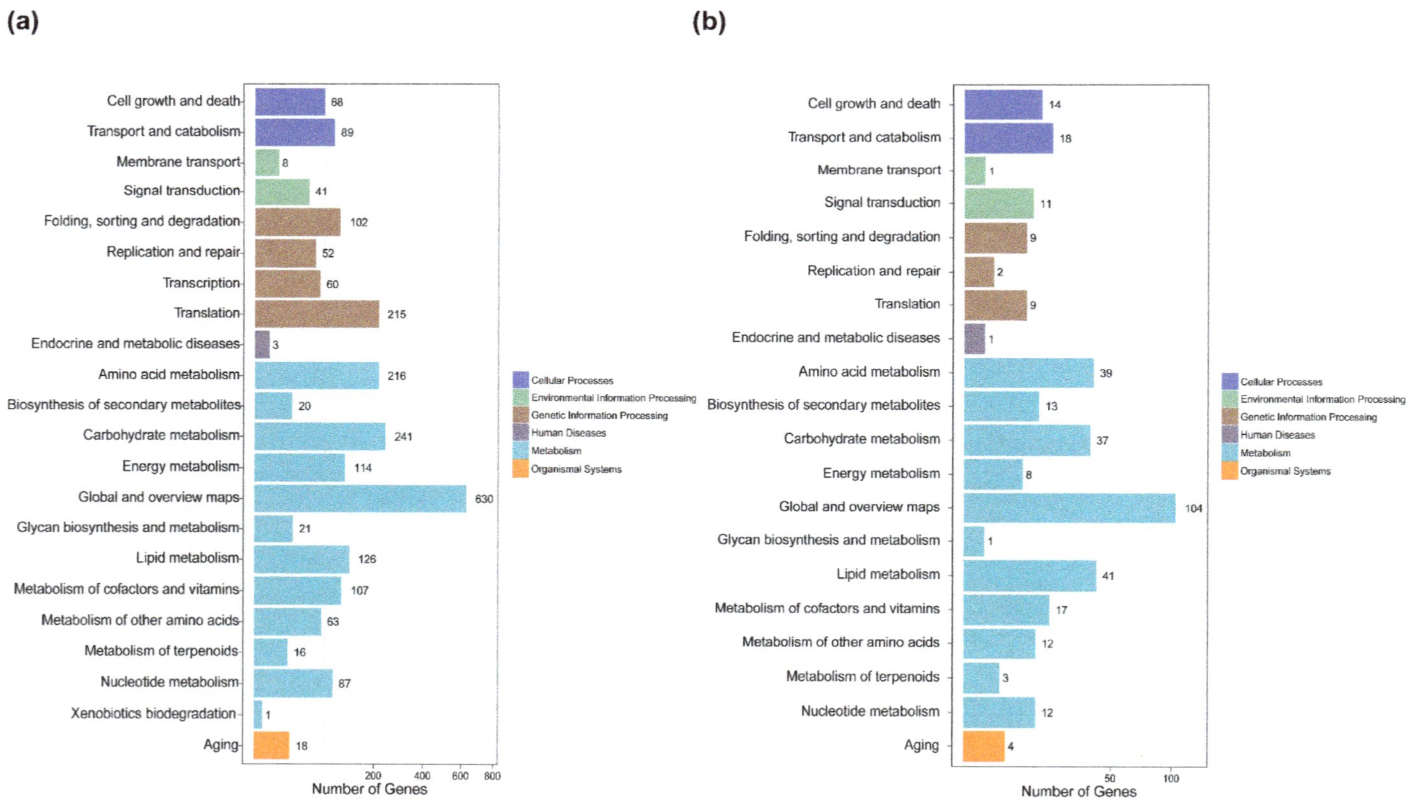

Figure 6. KEGG classification on DEGs for each comparison at 0 h vs. 5 h (**a**) and 5 h vs. 10 h (**b**). X-axis means the number of DEGs. Y axis represents second KEGG pathway terms. All second pathway terms are grouped in top pathway terms indicated with different colors.

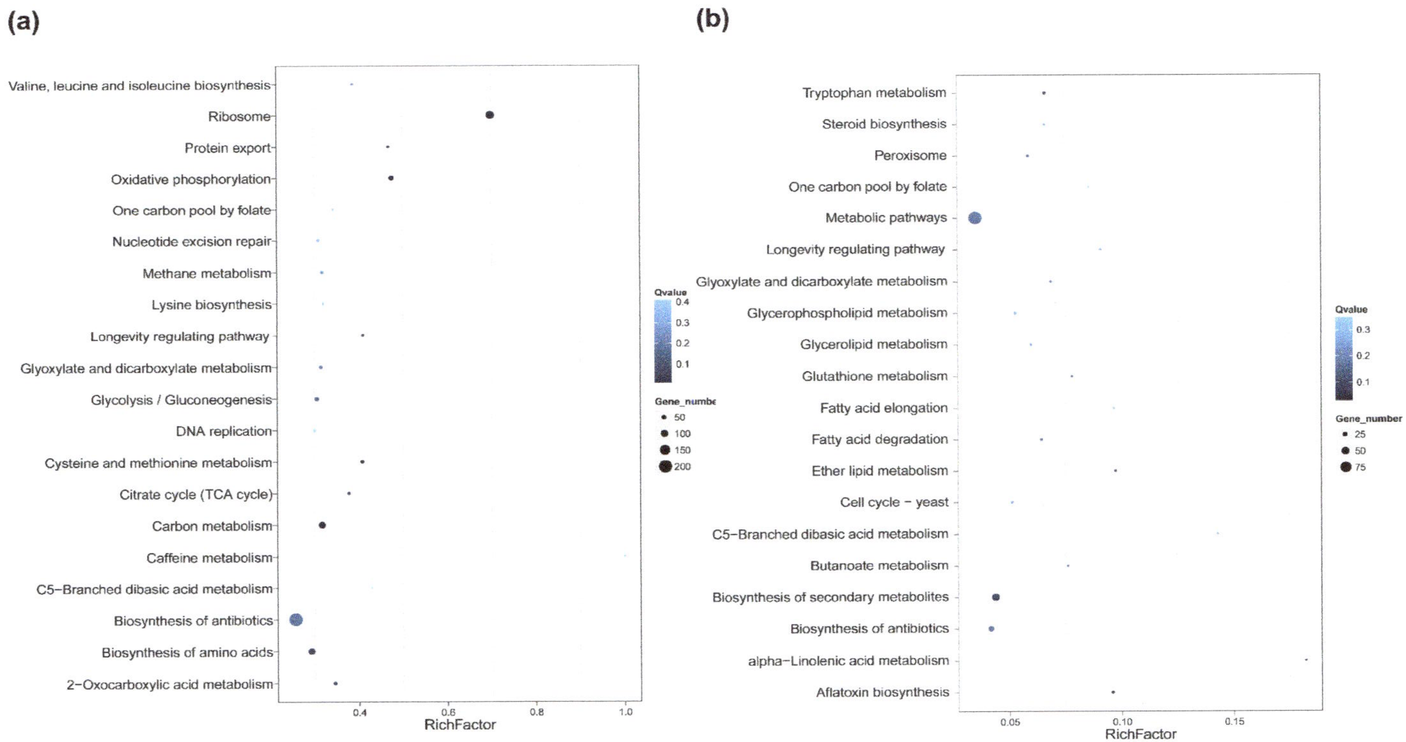

Figure 7. Statistics of pathway enrichment of DEGs in each comparison at 0 h vs. 5 h (**a**) and 5 h vs. 10 h (**b**). Rich factor is the ratio of differentially expressed gene numbers annotated in this pathway term to all gene numbers annotated in this pathway term. A greater rich factor means greater intensiveness. Q-value is corrected *p*-value ranging from 0 to 1, and a lower Q-value means greater intensiveness. We only displayed the top 20 enriched pathway terms.

Table 2. The differentially expressed genes grouped by GO, KEGG and enriched pathways of interest between 5 and 10 h.

Pathway	Gene ID	log2Ratio [a]	Up/Down	Probability	Description
Alpha-Linolenic acid metabolism	7917785	−1.02	Down	0.80373	3-ketoacyl-coA thiolase peroxisomal A precursor, mRNA
	7915336	−1.69	Down	0.80566	FMN binding oxidoreductase, putative, mRNA
	7912988	−1.95	Down	0.86628	NADH-dependent flavin oxidoreductase, putative, mRNA
	7912986	−2.05	Down	0.89858	Fatty-acyl coenzyme A oxidase (Pox1), putative, mRNA
	7910815	2.87	Up	0.929795	40S ribosomal protein S22, partial mRNA
Aflatoxin biosynthesis	7909985	−1.08	Down	0.810561786	PKS-like enzyme, putative, mRNA
	7910374	3.32	Up	0.863160112	short chain type dehydrogenase, putative, mRNA
	7912783	1.45	Up	0.842033907	toxin biosynthesis ketoreductase, putative, mRNA
	7914380	7.38	Up	0.83854578	benzoate 4-monooxygenase cytochrome P450, mRNA
	7911412	5.07	Up	0.865160849	cytochrome P450, putative, mRNA
	7911415	6.16	Up	0.941794345	short-chain dehydrogenase, putative, mRNA
	7911112	3.25	Up	0.826041173	cytochrome P450, putative, mRNA
	7915318	5.90	Up	0.834952351	O-methyltransferase family protein, mRNA
	7912683	8.03	Up	0.80137419	O-methyltransferase, putative, mRNA
	7915318	5.90	Up	0.834952351	O-methyltransferase family protein, mRNA
	7911961	1.56	Up	0.801874375	O-methyltransferase family protein, mRNA

[a] log2Ratio was determined as the log2 mean value of mRNA abundance at 5 h vs. 10 h.

Additionally, the aflatoxin biosynthesis pathway activated in the stage of germination (Table 2 and Figure S3), which means that the secondary metabolism was triggered with the germination process and became ready for aflatoxin biosynthesis.

2.6. Antioxidant System during Conidia Germination

The identification of the redox gene effect during conidia germination is of paramount importance. The essential oil has been reported to efficiently kill conidia of *A. flavus* via triggering reactive oxygen species and causing redox-balance damage. According to the results of RNA-Seq, the redox gene expression was determined by RT-PCR (Figure 8a) more specifically during conidia germination and separated into four different stages. Real-time qPCR results showed that redox gene mRNA levels of ss-cat and cat2 increased, while m-cat decreased as conidia germination progresses, which was also demonstrated in RNA-Seq results. With the coumalic acid and geraniol supplementation, conidia germination was inhibited as Figure 8b shows, and the mRNA abundance of ss-cat, cat, and cat2 increased.

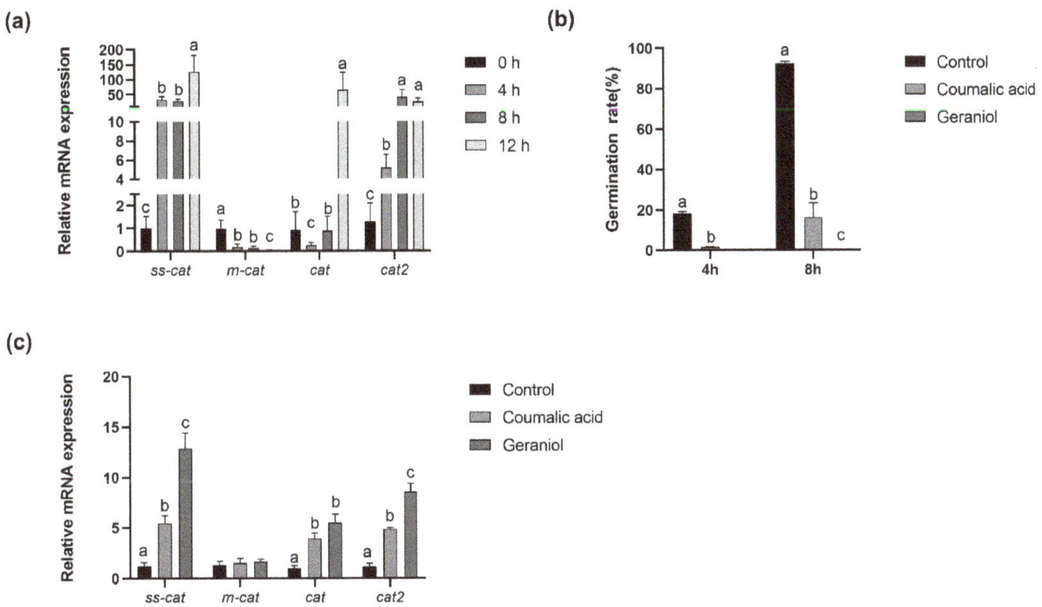

Figure 8. Relative mRNA abundance of redox genes during conidia germination of *A. flavus* (**a**). Germination rate with coumalic acid and geraniol supplement (**b**) during the germination of *A. flavus*. Effect on coumalic acid and geraniol supplement on relative mRNA abundance of redox genes after 8 h germination of *A. flavus* (**c**). Values are mean ± SEM, $n = 6$. Means without a common letter differ, $p < 0.05$.

3. Discussion

For the conidia of *A. flavus*, germination is the first crucial step from asexual propagule to vegetative mycelium growth and the production of aflatoxin, which causes contamination and the spoilage of food and feed. Therefore, an improved understanding of the conidia germination process, metabolism and key genes and pathways can provide significant contributions to studies focused on controlling aflatoxin contamination and improving food and feed safety.

The transitions of conidia germination are recognized in three different stages: dormant conidia, isotropic growth and polarized growth [21]. The generated hyphae are then separated into compartments by septa [22]. Each stage has its own unique morphological characteristics. A previous study found that dormant conidia are highly stress-tolerant structures [23], and they are able to survive and germinate under high-pressure conditions such as dehydration, extreme temperature, osmotic pressure variations in pH and UV due to the three layers of the cell wall and several inner characteristics [11,24]. Dormant conidia germinated when flexible nutrients such as sugars, inorganic salt and nitrogen source were supplemented in most *Aspergillus* strains. However, germination times depend on the different culture conditions and variations in different *Aspergillus* strains.

In our study, we chose 5 h after inoculation as the stage of isotropic growth and 10 h germination as the stage of polarized growth, respectively, using a series of microscopes and flow cytometry. While the morphology change was similar to other typical *Aspergillus* such as *A. niger* and *A. fumigatus*, during the isotropic stage, the cell's size was up to twice that of dormant conidia and the germ tube grew out from one side [25]. When the length of the tube was equal to the conidia's radius, this meant that conidia germination was successful. The morphological impact of germinating conidia on the surface ultrastructure of *A. flavus* spores was investigated by scanning electron microscope (SEM).

The goal to understand the transcriptome landscape of dormant and geminating conidia of the filamentous fungi *A. flavus* was achieved in this study. Presumably, our research is the first report to analyze the transcriptome levels of the dormant and germinating conidia within the aflatoxigenic *A. flavus* strain. In the current study, the RNA-Seq produced an average of 1.21 billion bp raw data size and 24.1 million raw reads for each treatment, and approximately 10,000 genes were characterized after filtering out low quality reads. The data indicated that the RNA expression level of dormant conidia is substantially different when compared to other stages of germination, each of which is characterized by a typical morphology. The transcriptome of conidia changed gradually before the stage of isotropic growth (swelling), in which the gene's expression had many variations. The correlation of the expression of the dormant and germinating conidia 5 h after inoculation (0.350) and 10 h after inoculation (0.348), as well as the correlation between the 5 h and 10 h time points, is 0.92, which provides evidence for these changes.

About 23,320 genes were expressed in vegetative growth in a control group of *A. flavus* in different water activity treatments, while transcripts of 33% of the genes were active in dormant conidia of *A. niger* [18] and a similar trend was found in *Aspergillus. fumigatus* [10]. Compared to vegetative hyphae and aerial structures, the complexity of conidial RNA is lower because these spores represent a single cell type [26]. In contrast, mycelium, vegetative hyphae and aerial structures are composed of different types of hyphae and cells. Previous studies have shown that the RNA profile has a few changes after one-year storage in the dormant conidia of *A. fumigatus,* and it was thought that mRNA was in a pre-packed pool stage for the translation and quick response of conidia germination [10]. For instance, a few compounds such as heat shock proteins, trehalose, mannitol and dehydrins in dormant conidia are key for maintaining the structures for surviving extreme conditions [11]. The transcripts of genes for encoding these related proteins were not only highly accumulated but the transcripts of genes related to the synthesis and degradation of compatible solutes were also unique in dormant spores. Similar research has shown that genes involved in the defense of the conidia cell wall (for example, the genes responsible for making hydrophobins and pigmentation [27,28]) are specific for dormant conidia in *A. niger* and *A. fumigatus*. Furthermore, some transcription-factor-related genes that are essential for spore formation and maturation were only found in dormant conidia but absent in germinating conidia [18].

In this study, to evaluate the changes between the breaking of dormancy and dormant conidia in *A. flavus*, RNA-Seq was performed. Compared to other research, significant transcriptional changes occurred over the first 2 h of germination in *A. niger* by using genome-wide microarrays, but the total gene number was only about 4000, which is far less than the 11,000 genes in our study [15]. Other research also revealed that the most significant changes occurred over the initial stages of conidia germination when compared to the subsequent stages of germination in *A. fumigatus* [9]. As a result of this observation, RNA-Seq technology was used to study this period of the breaking of dormancy in more detail and as a tool to validate the microarray's results. In our study, we found that significant changes occurred during the first stages of *A. flavus* conidia germination.

For GO class analysis, metabolic processes in biological process contain the largest number of DEGs between dormant conidia and conidia 5 h after inoculation. With the exception of the global and overview maps of KEGG enrichment analysis, the DEGs enriched in the nutrient metabolism pathway were the most considerable amount. From a metabolic perspective, the germination process involves a transitioning from a relatively quiescent, dormant state to a germinated state. There needs to be resumption and an increase in metabolic activities including respiration, DNA synthesis, mitosis, cell wall synthesis, RNA and protein biosynthesis throughout germination.

Protein synthesis is vital for germination in *A. fumigatus* and *A. niger* because the protein synthesis inhibitor cycloheximide prevents germ tube formation at moderate concentrations [13,15]. Both protein synthesis and polysome assembly are early events in germination and transcriptome research, with *A. niger* also supporting this conclusion [5].

In our study, pathway enrichment analyses revealed that the genes related to the ribosome and ribosome biogenesis significantly changed after conidia germination. In CZ culture medium, sodium nitrate was the only nitrogen source, and several genes involved in the nitrogen's metabolism were regulated at the onset of germination. For instance, *NR*, *RT* and *NIT-6* were responsible for converting nitrate into ammonia and increased their transcript levels upon germination. Then, L-amino acids were synthesized after a series of biological processes. L-amino acids are the building blocks of new proteins, and the data showed that transcripts encoding transcription factor *CpcA*, which monitors L-amino acid metabolism increased at the initial stages of germination [28], possibly act as signals for replenishing the pool of L-amino acids intracellularly, which involved the same tendency as conidia germination in *A. niger* [24].

For the energy process, the transcript-encoding enzymes of the tricarboxylic acid cycle (TCA), glycolysis/gluconeogenesis and pentose phosphate pathway were found to be highly abundant in the first stage of germinating conidia but were absent in dormant conidia. Fatty acids can act as a catalyst that starts the gluconeogenesis pathway because they can feed into it. The mRNA profile of genes in *Aspergillus* conidia indicates that gluconeogenesis may be significant for spore survival and germination through the use of stored lipids [15]. In our study, the highly abundant fatty acid degradation and metabolism at 5 h of germination also agreed with this conclusion. Furthermore, the fatty acid elongation and biosynthesis were also highly abundant in the swelling stage, which means lipid metabolism is crucial for the breaking of dormancy.

L-amino acids are also possible substrates for gluconeogenesis after germination and the transcriptome suggested that the proteasome is an organelle that could be functional. Additionally, in contrast to *A. niger* conidia germination, there was a lower abundance of transcripts encoding the proteasome in the 5 h germinating conidia compared to the dormant conidia [18].

In translation, the sequence of codons on mRNA directs the synthesis of a polypeptide chain. This process takes place on the ribosome and the movement of tRNA and mRNA through the ribosome is a complicated process that combines high speeds with high accuracy [29]. The ribosome, a large ribonucleoprotein particle, comprises two subunits (large and small) in all species. In our study, most DEGs related to the ribosome (see details in Table 3 and Figure S1) were upregulated between the first two stages, which means translation activity was highly frequent between these changes. Based on our data, the lipid metabolism pathway was a key pathway for the germ-tube stage, sphingolipids, a type of lipid, are major components of fungal plasma membranes and also an inhibition target that prevents polarized growth in *Aspergillus. nidulans* [30]. Alpha-linolenic acid metabolism was the most influenced pathway in the lipid metabolism pathway, whereby alpha-linolenic acid reduced growth and aflatoxin synthesis after several hours [31]. This finding also supported our research, whereby all four DEGs in this pathway were down-regulated in polarized growth stage.

Table 3. The differentially expressed genes grouped by GO, KEGG and enriched pathways of interest between 0 and 5 h.

Pathway	Gene ID	log2Ratio [a]	Up/Down	Probability	Description
Ribosome	7913639	3.02	Up	0.905636	ribosomal protein YmL41, partial mRNA
	7910217	2.97	Up	0.929728	60S ribosomal protein L31, partial mRNA
	7921558	2.87	Up	0.929755	60S ribosomal protein L1, partial mRNA
	7913089	2.87	Up	0.929914	40S ribosomal protein S20, partial mRNA
	7910815	2.87	Up	0.929795	40S ribosomal protein S22, partial mRNA
Redox genes	7912906	7.00	Up	0.974886	catalase Cat, mRNA
	7920700	−1.68	Down	0.904521	spore-specific catalase CatA, mRNA
	7917068	6.11	Up	0.950608	mycelial catalase Cat1, mRNA
	7918464	1.84	Up	0.850788	bifunctional catalase-peroxidase Cat2, mRNA

[a] log2Ratio was determined as the log2 mean value of mRNA abundance of 0 h vs. 5 h.

With conidia germination, respiration become more active, and the antioxidant system simultaneously become more effective [13,18,32]. In this regard, conidia germination affected four DEGs involved in the antioxidant process (Figure 8, Table 3). Notably, the *ss-cat* (spore-specific catalase) and *cat2* (bifunctional catalase-peroxidase) genes were up-regulated, while the *m-cat* (mycelial catalase) gene for protein was down-regulated, and the *cat* (catalase) gene was down-regulated and then up-regulated. These changes in gene abundance could result in mitigating oxidative stress during conidia germination. However, the different gene expression changes need to be explored in the future. Numerous studies have shown that the *cat* (catalase) gene plays an important role in fungal development, aflatoxin biosynthesis and virulence [33]; mycelial catalases transiently protect the fungus from external conditions [34]. However, few studies focus on the antioxidant system during conidia germination, and the function of these genes requires further research.

Coumalic acid and geraniol found in the essential oil of fruit and herbs have been suggested to represent a new class of agents to control *A. flavus* and aflatoxin contamination. The two materials have been reported to inhibit the germination of resting spores of some pathogens by interrupting the antioxidant balance system [35–37]. Consistent with previous studies, coumalic acid and geraniol exhibited a potent inhibitory effect on *A. flavus* conidia germination and the *ss-cat*, *cat*, and *cat2* genes were up-regulated at 5 h of germination via the induced antioxidant system imbalance (Figure 8). The changes in these genes might help us figure out the mechanism of *A. flavus* conidia germination. Most importantly, redox genes could be a potential target to inhibit *A. flavus* conidia germination.

In conclusion, the present study found that the many changes in the transcriptome were not correlated with distinct morphological changes during germination. In addition, DEGs related to aflatoxin synthesis were found during polarized growth, which means that the transcription process was triggered in an early stage. In general, RNA-Seq was used to uncover transcriptome changes at the conidia germination of *A. flavus*. Translation, amino acid metabolism and carbohydrate metabolism were the most active pathway in breaking conidia germination. Moreover, lipid metabolism, amino acid metabolism and carbohydrate metabolism were the top three pathways during germ-tube growth. Additionally, the antioxidant system plays a crucial role in conidia germination, and *ss-cat*, *cat* and *cat2* are essential redox genes. However, the further validation of the exact functions and mechanisms of these key DEGs in conidia germination needs to be further studied and might potentially be beneficial in preventing aflatoxin contamination.

4. Materials and Methods

4.1. Culturing Conditions and Sampling

Aspergillus flavus NRRL 3357 was obtained from Prof. Zhumei He (Sun Yat-Sen University, Guangzhou, China) [38]. The strain was grown on Potato Dextrose Agar (PDA) for 7 days at 30 °C to develop mature spores. Spores were then harvested with sterile 0.05% (*w*/*v*) Tween 80 solution. The spore suspension was filtered through 3 sterile layers of lens paper and kept on ice until further processing on the same day, and the spore population was quantified using a hemocytometer. For spore germination, 20 mL of 10^6 mL^{-1} spores was inoculated in 200 mL liquid Czapek–Dox (CZ) Medium. Three replicates were shaken at 150 rpm at 30 °C for each RNA isolation. At each time point, samples were pooled and centrifuged at 5 °C for 10 min at 3000× *g*. The pellet was frozen in liquid nitrogen for later RNA isolation. Coumalic acid and geraniol were dissolved in ethanol into a 100 mg/mL stock solution, protected from light, and stored at 4 °C. The final concentration of the coumalic acid treatment groups was 200 mg/L and geraniol was 100 mg/L.

4.2. Microscopy

For scanning electron microscopy (SEM) analysis, 1 × 10^6 conidia of *A. flavus* were harvested by centrifugation at 3000× *g* and washed with PBS (phosphate buffered saline, pH 7.4) twice. Then, conidia were fixed in 2.5% glutaraldehyde in PBS for 2 h at room temperature. The conidia were washed with PBS for 3 times, 15 min each, and then the conidia were post-fixed in 1% osmium tetroxide for 1–2 h at room temperature. After that, conidia were washed in PBS for 3 times. The dehydration of samples was achieved by transferring by increasing the concentration of (30–100%) ethanol solutions, and the samples were dried with Critical Point Dryer [39]. The samples were then attached to metallic stubs using stickers and sputter-coated with gold for 30 s. The observations were made on a HITACHI Regulus 8100 SEM (Tokyo, Japan).

4.3. Flow Cytometry of Spores

Flow cytometry was used to measure the size of spores (1×10^5) over the first few hours of germination when the conidia were swelling. Liquid CZ medium was inoculated with *A. flavus* conidia at a concentration of 10^6/mL and shaken at 150 rpm at 28 °C. The samples were collected 5 and 10 h after inoculation. The supernatant was removed, and conidia were washed 3 times with 1 mL Tween 80 (0.01% *v*/*v*) and resuspended in 0.5 mL Tween 80. The sample was then analyzed using flow cytometry (Beckman-CytoFLEX Coulter, Brea, CA, USA). FlowJo software was used to determine the forward scatter (FSC) parameter for each sample, which is a measure of conidial size [25]. The same number of dormant conidia was analyzed as well.

4.4. RNA Extraction and RNA-Seq

Total RNA was extracted from conidia using a TRIzol and chloroform RNA extraction protocol, as previously described [5]. Three replicate RNA-Seq libraries were prepared from dormant conidia at 5 h and 10 h after the inoculation of *A. flavus*. A total of the nine libraries were sequenced separately using BGISEQ-500 sequencer. Raw sequencing reads were cleaned by removing adaptor sequences, reads containing ply-N sequences and low-quality reads. Approximately 24,006,405 clean reads were mapped to the Nipponbare reference genome using HISAT [40]/Bowtie [41] tools. After data were mapped, normalization was performed and then FPKM (fragments per kilobase per million mapped reads) was calculated using RESM software [42]. As previously described [43], a false discovery rate (FDR) < 0.01 and absolute value of log2 ratio ≥ 1 were used to identify differentially expressed genes in dormant conidia versus 5 h and 5 h versus 10 h samples.

4.5. Real-Time Quantitative PCR

Total RNA from 6 individual *A. Flavus* spore samples in each treatment (0 h, 4 h, 8 h and 12 h) and 2 essential oil supplement groups (after 8 h inoculation) were isolated, and the quality and quantity of RNA were analyzed by using Thermo NanoDrop (Thermo, Waltham, MA, USA). To estimate the accuracy of transcriptome results and for further investigation, 4 DEGs, *ss-cat* (spore-specific catalase), *cat2* (bifunctional catalase-peroxidase), *m-cat* (mycelial catalase) and *cat* (catalase), were selected using Real-time quantitative PCR (RT-qPCR). RT-qPCR was conducted on a Bio-Rad CFX384 Real-Time PCR System (Bio-Rad, Hercules, CA, USA) with TB Green® Premix Ex Taq™ II (Tli RNaseH Plus) (Takara, Dalian, China). The relative amounts of mRNAs were normalized with the housekeeping gene *GAPDH* and were analyzed by the $2^{-\Delta\Delta Ct}$ method.

4.6. Statistical Analysis

For spore germination rates and transcriptomic data, statistical analyses were performed using Graphpad Prism (San Diego, CA, USA) for Windows (version 8.00). The data were expressed as mean ± SEM (standard error of mean). Differential effects were analyzed by one-way analysis of variance (ANOVA). A p value < 0.05 was considered significant (*), and p value < 0.001 was considered extremely significant (***).

4.7. Data Submission

All the amplicon sequencing datasets in this study were submitted to NCBI Sequence Read Archive (SRA) under accession number PRJNA698788.

Supplementary Materials: The following supporting information can be downloaded at https://www.mdpi.com/article/10.3390/toxins14080560/s1, Figure S1: DEGs in ribosome pathway in *A. flavus* between 0/5 h. Figure S2: DEGs in alpha-linolenic acid metabolism pathway in *A. flavus* between 5/10 h. Figure S3: DEGs in aflatoxin biosynthesis pathway in *A. flavus* between 5/10 h. Excel S1: All DEGs between 0 h vs. 5 h. Excel S2: All DEGs between 5 h vs. 10 h.

Author Contributions: D.Q. and S.W. designed the research study. C.L., S.J. and S.A.R. performed research, analyzed data and wrote the paper. All authors have read and agreed to the published version of the manuscript.

Funding: This project was funded by the National Natural Science Foundation Project of China (Project no. 31772635).

Institutional Review Board Statement: Not applicable.

Informed Consent Statement: Not applicable.

Data Availability Statement: The data presented in this study are available in this article and supplementary materials.

Conflicts of Interest: The authors declare no conflict of interest.

References

1. Tian, J.; Ban, X.; Zeng, H.; He, J.; Huang, B.; Wang, Y. Chemical composition and antifungal activity of essential oil from *Cicuta virosa* L. var. *latisecta* Celak. *Int. J. Food Microbiol.* **2011**, *145*, 464–470. [CrossRef] [PubMed]
2. Liu, J.; Sun, L.; Zhang, N.; Zhang, J.; Guo, J.; Li, C.; Rajput, S.A.; Qi, D. Effects of Nutrients in Substrates of Different Grains on Aflatoxin B1 Production by *Aspergillus flavus*. *Biomed. Res. Int.* **2016**, *2016*, 7232858. [CrossRef] [PubMed]
3. WHO. IARC Monographs on the Evaluation of Carcinogenic Risks to Humans. Volume 56. Some Naturally Occurring Substances: Food Items and Constituents, Heterocyclic Aromatic Amines and Mycotoxins. *Cancer Causes Control* **1994**, *5*, 89–90.
4. Egal, S.; Hounsa, A.; Gong, Y.Y.; Turner, P.C.; Wild, C.P.; Hall, A.J.; Hell, K.; Cardwell, K.F. Dietary exposure to aflatoxin from maize and groundnut in young children from Benin and Togo, West Africa. *Int. J. Food Microbiol.* **2005**, *104*, 215–224. [CrossRef]
5. Novodvorska, M.; Hayer, K.; Pullan, S.T.; Wilson, R.; Archer, D.B. Transcriptional landscape of *Aspergillus niger* at breaking of conidial dormancy revealed by RNA-Sequencing. *BMC Genom.* **2013**, *14*, 246. [CrossRef]
6. Wang, B.; Lv, Y.; Li, X.; Lin, Y.; Deng, H.; Pan, L. Profiling of secondary metabolite gene clusters regulated by LaeA in *Aspergillus niger* FGSC A1279 based on genome sequencing and transcriptome analysis. *Res. Microbiol.* **2017**, *169*, 67–77. [CrossRef]
7. Brown, G.D.; Denning, D.W.; Gow, N.A.R.; Levitz, S.M.; Netea, M.G.; White, T.C. Hidden Killers: Human Fungal Infections. *Sci. Transl. Med.* **2012**, *4*, 165rv113. [CrossRef]
8. Verweij, P.P.E.; Zhang, J.; Debets, A.J.M.; Meis, J.F.; Melchers, W.J.G. In-host adaptation and acquired triazole resistance in *Aspergillus fumigatus*: A dilemma for clinical management. *Lancet Infect. Dis.* **2016**, *16*, e251–e260. [CrossRef]
9. Nir, O.; May, G.S. The molecular mechanisms of conidial germination. *FEMS Microbiol. Lett.* **2001**, *199*, 153–160. [CrossRef]
10. Lamarre, C.; Sokol, S.; Debeaupuis, J.P.; Henry, C.; Lacroix, C.; Glaser, P.; Coppée, J.-Y.; François, J.M.; Latgé, J.-P. Transcriptomic analysis of the exit from dormancy of *Aspergillus fumigatus* conidia. *BMC Genom.* **2008**, *9*, 417. [CrossRef]
11. Baltussen, T.J.H.; Zoll, J.; Verweij, P.E.; Melchers, W.J.G. Molecular Mechanisms of Conidial Germination in *Aspergillus* spp. *Microbiol. Mol. Biol. Rev.* **2019**, *84*, e00049-19. [CrossRef]
12. Thanh, N.V.; Rombouts, F.M.; Nout, M.J.R. Effect of individual amino acids and glucose on activation and germination of *Rhizopus oligosporus* sporangiospores in tempe starter. *J. Appl. Microbiol.* **2005**, *99*, 1204–1214. [CrossRef] [PubMed]
13. Taubitz, A.; Bauer, B.; Heesemann, J.; Ebel, F. Role of respiration in the germination process of the pathogenic mold *Aspergillus fumigatus*. *Curr. Microbiol.* **2007**, *54*, 354–360. [CrossRef] [PubMed]
14. Leeuwen, M.R.V.; Doorn, T.M.V.; Golovina, E.; Stark, J.; Dijksterhuis, J. Water- and Air-Distributed Conidia Differ in Sterol Content and Cytoplasmic Microviscosity. *Appl. Environ. Microbiol.* **2009**, *76*, 366–369. [CrossRef] [PubMed]
15. Osherov, N.; May, G.S. Conidial germination in *Aspergillus nidulans* requires RAS signaling and protein synthesis. *Genetics* **2000**, *155*, 647–656. [CrossRef] [PubMed]
16. Van Leeuwen, M.R.; Smant, W.; De Boer, W.; Dijksterhuis, J. Filipin is a reliable in situ marker of ergosterol in the plasma membrane of germinating conidia (spores) of *Penicillium discolor* and stains intensively at the site of germ tube formation. *J. Microbiol. Methods* **2008**, *74*, 64–73. [CrossRef] [PubMed]
17. Mellon, J.E.; Dowd, M.K.; Cotty, P.J. Substrate Utilization by *Aspergillus flavus* in Inoculated Whole Corn Kernels and Isolated Tissues. *J. Agric. Food Chem.* **2005**, *53*, 2351–2357. [CrossRef]
18. van Leeuwen, M.R.; Krijgsheld, P.; Bleichrodt, R.; Menke, H.; Stam, H.; Stark, J.; Wosten, H.A.; Dijksterhuis, J. Germination of conidia of *Aspergillus niger* is accompanied by major changes in RNA profiles. *Stud. Mycol.* **2013**, *74*, 59–70. [CrossRef]
19. Szilagyi, M.; Miskei, M.; Karanyi, Z.; Lenkey, B.; Pocsi, I.; Emri, T. Transcriptome changes initiated by carbon starvation in *Aspergillus nidulans*. *Microbiology* **2013**, *159*, 176–190. [CrossRef]

20. Zhang, F.; Guo, Z.; Zhong, H.; Wang, S.; Yang, W.; Liu, Y.; Wang, S. RNA-Seq-based transcriptome analysis of aflatoxigenic *Aspergillus flavus* in response to water activity. *Toxins* **2014**, *6*, 3187–3207. [CrossRef]
21. Baltussen, T.J.H.; Coolen, J.P.M.; Zoll, J.; Verweij, P.E.; Melchers, W.J.G. Gene co-expression analysis identifies gene clusters associated with isotropic and polarized growth in *Aspergillus fumigatus* conidia. *Fungal Genet. Biol.* **2018**, *116*, 62–72. [CrossRef] [PubMed]
22. Borkovich, K.A.; Ebbole, D.J. Hyphal Growth and Polarity. In *Cellular and Molecular Biology of Filamentous Fungi*; ASM Press: Washington, DC, USA, 2010; pp. 238–259. [CrossRef]
23. Lestrade, P.P.A.; Meis, J.F.; Melchers, W.J.G.; Verweij, P.E. Triazole resistance in *Aspergillus fumigatus*: Recent insights and challenges for patient management. *Clin. Microbiol. Infect.* **2018**, *25*, 799–806. [CrossRef] [PubMed]
24. Dijksterhuis, J. Fungal spores: Highly variable and stress-resistant vehicles for distribution and spoilage. *Food Microbiol.* **2019**, *81*, 2–11. [CrossRef]
25. Pereira de Souza, C.C.; Moreira Prado, G.; da Conceicao Freitas, R.C.; Silva Santos Guimaraes, P.; Calegario de Oliveira, L.; Eustaquio Alvim Brito-Melo, G.; de Figueiredo Conte Vanzela, A.P. Analysis of *Aspergillus nidulans* germination, initial growth and carbon source response by flow cytometry. *J. Basic Microbiol.* **2011**, *51*, 459–466. [CrossRef] [PubMed]
26. Krijgsheld, P.; Bleichrodt, R.; Veluw, G.J.V.; Wang, F.; Müller, W.H.; Dijksterhuis, J.; Wsten, H.A.B. Development in *Aspergillus*. *Stud. Mycol.* **2013**, *74*, 1–29. [CrossRef]
27. Battaglia, M.; Olvera-Carrillo, Y.; Garciarrubio, A.; Campos, F.; Covarrubias, A.A. The Enigmatic LEA Proteins and Other Hydrophilins. *Plant Physiol.* **2008**, *148*, 6–24. [CrossRef]
28. Sasse, C.; Bignell, E.M.; Hasenberg, M.; Haynes, K.; Gunzer, M.; Braus, G.H.; Krappmann, S. Basal expression of the *Aspergillus fumigatus* transcriptional activator CpcA is sufficient to support pulmonary aspergillosis. *Fungal Genet. Biol.* **2008**, *45*, 693–704. [CrossRef]
29. Ramakrishnan, V. Ribosome Structure and the Mechanism of Translation. *Cell* **2002**, *108*, 557–572. [CrossRef]
30. Pearson, C.L.; Xu, K.; Sharpless, K.E.; Harris, S.D. MesA, a Novel Fungal Protein Required for the Stabilization of Polarity Axes in *Aspergillus nidulans*. *Mol. Biol. Cell* **2004**, *15*, 3658. [CrossRef]
31. Yan, S.; Liang, Y.; Zhang, J.; Chen, Z.; Liu, C.-M. Autoxidated linolenic acid inhibits aflatoxin biosynthesis in *Aspergillus flavus* via oxylipin species. *Fungal Genet. Biol.* **2015**, *81*, 229–237. [CrossRef]
32. Chitarra, G.S.; Abee, T.; Rombouts, F.M.; Dijksterhuis, J. 1-Octen-3-ol inhibits conidia germination of *Penicillium paneum* despite of mild effects on membrane permeability, respiration, intracellular pH, and changes the protein composition. *FEMS Microbiol. Ecol.* **2005**, *54*, 67–75. [CrossRef] [PubMed]
33. Zhu, Z.; Yang, M.; Bai, Y.; Ge, F.; Wang, S. Antioxidant-related catalase CTA1 regulates development, aflatoxin biosynthesis, and virulence in pathogenic fungus *Aspergillus flavus*. *Environ. Microbiol.* **2020**, *22*, 2792–2810. [CrossRef] [PubMed]
34. Paris, S.; Wysong, D.; Debeaupuis, J.P.; Shibuya, K.; Philippe, B.; Diamond, R.D.; Latgé, J.P. Catalases of *Aspergillus fumigatus*. *Infect. Immun.* **2003**, *71*, 3551–3562. [CrossRef] [PubMed]
35. Ohi, M.; Kitamura, T.; Hata, S. Stimulation by caffeic acid, coumalic acid, and corilagin of the germination of resting spores of the clubroot pathogen *Plasmodiophora brassicae*. *Biosci. Biotechnol. Biochem.* **2003**, *67*, 170–173. [CrossRef]
36. Geraci, A.; Di Stefano, V.; Di Martino, E.; Schillaci, D.; Schicchi, R. Essential oil components of orange peels and antimicrobial activity. *Nat. Prod. Res.* **2017**, *31*, 653–659. [CrossRef] [PubMed]
37. Sharma, R.; Rao, R.; Kumar, S.; Mahant, S.; Khatkar, S. Therapeutic Potential of Citronella Essential Oil: A Review. *Curr. Drug Discov. Technol.* **2019**, *16*, 330–339. [CrossRef]
38. Nierman, W.C.; Yu, J.; Fedorova-Abrams, N.D.; Losada, L.; Payne, G.A. Genome Sequence of *Aspergillus flavus* NRRL 3357, a Strain That Causes Aflatoxin Contamination of Food and Feed. *Genome Announc.* **2014**, *3*, e00168-15. [CrossRef]
39. Shishodia, S.K.; Tiwari, S.; Hoda, S.; Vijayaraghavan, P.; Shankar, J. SEM and qRT-PCR revealed quercetin inhibits morphogenesis of *Aspergillus flavus* conidia via modulating calcineurin-Crz1 signalling pathway. *Mycology* **2020**, *11*, 118–125. [CrossRef]
40. Kim, D.; Langmead, B.; Salzberg, S.L. HISAT: A fast spliced aligner with low memory requirements. *Nat. Methods* **2015**, *12*, 357–360. [CrossRef]
41. Langmead, B.; Trapnell, C.; Pop, M.; Salzberg, S.L. Ultrafast and memory-efficient alignment of short DNA sequences to the human genome. *Genome Biol.* **2009**, *10*, R25. [CrossRef]
42. Li, B.; Dewey, C.N. RSEM: Accurate transcript quantification from RNA-Seq data with or without a reference genome. *BMC Bioinform.* **2011**, *12*, 323. [CrossRef] [PubMed]
43. Benjamini, Y.; Drai, D.; Elmer, G.; Kafkafi, N.; Golani, I. Controlling the false discovery rate in behavior genetics research. *Behav. Brain Res.* **2001**, *125*, 279–284. [CrossRef]

toxins

MDPI

Article

Aflatoxin B_1 Degradation by Ery4 Laccase: From In Vitro to Contaminated Corn

Martina Loi [1,*], Silvana De Leonardis [2], Biancamaria Ciasca [1], Costantino Paciolla [2], Giuseppina Mulè [1] and Miriam Haidukowski [1,*]

[1] Institute of Sciences of Food Production, National Research Council of Italy (CNR), Via Amendola 122/O, 70126 Bari, Italy
[2] Department of Biosciences, Biotechnology and Environment, Università degli Studi di Bari Aldo Moro, Via E. Orabona 4, 70125 Bari, Italy
* Correspondence: martina.loi@ispa.cnr.it (M.L.); miriam.haidukowski@ispa.cnr.it (M.H.); Tel.: +39-0805929513 (M.H.)

Abstract: Aflatoxins (AFs) are toxic secondary metabolites produced by *Aspergillus* spp. and are found in food and feed as contaminants worldwide. Due to climate change, AFs occurrence is expected to increase also in western Europe. Therefore, to ensure food and feed safety, it is mandatory to develop green technologies for AFs reduction in contaminated matrices. With this regard, enzymatic degradation is an effective and environmentally friendly approach under mild operational conditions and with minor impact on the food and feed matrix. In this work, Ery4 laccase, acetosyringone, ascorbic acid, and dehydroascorbic acid were investigated in vitro, then applied in artificially contaminated corn for AFB_1 reduction. AFB_1 (0.1 µg/mL) was completely removed in vitro and reduced by 26% in corn. Several degradation products were detected in vitro by UHPLC-HRMS and likely corresponded to AFQ_1, epi-AFQ_1, AFB_1-diol, or AFB_1dialehyde, AFB_{2a}, and AFM_1. Protein content was not altered by the enzymatic treatment, while slightly higher levels of lipid peroxidation and H_2O_2 were detected. Although further studies are needed to improve AFB_1 reduction and reduce the impact of this treatment in corn, the results of this study are promising and suggest that Ery4 laccase can be effectively applied for the reduction in AFB_1 in corn.

Keywords: aflatoxin B_1; laccase; corn; bioremediation; degradation products; hydrogen peroxide; ascorbic acid; dehydroascorbic acid; AFQ_1; AFB_{2a}; AFB_1-diol

Key Contribution: Aflatoxin degradation in vitro and in corn flour was assessed. Degradation products were detected by UHPLC-HRMS. In addition, the protein content and oxidative status of the matrix after the enzymatic treatment were evaluated. Significant improvement in the safety and a minimum impairment of the oxidative status were observed, proving that the laccase treatment was a promising aflatoxin reducing treatment.

Citation: Loi, M.; De Leonardis, S.; Ciasca, B.; Paciolla, C.; Mulè, G.; Haidukowski, M. Aflatoxin B_1 Degradation by Ery4 Laccase: From In Vitro to Contaminated Corn. *Toxins* **2023**, *15*, 310. https://doi.org/10.3390/toxins15050310

Received: 31 March 2023
Revised: 17 April 2023
Accepted: 24 April 2023
Published: 27 April 2023

1. Introduction

Aflatoxins (AFs) are secondary toxic metabolites produced by *Aspergillus* spp., which can contaminate food and feed worldwide [1]. AFs include more than 20 different furanocoumarin derivatives with carcinogenic, teratogenic, mutagenic, nephrotoxic, and hepatotoxic properties [2,3]. AFB_1 is the most potent carcinogen known (Group 1 carcinogen) and the most occurring mycotoxin reported by the Rapid Alert System for Food and Feed [4]. AFs are chemically stable compounds, and currently their post-harvest reduction is performed only by physical methods, i.e., by sorting and adsorption. Thus far, effective AFs degradation can be achieved only by means of strong oxidants from physical (plasma, photolysis, photocatalysis), chemical (ammoniation), or biological (oxidoreductase enzymes) origin [5,6].

Enzymes represent an effective yet mild and environmentally friendly method to reduce AFs. So far, AFs enzymatic degradation has been achieved by using oxidoreductases, such as laccases, peroxidases, or so-called "aflatoxin oxidases" [7,8]. In particular, laccases (LCs, benzenediol: oxygen oxidoreductase, EC 1.10.3.2) are copper containing enzymes, able to oxidize simple and substituted phenolic compounds, thiols, anilines, amines, and complex aromatic compounds to the corresponding quinones, concurrently to the four-electron reduction in oxygen to water [9]. The catalytic activity of LCs can be further broadened to compounds which cannot be oxidized due to their high redox potential or steric hindrance thanks to the use of redox mediators. The use of these compounds allows for fine-tuning of the oxidation process and degrade a wide range of chemically unrelated compounds, such as mycotoxins [10]. Among redox mediators, the use of natural antioxidant compounds, such as phenols, has attracted attention because they are regarded as safe and can be used to improve existing industrial processes or develop new ones for the production of high value products [11]. Although the enzymatic degradation has been proven to be an effective method for mycotoxins reduction in feed, its application in food still has to be investigated. In Europe, Regulation 786/2015 defines the acceptability criteria for detoxification processes applied to products intended for animal feed [12].

A detoxification process implies that the toxin is converted to a less toxic, possibly safe, compound. Oxidases convert AFB_1 into hydroxylated metabolite AFQ_1, or to the 8,9-epoxy-AFB_1, which spontaneously converts to 8,9-dihydroAFB_1. Other hypothesized products derive from hydrolysis of the lactone ring followed by its opening (i.e., AFD_1), from addition of water to the double bond of the terminal furan (AFB_{2a}), or from demethylation (AFP_1) [13]. These compounds have been found in vivo as a result of cytochrome detoxification in the liver [14].

Other than safety and efficacy, another mandatory requisite is that the method must not adversely affect the characteristics and the nature of the feed. Although food detoxification is not authorized yet, similar, if not more stringent, criteria will be likely set for food detoxification procedures in the near future.

Corn is one of the main staple food commodities worldwide and performs a central role in global agro-food systems. Contamination of corn grain with AFs is a concerning issue, especially in developing countries, where the majority of the product is self-produced by smallholder farmers in rural subsistence farming communities [15]. Despite being an important component of the human diet, corn is one of the main ingredients of livestock feed, it has multiple industrial uses, and its by-products find application in the energetic supply chain [16–18].

The application of an enzymatic degradation step within the common corn processing should encompass the addition of a buffered solution to easily convey the enzyme and natural redox mediators. Water addition is already included in both dry and wet milling processes.

Dry milling is the main industrial process used in the corn supply chain to separate the pericarp, the endosperm, and the germ; and obtain hominy grits, corn flours and feed meals [19,20]. It may encompass the tempering step, in which water is added to faster separate corn tissues and obtain fractions with low fat content, suitable for the manufacture of extruded products. In wet milling process, the kernels are steeped in SO_2 and lactic acid solution for 24–48 h to facilitate the separation of kernel's components [21]. In a complex matrix, such as corn flour, the addition of exogenous antioxidants could be investigated to support mediator reconversion and reduce the oxidative damage induced by the laccase mediator systems (LMS). Vitamin C (L-ascorbic acid, ascorbate, and ASC) is the most abundant water-soluble compound widely used as antioxidant in food and feed products. Its oxidation product, dehydroascorbic acid (DHA), in the apoplast, is readily taken up by the plasma membrane and reduced to ascorbate in the cytosol [22]. In this regard, either the direct or indirect (by reduced DHA) addition of vitamin C could be beneficial in supporting the enzymatic AFB_1 reduction.

Therefore, in this work, an enzymatic treatment for AFB_1 reduction was investigated in vitro using different LMSs, including acetosyringone (AS), a naturally occurring phenol, ASC, and DHA; in vitro degradation products were also identified. Then, AFB_1 reduction was assayed in corn to assess enzyme performance in the real matrix. Additionally, to monitor the oxidative status, the effect of the different treatments in terms of protein content, lipid peroxidation, and H_2O_2 was also assessed.

2. Results

2.1. *Aflatoxin B_1 Degradation in Buffer Solution Using Different LMSs*

In a previous work, the efficacy of different LMSs for AFB_1 was screened in a 72 h-in vitro assay. The maximum degradation of 1 μg/mL of toxin was 73%, obtained using AS as redox mediator [10]. This LMS was selected for further investigations to improve AFB_1 degradation.

Therefore, in this study, AFB_1 degradation (0.1 μg/mL) was evaluated over time using different LMS, namely Ery4 with AS, also in combination with ASC or DHA at 1 and 10 mM. Degradation, expressed as percentage with respect to the control not containing LC, is shown in Table 1.

Table 1. Time course in vitro degradation of aflatoxin B_1 (0.1 μg/mL) using Ery4 laccase (5 U/mL), acetosyringone (AS) in combination with dehydroascorbic acid (DHA) 1 or 10 mM.

Time (h)	Ery4 + AS	Ery4 + AS + DHA 1 mM	Ery4 + AS + DHA 10 mM
1	100	84.1 ± 3.5	6.1 ± 1.0
2	100	90.3 ± 0.8	8.6 ± 6.5
3	100	93.2 ± 2.7	8.4 ± 4.2
6	100	95.7 ± 1.9	11.8 ± 0.3
24	100	100	16.6 ± 1.6
48	100	100	20.3 ± 1.8

AFB_1 was completely removed from the buffer by Ery4 + AS even after only 1 h. The addition of ASC and DHA was deleterious, especially at higher concentrations. No degradation was observed using ASC. When used at 1 mM, DHA slowed the enzymatic degradation, and AFB_1 was completely removed only after 24 h. DHA 10 mM inhibited AFB_1 degradation, which reached only 20.3 ± 1.8% after 48 h.

2.2. *In Vitro Study of Aflatoxin B_1 Degradation Products*

To further study the ability of Ery 4 laccase to degrade AFB_1 in the presence of the mediator AS, an UPLC-HRMS analysis was carried out. For this purpose, full-scan/variable data-independent acquisitions in positive ion mode of control samples containing Ery4 5 U/mL and AS 10 mM in sodium acetate buffer 1 mM (pH5) (C_Ery4_AS) and treated samples with AFB_1 (1 μg/mL) incubated with Ery4 laccase (5 U/mL), and AS 10 mM in sodium acetate buffer 1 mM, pH 5, for 24 h (AF_Ery4_AS) were acquired. The comparison between the control and the AFB_1-treated sample confirmed a decrease of 55% of AFB_1 content and the formation of additional peaks after enzymatic treatment, which could be attributed to oxidation products of AFB_1. Proposed reaction products, chemical structure and formulas are presented in Figure 1.

A measured mass of 347.0761, which was attributable to a molecular formula $C_{17}H_{14}O_8$ corresponding to the ion $[M+H]^+$, showed one peak eluting at 21.5 min (mass error: 1.6 ppm) and two overlapping peaks at 23.6 min (mass error: 1.3 ppm) and 24.7 min (mass error: 1.3 min) (Figure 2). A difference of 34 mass units compared to aflatoxin B_1 indicated the presence of two hydroxyl groups; therefore, the following molecular formula could be attributed to AFB_1 8,9-dihydrodiol or to AFB_1 dialdehyde. Considering the polarity of these compounds, the peak at 21.5 was assumed to be relative to dihydrodiol or dialdehyde. The $[M+H]^+$ molecular ion at 331.0812, which was attributable to a molecular formula $C_{17}H_{14}O_7$, showed one main peak at 22.0 min (mass error: 1.2 ppm) and could be related to AFB_{2a} or product 1 (P1) (Figure 1). Finally, the $[M+H]^+$ molecular ion at 329.0656, which was attributable to a molecular formula of $C_{17}H_{13}O_7$, corresponded to two main peaks, eluting at 22.8 min (mass error: 2.2 ppm) and 23.6 min (mass error: 1.9 ppm); one less abundant peak eluted at 24.8 min (mass error: 3.2 ppm). These peaks could be related to AFQ_1, epi AFQ_1, AFB_1-8,9-epoxyde, or AFM_1.

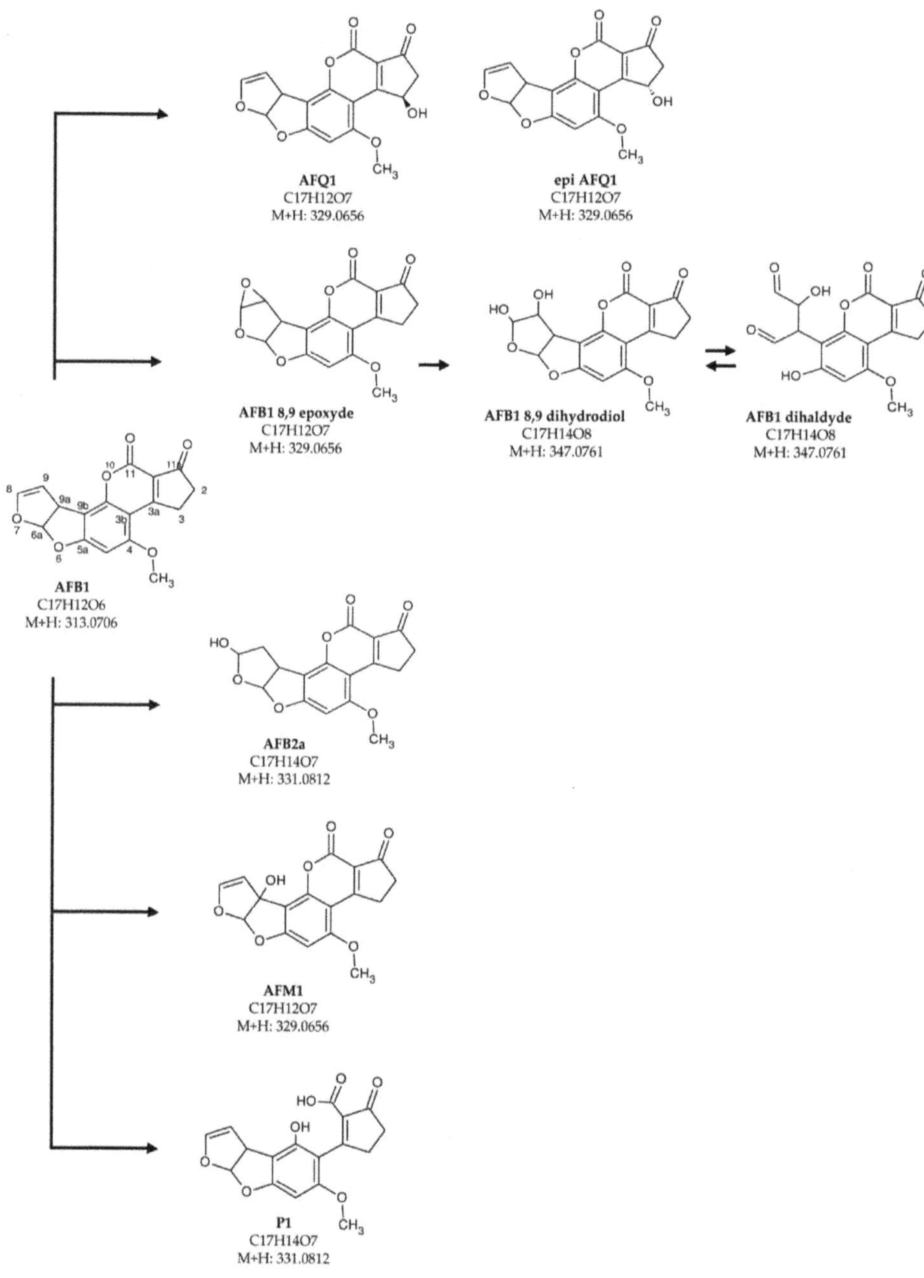

Figure 1. Proposed AFB_1 degradation products.

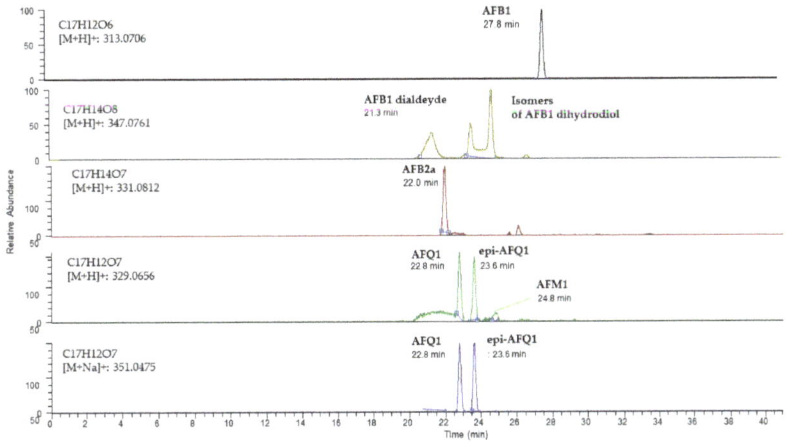

Figure 2. UHPLC-HRMS chromatogram of treated sample with AFB_1 (1 µg/mL) incubated with Ery4 laccase (5 U/mL) and AS 10 mM in sodium acetate buffer 1 mM, pH5, for 24 h (AF_Ery4_AS). Peaks attributable to AFB1 and LMS oxidation products (AFB2a, AFQ_1, epi AFQ_1, AFM_1, AFB_1 dialdehyde and isomers of AFB_1 dihydrodiol are shown. Resolution: 70,000 full width at half maximum; extraction window tolerance 5 ppm.

Identity confirmation of the putative product of the enzymatic reaction was performed by matching the detected fragments with MS2 spectra reported in the literature (if available), as shown in Table 2. In the case of precursor at 329.0656, fragments obtained in AF_Ery4_AS sample were reported in Figure 3. MS/MS spectra of the first two peaks (22.8 min and 23.6 min) showed some characteristic fragments of AFQ_1, such as the peak of m/z 311.0547, originated by the loss of water (neutral loss of 18 a.m.u.), and fragments of m/z 283.0606, 206.0673, and 141.0180.

Table 2. Precursor ion, exact mass, retention time, and fragments of proposed AFB_1 degradation products.

Proposed Product	Molecular Formula	Mass Exact $[M+H]^+/[M+Na]^+$	Error (ppm)	Retention Time (min)	Fragments
AFB_1-dialdehyde/ AFB_1-dihydrodiol	$C_{17}H_{14}O_8$	347.0761	1.6	21.3	No data
AFB_{2a}/P1	$C_{17}H_{14}O_7$	331.0812	1.2	22.0	303.0861, 299.0550, 284.0316, 267.0288, 239.0338
AFQ_1	$C_{17}H_{12}O_7$	329.0656/351.0475	2.2	22.8	311.0343, 283.0601, 259.0601, 247.0602
Unidentified peak 1	$C_{17}H_{14}O_8$	347.0761	1.3	23.6	No data
epi-AFQ_1	$C_{17}H_{12}O_7$	329.0656/351.0475	1.9	23.6	311.0343, 283.0601, 259.0601, 247.0603
Unidentified peak 2	$C_{17}H_{14}O_8$	347.0761	1.3	24.7	No data
AFM_1	$C_{17}H_{12}O_7$	329.0656/351.0475	3.2	24.8	329.0656, 301.07, 273.05
Unidentified peak3	$C_{17}H_{14}O_8$	347.0761	2.3	26.7	No data
AFB_1	$C_{17}H_{12}O_6$	313.0706	1.0	27.8	285.0575, 270.0522, 243.0652, 201.0912

Peak eluting at 24.81 min presented different relative abundances of fragments 329.0652 and 301.0706. In addition, the fragment ion at 273.0757 $[M - 74 + H]^+$ was shown. These fragments are characteristic of AFM_1 [9,23].

In the case of the precursor at 331.0812, fragments at m/z 303.0861 $[M\text{-}CO + H]^+$, 284.0316, 267.0288, and 239.0338 were shown. No fragments were detected for the precursor at 347.0812.

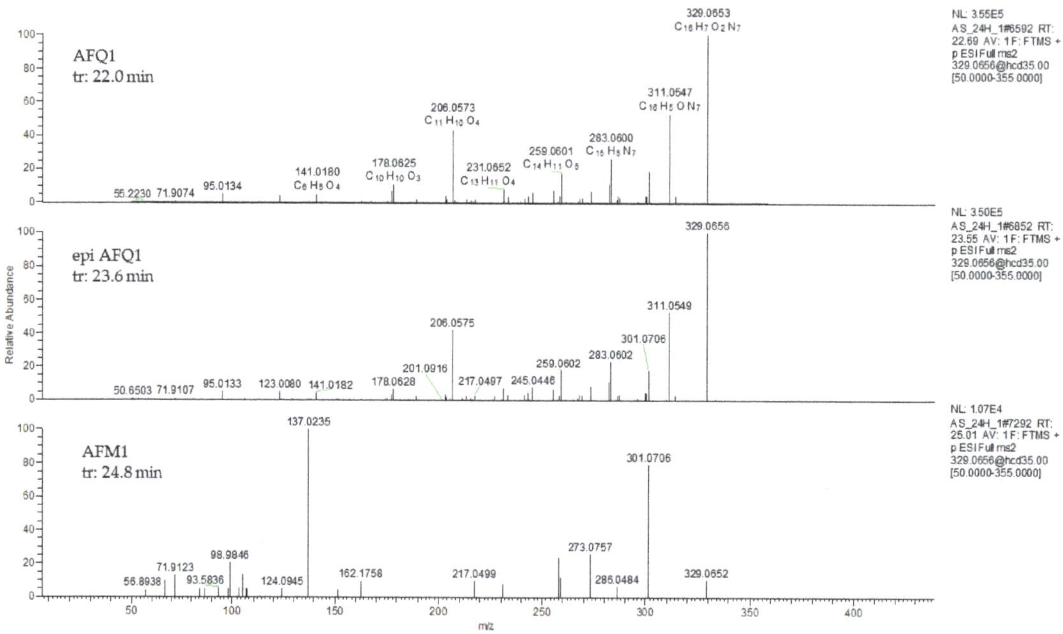

Figure 3. Parallel reaction monitoring (PRM) spectra (collision energy 35 eV) of 329.0656 in treated sample with AFB1 (1 μg/mL) incubated with Ery4 laccase (5 U/mL) and AS 10 mM in sodium acetate buffer 1 mM, pH 5, for 24 h (AF_Ery4_AS).

A rough estimation on the basis of peak area ratios indicated that among the identified products, the most prevalent one was AFQ_1 (41.2%), followed by AFB_{2a}/P1 (29.6%), AFB_1-dihydrodiol/AFB_1dialdehyde (14.8%), and AFM_1 (3.7%). AFB_{2a} may also be formed spontaneously in acidic conditions, in agreement with other literature data [13].

2.3. *Aflatoxin B_1 Degradation in Corn*

Following the results obtained in vitro, only three LMSs (Ery4, AS and DHA) were tested in artificially contaminated corn flour (50 μg/kg AFB_1). After the reaction, samples were centrifuged, and both the supernatant and pellets were analyzed. No AFB_1 was detected in the supernatant, while appreciable degradation could be observed in the pellets (Figure 4). AFB_1 degradation levels were lower with respect to the **in vitro** trials, although Ery4+AS was confirmed to be the most efficient LMS. While no difference could be observed when DHA 1 mM was added, a clear inhibiting effect was exerted by DHA 10 mM, leading to ineffective degradation.

2.4. *Protein Content*

As shown in Figure 5, the enzymatic treatment did not alter the total protein content, calculated as a sum of water-soluble, ethanol soluble, and insoluble fractions. Conversely, statistically significant differences were shown in samples containing DHA. In particular, a dose dependent reduction was observed irrespectively of the presence of Ery4 and AS, highlighting that protein reduction could be ascribed to DHA addition rather than to LMS.

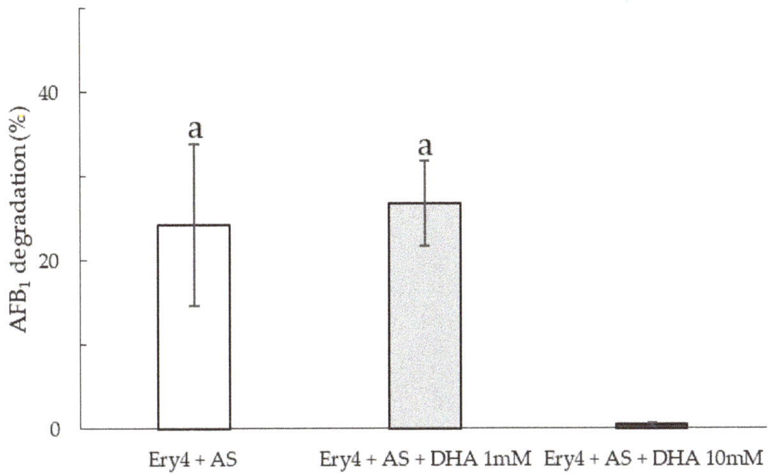

Figure 4. Aflatoxin B_1 degradation (%) in corn samples treated with Ery4 (5 U/mL) acetosyringone (AS, 10 mM) 10 mM and dehydroascorbic acid (DHA) at 1 and 10 mM. Different lowercase letters above columns indicate significant differences between treatments ($p < 0.05$).

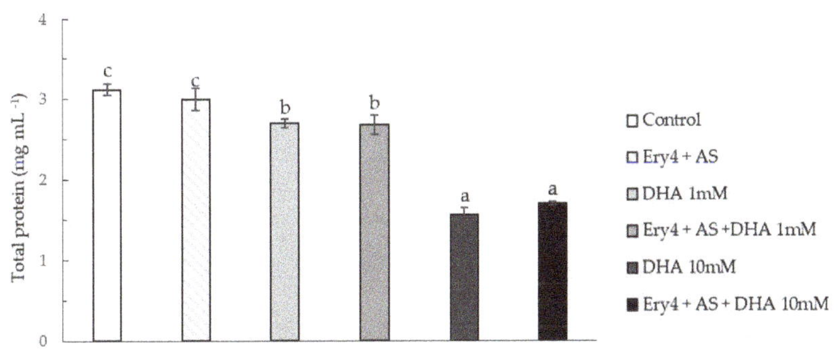

Figure 5. Total protein content (mg mL^{-1}) in untreated corn samples (Control) and samples treated with Ery4 (5 U/mL), acetosyringone (AS, 10 mM), and dehydroascorbic acid (DHA) at 1 and 10 mM. Different lowercase letters above columns indicate significant differences between treatments ($p < 0.05$).

2.5. Lipid Peroxidation and H_2O_2 Content

The oxidative status of both supernatant and pellets was analyzed in terms of H_2O_2 content and MDA levels. The enzymatic treatment had a detrimental effect on H_2O_2 content, both in the pellet (Figure 6A) and in the supernatant (Figure 6B; a synergistic oxidative effect was observed in samples treated with DHA 1 mM, as H_2O_2 levels further increased up to 95,65 ± 0.79 mmol/mL. Conversely, lower H_2O_2 values were registered in samples containing DHA 10 mM (66.5 ± 0.20 mmol/mL).

As reported for H_2O_2, higher MDA content was shown in the supernatants rather than in the pellets. In this latter case, only samples containing DHA 10 mM showed statistically significant increased level. In the supernatants, the oxidative effect of Ery4 + AS enzymatic treatment was more pronounced, and the synergic effect of DHA could be observed only at 10 mM concentration.

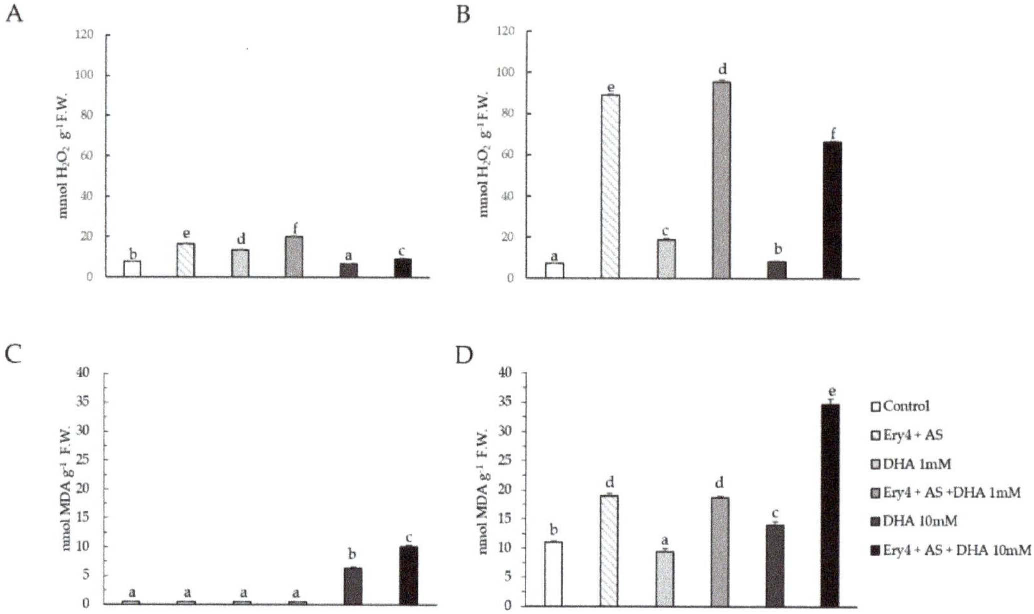

Figure 6. Hydrogen peroxide (Panel (**A**), pellet; Panel (**B**), supernatant) and lipid peroxidation (Panel (**C**), pellet; Panel (**D**), supernatant) content of corn samples using Ery4 (5 U/mL) acetosyringone (AS) 10 mM, and dehydroascorbic acid (DHA) at 1 and 10 mM. Data were expressed as mmol or nmol per fresh weight (F.W.). Different lowercase letters above columns indicate significant differences between treatments ($p < 0.05$).

3. Discussion

Mycotoxins degradation via LMS has been explored with several mediators of natural and synthetic origin [10]. Natural phenols, such as AS, were applied as promising mediators for bioremediation, with potential application in the food industry [24].

AS is a syringic acid derivative found as phenolic humic constituents in natural organic matter [25]. AS, together with other structurally related compounds, have been reported to be efficient mediators for the reduction in organic pollutants, dyes, and mycotoxins [10,26–28].

AS has a redox potential of 0.580 V, which is not among the highest potential reported for LC mediators. Nonetheless, the mediator efficacy does not only depend upon the redox potential but also on the rate of oxidation by LC, stability of the oxidized form of the mediator, its capacity of being recycled, and not to inhibit LC active site [29]. AS's good mediator activity is due to the presence of 2,6-dimethoxy electron-donating groups that give stable phenoxy radicals with a relative long half-life and low free radical activity [30,31].

AS oxidation was reported to proceed via electron transfer and hydrogen atom abstraction mechanism to give a phenoxy radical. This radical intermediate is also stabilized by the acetyl group in orto position, where a further electron delocalization takes place. Additionally, AS oxidation intermediates can still be oxidized by LC as long as it has a phenolic group that can be oxidized [30]. Due to the radical nature of the oxidation mechanism, the addition of a natural antioxidant, ASC, was evaluated for AS reconversion. Moreover, due to the existing reconversion route of ASC from DHA in plasma membrane, DHA supplementation was also assayed.

ASC is a pivotal antioxidant compound and a key element for the metabolism of almost all living organisms. It is a dibasic acid with an enediol group on C2 and C3 of a heterocyclic lactone ring, and at physiological pH, the hydroxyl group at C3 is deprotonated, giving a monovalent anion, ASC [32,33].

The ASC is the only reductant present at a significant level in the apoplast, with a redox potential ranging from +0.40 to +0.50 V [34]. When both electrons of the enediol group of ASC are donated, ASC can be oxidized in this compartment to DHA by ASC oxidases [22,35]. Conversely, when in excess, DHA can be transported via the cell membrane through a carrier mediated uptake and reduced again to ASC. This is part of the cellular redox gradient across the plasma membrane, connecting intra- and extra-cellular environments. The redox environment of the cell is determined by the global poise of its oxidation/reduction systems and may contribute to regulating the effectiveness of the LMS. Indeed, there is a complex link between redox state and simplistic and apoplastic metabolism [36], which is also determined by ROS level, produced at either physiological or toxic levels [35].

The addition of ASC completely inhibited AFB_1 degradation. Similarly, DHA negatively impacted AFB_1 degradation, proving that DHA does not participate in LMS and possibly inhibits LC at high concentration.

ASC was reported to non-competitively inhibit LC from *Botritis cynerea* [37]. Accordingly, in our study, ASC reduced the rate of AFB_1 degradation, possibly by inhibiting LC or scavenging AS reactive radicals before toxin degradation. To our knowledge, no report is available on how DHA affects LC activity. DHA can undergo further irreversible degradation, such as hydrolyzation to 2,3-diketo-L-gulonate, or oxidation to a range of products, such as L-threonic acid, oxalic acid, and their esters; therefore, it may contribute to radical quencing [38].

In the present study, we wanted also to investigate the effectiveness of the detoxification process of AFB_1 by LMS under the optimal degradation condition. To this purpose, a UHPLC-HRMS analysis was carried out to investigate the degradation products by LMS in presence of AS 10 mM after 24 h incubation and 1 μg/mL in **in vitro** samples. To date, neither the mechanism of the laccase-catalyzed degradation of AFB_1 nor the degradation products have been fully disclosed; however, a review on the application of both bacterial and fungal laccase enzyme in AFB_1 degradation was reported by Okawara and colleagues [39]. LCs act on AFB_1 in two ways; on the terminal furan ring of AFB_1, leading to the formation of AFB_1-8,9 epoxide, which is further converted to AFB_1-8,9 dihydrodiol or may directly open the lactone ring by introducing hydroxyl groups at the carbon 10 and 11 positions in AFB_1 (product P1 Figure 1).

Trametes versicolor laccase, Lac2 from *Pleurotus pulmonarius* and the Ery4 from *P. eryngii* were demonstrated to degrade AFB_1 via the mediation of natural phenolic compounds such as AS, syringaldehyde, ferulic acid, etc., or synthetic compounds; however, the degradation products have not been reported. The oxidation of AFB_1 in AFQ_1 was reported in degradation study on CotA laccase from *Bacillus licheniformis* [7] and on Lac2 produced by *Cerrena unicolor* 6884 [40]. The latter study also reported the presence of AFQ_1 epimer (epi AFQ_1). In these studies, the presence of AFQ_1 and epi AFQ_1 products was justified assuming the action of LMS on the lactone ring of AFB_1, possibly by hydrogen atom transfer followed by addition of water to C3. In our study, based on the structure, relative polarity and fragment ions, the epi AFQ_1, AFQ_1, and AFM_1 were identified as oxidation products, corresponding to the peaks at 22.8, 23.6, and 24.8 min. The presence of AFQ_1 is in agreement with the several reports of AFB_1 degradation with LCs [40,41], peroxidases [8], or other oxidases [7].

LC and oxidases were also reported to convert AFB_1 into the toxic 8,9-AFB_1 epoxide [13,39]. Nonetheless, this compound has a fast rate, non-enzymatic conversion to AFB_1-diol in water [42], thus it is hardly detectable by UPLC-HRMS. Based on 8,9-AFB_1 epoxide hydrolysis and kinetics of rearrangement of the dihydrodiol, and considering LMS mechanism, retention times, and polarity of detected compounds in UPLC-HRMS, the ion at m/z 347.0812 corresponding to a molecular formula $C_{17}H_{14}O_8$ could be likely addressed as AFB_1-dialdehyde or AFB_1-diol isomers.

In addition, a measured mass of m/z 331.0812, which was attributable to the molecular formula $C_{17}H_{14}O_7$, corresponding to the ion $[M+H]^+$, with a mass accuracy of 1.2 ppm, was detected. Two candidate compounds were in agreement with this formula (AFB_{2a}

and P1). Despite LCs have been reported to degrade AFB_1 into P_1, according to polarity compounds, the measured mass of m/z 331.0812 at retention time 22.0 min could be more likely AFB_{2a}.

All detected degradation products show a higher polarity and a higher excretion rate via urine and faeces, thus, lower toxicity than AFB_1 [43]. Some of the products found lack of the reactive C8–C9 double bond and possess reduced mutagenicity. Nonetheless, they retain the ability to form Shiff bases with primary amines in proteins, leading to adducts responsible for residual cytotoxicity.

The same degradation trend was registered in vitro and in corn samples, though the differences were evened out, likely due to the matrix effect. Indeed, the interaction between the toxin and the active mediator may be hindered by proteins, carbohydrates, and lipids in corn flour, resulting in a lower efficacy. Competition of food components for the enzyme, enzyme adsorption to food components, and higher viscosity may also contribute to reducing the efficacy of LMS in corn flour.

The enzymatic treatment had slightly impacted the oxidative status of the matrix, while more significant effects were observed in the supernatants. The increased MDA content in corn sample pellets treated with higher DHA-concentration indicated the presence of increased lipid peroxidation of the biological membranes. This reflects the fact that the DHA is toxic in cell if is largely accumulated [44] and may activate induced systemic resistance via ROS production and salicylic acid pathway activation [45]. On the other hand, the decreased hydrogen peroxide level observed, at least for the corn pellet samples treated with the highest DHA concentration, indicates that H_2O_2 could oxidize the biological membranes, as supported by the higher lipid peroxidation in these samples.

Overall, these results suggest that the transport mechanism for DHA via the plasma membrane with its reconversion to ASC would appear not to be present, at least for the corn kernel.

H_2O_2 is the most commonly studied ROS due to its stability and capability of penetrate through cellular membranes, and it has been recognized as a subcellular signaling molecule. Plants can well tolerate relatively high H_2O_2 (up to 10^2–2×10^5 μM), and its endogenous concentration was reported to range from nanomoles to several hundred micromoles [46,47]. Thus, H_2O_2 levels found in this study, although significantly higher in the LMS treated samples, were still in the tolerable ranges reported in the literature for plant cells [47].

The decreased total protein content in the DHA-treated samples underlines the presence of an action of DHA on the protein structure and an interference with the dye response. Particularly, corn proteins are rich in prolamins, which are thiol containing proteins. Indeed, a link between reduction DHA and oxidation of thiol group has been found [48]. Consequently, this event could have a negative impact on the protein folding due to the interaction of carbonyl groups of the DHA with amino acid residues. Indeed, DHA irreversibly inhibits some enzymes, such as human type I hexokinase, that shows a smaller number of cysteine residues [49,50].

A wide number of reports of in vitro enzymatic AFs degradation are available in the literature [10,51,52]. Conversely, fewer studies have been conducted on food matrices. Enzymatic degradation has been explored in food or feed for AFs, zearalenone, thricothecenes, and fumonisin [53–56], although they did not focus on the evaluation of the characteristics of the food matrix after the treatment. To our knowledge, this is the first study that evaluates the oxidative status of corn flour after the application of an enzyme-based degradation treatment.

In order to be applied in feed matrices, mycotoxin reduction methods must not alter the characteristics of the matrix. Therefore, the evaluation of the effects exerted by any reduction treatment must be assessed. So far, this is the first time that the effects on protein content and the oxidative status of corn flour after an enzymatic reduction treatment were studied. Indeed, the work performed by Dini and colleagues [57] only focus on aflatoxins enzymatic degradation in corn flour, obtaining similar results (30% of reduction) with the same level of contamination. Aflatoxin degradation was studied in other matrices, such as milk and beer, with promising results [8,10,58–61]. Nonetheless, few studies investigated the effect of the enzymatic treatment on the protein content and quality, antioxidant activity and technological properties though in a liquid matrix, such as milk [60,61].

4. Conclusions

Different LMSs were tested in vitro and in corn flour with the aim of reducing AFB_1 contamination. Complete degradation was achieved in vitro with Ery4 and AS; the addition of ASC completely inhibits the degradation, while DHA decreased AFB_1 degradation in a dose-dependent manner. The same behavior was observed in corn, even though the rate of degradation was reduced of one fourth due to matrix effect. Several degradation products characterized by lower toxicity were found in vitro by UHPLC-HRMS, namely AFQ_1, epi-AFQ_1, AFB_1-diol or AFB_1dialehyde, AFB_{2a}, and AFM_1.

The protein content was not altered by the sole enzymatic treatment, while it was lowered by DHA in a dose dependent manner. Conversely, LMS treatment affected the oxidative status of corn flour. Increased lipid peroxidation and H_2O_2 content were registered in enzyme- treated samples irrespectively of the amount of DHA added.

Even though further studies are needed to reduce matrix effect and assess the technological impact of this reduction methods, the results of this study are promising and suggest that AFB_1 can be reduced completely in vitro and by 26% in corn flour. Therefore, since only slight oxidation occurred in corn flour, minimum impairment of the nutritional or technological properties could be expected by this treatment, but with significant improvement in its safety.

5. Materials and Methods

5.1. Chemicals, Reagents, and Corn Kernels

Analytical-grade acetonitrile (ACN), methanol (MeOH), and toluene (for HPLC purpose) were purchased from Mallinckrodt Baker (Milan, Italy). Ultrapure water was produced by a Millipore Milli-Q system (Millipore, Bedford, MA, USA). Filter paper and Glass microfiber filters (GF/A) were purchased from Whatman (Maidstone, UK).

Standard of aflatoxin B_1, 2-azino-di-[3-ethylbenzo-thiazolin-sulphonate] (ABTS), syringaldehyde, and acetosyringone were obtained from Sigma Aldrich (Milan, Italy). Immunoaffinity columns AflaTest® Wide Bore were obtained from Vicam L.P. (Watertown, MA, USA).

Organic corn kernels (*Zea Mais* L.) were purchased from Bioseme s.c.a.r.l.

5.2. Preparation of Standards

Standard solution of AFB_1 was prepared by dissolving the solid commercial toxin in toluene/acetonitrile (9:1, v/v) to a concentration of 10 µg/mL. The exact concentration of AFB_1 was determined according to AOAC Official Method 971.22 [62]. Aliquots of the solution were transferred to 4 mL amber silanized glass vials and evaporated to dryness under a stream of nitrogen at 50 °C. The residue was dissolved with water/methanol (60:40, v/v) to obtain final concentrations in a range of 0.5 to 50 ng/mL of aflatoxin B_1. Standard solutions were stored at −20 °C and warmed to room temperature before use.

5.3. Laccase Production and Purification

The recombinant Ery4 laccase was produced from *Saccharomyces cerevisiae* ITEM 17,289 of the Agri-Food Microbial Fungi Culture Collection of the Institute of Sciences of Food (http://www.ispa.cnr.it/Collection, accessed on 25 October 2022). Laccase purification was performed by concentration/ultrafiltration of the cultured media with Tris HCl 50 mM, pH 8, and anion exchange chromatography, as reported in Loi et al. [61].

5.4. Laccase Activity Assay

The enzymatic activity was assessed by the ABTS colorimetric assay using a spectrophotometer (Ultraspec 3100pro, Amersham Pharmacia Biotech Italia, Cologno Monzese,

Italy). [7]. The reaction was performed in 100 mM sodium acetate pH 4.5, 5 mM ABTS and an appropriate amount of enzyme solution in a final volume of 1 mL. The oxidation of ABTS was determined after 10 min at 420 nm (ε420 = 36,000 $M^{-1}cm^{-1}$). One unit was defined as the amount of enzyme which oxidized 1 μmol of substrate per min.

5.5. *Aflatoxin B_1 Degradation In Vitro*

AFB_1 degradation (0.1 μg/mL) was assessed in sodium acetate buffer (1 mM, pH 5) using 5 U/mL of Ery4 laccase and AS 10 mM. ASC and DHA were also tested at two concentrations (1 or 10 mM). Aliquots were incubated at 25 °C and withdrawn after 1 h, 2 h, 3 h,6 h, 24 h, and 48 h, respectively, then immediately added with methanol (1:1 *v*/*v*) and stored at −20 °C until analysis.

5.6. *In Vitro Study of Aflatoxin B_1 Degradation Products*

In order to analyze AFB_1 degradation products, a degradation assay was performed as described in Section 5.4, but with higher amount of toxin (1 μg/mL). Controls and samples containing Ery4 were analyzed after 24 h of static incubation at 25 °C.

5.7. *UHPLC-HRMS Analysis*

The UHPLC-HMRS analysis was performed on a Q-Exactive Plus mass spectrometer equipped with a heated electrospray ion source (HESI II) coupled to an Ultimate 3000 UHPLC system (all from Thermo Fisher Scientific, San Jose, CA, USA).

The LC column was a Gemini C18 column (150 mm × 2 mm, 5-μm particles) (Phenomenex, Torrance, CA, USA) preceded by a Gemini C18 guard column (4 mm × 2 mm). The mass spectrometer operated in full scan mode combined with 5 MS2 events (all related instrumental parameters can be found in Ciasca et al. (2020) [63]. In addition, putative compound was identified by target MS/MS analysis (parallel reaction monitoring (PRM) mode). Settings for PRM data acquisition were as follows: resolution, 70,000 fwhm; microscans, 1; AGC target, 5 × 10^5; maximum injection time, 200 ms; isolation window, 0.5 m/z; nor-malized collision energy (NCE), 35 eV; spectrum data type, centroid. The inclusion list contained the monoisotopic masses of main significant features. The system was controlled by the Xcalibur (version 3.1), Chromeleon MS Link 6.8, and Q-Exactive Tune 2.8 software package.

5.8. *Aflatoxin B_1 Degradation in Corn*

Corn kernels were finely ground (≤500 μm of diameter) by a Model Retsch ZM 200 laboratory mill (Retsch, Haan, Germany) and spiked with 50 μg/kg of AFB_1. The sample was left all night to allow solvent evaporation prior to perform the degradation test.

The enzymatic reactions were performed using 2 g of corn flour in 15 mL tubes with 6 mL of sodium acetate buffer containing Ery4 (5 U/mL) and AS 10 mM. The effect of DHA was also evaluated together with Ery4 and AS at two different concentrations, namely 1 and 10 mM. Samples were incubated at 25 °C under shaking 150 rpm for 3 h.

5.9. *Aflatoxin Extraction and Chemical Analyses*

5.9.1. Corn Samples Clean-Up

After incubation, all sample tubes were centrifuged at 15,000 rpm for 10 min, giving a supernatant (buffer) and a pellet (corn flour); then, AFB_1 was quantified. AFB_1 analyses were performed according to the AOAC Official Method 991.31 [64], based on immunoaffinity column clean-up and toxin determination by HPLC/FLD with post-column photochemical derivatization (UVE™, LCTech GmbH, Dorfen, Germany).

Briefly, the pellet plus 0.5 g of NaCl was extracted with 8 mL of methanol/water (70:30, *v*/*v*) by 60 min shaking. After filtration (filter paper, Whatman n. 4), 4 mL was diluted with 8 mL water and filter (glass microfiber filter, Whatman GF/A). The supernatant was filter through glass microfiber filter. A total of 6 mL of pellet extract fraction and 3 mL supernatant extract were purified through Afla Test™ WB immunoaffinity column. The column was washed with 10 mL water, then eluted with 1 mL methanol. Afterwards the extracts were diluted with 1 mL of water.

5.9.2. HPLC Analyses

Analyses were performed on a HPLC apparatus with a full loop injection system; 100 μL of each sample were injected. The fluorometric detector was set at wavelengths of 365 nm (excitation) and 435 nm (emission). The mobile phase consisted of a mixture of water/acetonitrile (70:30, *v*/*v*), and the flow rate was 1.0 mL/min. The temperature of the column was maintained at 40 °C. AFB_1 was quantified by measuring peak areas at the retention time of aflatoxin standard solutions and comparing these areas with the relevant calibration curve. With this mobile phase, the retention time was about 12 min. The limit of quantification (LOQ) was 2 μg/kg for pellet and 1 μg/kg for supernatant based on a signal to noise ratio of 10:1, and the limit of detection (LOD) were 1 μg/kg for pellet and 0.5 μg/kg for supernatant based on a signal to noise ratio of 3:1.

5.10. Lipid Peroxidation and H_2O_2 Content

Lipid peroxidation was measured in terms of malondialdehyde (MDA) concentrations, following the method reported by Villani and colleagues [65]. Absorbance was measured at 532 and 600 nm, and MDA content was calculated and expressed as nmol g^{-1} fresh weight.

The homogenate was filtered through four layers of cheesecloth to remove cellular debris and then centrifuged at 18,000× *g* for 20 min at 4 °C. The H_2O_2 content was measured as reported by Lanubile et al. [66]. A supernatant aliquot of the reaction mixture was read at 436 nm, and its absorbance was compared to the extinction coefficient of an H_2O_2 standard.

5.11. Protein Content

After the enzymatic treatment, samples were added with NaCl 0.4 M and 0.4% (*v*:*v*) of protease inhibitor cocktail (Sigma Aldrich, Milan, Italy) and incubated for additional 20 min. Then, samples were centrifuged at 10,000 rpm for 20 min, and the pellet and supernatant were separated.

The supernatant was dialyzed against H_2O for 3 h to obtain the first water-soluble protein fraction. The pellet was resuspended in a solution containing EtOH 70% and 2-mercaptoethanol 0.01 M and incubated for 20 min. After centrifugation at 10,000 rpm for 20 min, an ethanol soluble fraction was obtained, while the pellet was further extracted using PBS 0.1 M, pH 7.4, SDS 2.5%, and NaCl 0.01 M to obtain the alcohol-insoluble protein fraction. The three protein fractions were quantified using Bradford method [67].

5.12. Statistical Analyses

Data are the means ± standard deviation of at least three independent biological replicates. One-factor analysis of variance (ANOVA), followed by Tukey's HSD test, was performed on means. Differences between samples and relative control were considered significant for a $p < 0.05$.

Author Contributions: Conceptualization, M.L., G.M. and C.P.; investigation, M.L., S.D.L., B.C. and M.H.; writing—original draft preparation, M.L., G.M., C.P., B.C. and M.H; writing—review and editing, M.L., G.M., C.P., B.C. and M.H.; visualization, M.L., S.D.L. and B.C.; supervision, M.H., G.M. and C.P. All authors have read and agreed to the published version of the manuscript.

Funding: This research received no external funding.

Informed Consent Statement: Not applicable.

Data Availability Statement: Not applicable.

Conflicts of Interest: The authors declare no conflict of interest.

References

1. Jallow, A.; Xie, H.; Tang, X.; Qi, Z.; Li, P. Worldwide aflatoxin contamination of agricultural products and foods: From occurrence to control. *Compr. Rev. Food Sci. Food Saf.* **2021**, *20*, 2332–2381. [PubMed]
2. Aflatoxins. *IARC Monographs*; International Agency for the Research on Cancer: Lyon, France, 2012; Volume 100F.
3. Wang, Y.; Liu, F.; Zhou, X.; Liu, M.; Zang, H.; Liu, X.; Shan, A.; Feng, X. Alleviation of Oral Exposure to Aflatoxin B_1-Induced Renal Dysfunction, Oxidative Stress, and Cell Apoptosis in Mice Kidney by Curcumin. *Antioxidants* **2022**, *11*, 1082. [CrossRef] [PubMed]
4. RASFF Annual Report 2021. Available online: https://food.ec.europa.eu/system/files/2022-07/acn_annual-report_2021-final.pdf (accessed on 2 December 2022).
5. Loi, M.; Logrieco, A.F.; Pusztahelyi, T.; Hornok, L.; Pócsi, I. Advanced Mycotoxin Control and Decontamination Techniques In View Of An Increased Aflatoxin Risk In Europe Due To Climate Change. *Front. Microbiol.* **2023**, *13*, 5258. [CrossRef] [PubMed]
6. Guo, Y.; Zhao, L.; Ma, Q.; Ji, C. Novel strategies for degradation of aflatoxins in food and feed: A review. *Food Res. Int.* **2021**, *140*, 109878. [CrossRef] [PubMed]
7. Guo, Y.; Qin, X.; Tang, Y.; Ma, Q.; Zhang, J.; Zhao, L. CotA laccase, a novel aflatoxin oxidase from *Bacillus licheniformis*, transforms aflatoxin B_1 to aflatoxin Q_1 and epi-aflatoxin Q_1. *Food Chem.* **2020**, *325*, 126877. [CrossRef]
8. Loi, M.; Renaud, J.B.; Rosini, E.; Pollegioni, L.; Vignali, E.; Haidukowski, M.; Sumarah, M.W.; Logrieco, A.F.; Mule, G. Enzymatic transformation of aflatoxin B_1 by Rh_DypB peroxidase and characterization of the reaction products. *Chemosphere* **2020**, *250*, 126296. [CrossRef] [PubMed]
9. Loi, M.; Glazunova, O.; Fedorova, T.; Logrieco, A.F.; Mulè, G. Fungal Laccases: The Forefront of Enzymes for Sustainability. *J. Fungi* **2021**, *7*, 1048. [CrossRef]
10. Loi, M.; Fanelli, F.; Cimmarusti, M.T.; Mirabelli, V.; Haidukowski, M.; Logrieco, A.F.; Caliandro, R.; Mule, G. In vitro single and combined mycotoxins degradation by Ery4 laccase from Pleurotus eryngii and redox mediators. *Food Control* **2018**, *90*, 401–406. [CrossRef]
11. Moreno, A.D.; Ibarra, D.; Eugenio, M.E.; Tomás-Pejó, E. Laccases as versatile enzymes: From industrial uses to novel applications. *J. Chem. Technol. Biotechnol.* **2020**, *95*, 481–494. [CrossRef]
12. Commission Regulation (EU) 2015/786 of 19 May 2015 Defining Acceptability Criteria for Detoxification Processes Applied to Products Intended for Animal Feed as Provided for in Directive 2002/32/EC of the European Parliament and of the Council. Available online: https://eur-lex.europa.eu/eli/reg/2015/786 (accessed on 14 November 2022).
13. Wang, L.; Huang, Q.; Wu, J.; Wanying, W.; Jiang, J.; Yan, H.; Huang, J.; Sun, Y.; Deng, Y. The metabolism and biotransformation of AFB1: Key enzymes and pathways. *Biochem. Pharmacol.* **2022**, *199*, 115005. [CrossRef]
14. Deng, J.; Zhao, L.; Zhang, N.Y.; Karrow, N.A.; Krumm, C.S.; Qi, D.S.; Sun, L.H. Aflatoxin B1 metabolism: Regulation by phase I and II metabolizing enzymes and chemoprotective agents. *Mutat. Res./Rev. Mutat. Res.* **2018**, *778*, 79–89. [CrossRef] [PubMed]
15. Mahuku, G.; Nzioki, H.S.; Mutegi, C.; Kanampiu, F.; Narrod, C.; Makumbi, D. Pre-harvest management is a critical practice for minimizing aflatoxin contamination of maize. *Food Control* **2019**, *96*, 219–226. [CrossRef] [PubMed]
16. Erenstein, O.; Jaleta, M.; Sonder, K.; Mottaleb, K.; Prasanna, B.M. Global maize production, consumption and trade: Trends and R&D implications. *Food Secur.* **2022**, *14*, 1295–1319. [CrossRef]
17. Mohammadi Shad, Z.; Venkitasamy, C.; Wen, Z. Maize distillers dried grains with solubles: Production, properties, and potential uses. *Cereal Chem.* **2021**, *98*, 999–1019. [CrossRef]
18. Mohanty, S.K.; Swain, M.R. Bioethanol production from maize and wheat: Food, fuel, and future. In *Bioethanol Production from Food Crops*; Ray, R.C., Ramachandran, S., Eds.; Academic Press: London, UK, 2019; pp. 45–59. [CrossRef]
19. Scarpino, V.; Vanara, F.; Sulyok, M.; Krska, R.; Blandino, M. Fate of regulated, masked, emerging mycotoxins and secondary fungal metabolites during different large-scale maize dry-milling processes. *Food Res. Int.* **2021**, *140*, 109861. [CrossRef]
20. Vanara, F.; Scarpino, V.; Blandino, M. Fumonisin distribution in maize dry-milling products and by-products: Impact of two industrial degermination systems. *Toxins* **2018**, *10*, 357. [CrossRef]
21. Deepak, T.S.; Jayadeep, P.A. Prospects of Maize (Corn) Wet Milling By-Products as a Source of Functional Food Ingredients and Nutraceuticals. *Food Technol. Biotechnol.* **2022**, *60*, 109–120. [CrossRef]
22. Parsons, H.T.; Fry, S.C. Oxidation of dehydroascorbic acid and 2,3-diketogulonate under plant apoplastic conditions. *Phytochemistry* **2012**, *75*, 41–49. [CrossRef]
23. Vega, V.A.; Young, M.; Todd, S. Laboratory Information Bulletin: Quantitation of Aflatoxin M_1 in Bovine Milk by Liquid Chromatography with Fluorescence Detection. *J. AOAC Int.* **2016**, *99*, 174–179. [CrossRef]
24. Albuquerque, B.R.; Heleno, S.A.; Oliveira, M.B.P.; Barros, L.; Ferreira, I.C. Phenolic compounds: Current industrial applications, limitations and future challenges. *Food Funct.* **2021**, *12*, 14–29. [CrossRef]
25. Song, Y.; Jiang, J.; Qin, W.; Li, J.; Zhou, Y.; Gao, Y. Enhanced transformation of organic pollutants by mild oxidants in the presence of synthetic or natural redox mediators: A review. *Water Res.* **2021**, *189*, 116667. [CrossRef] [PubMed]
26. Mani, P.; Fidal Kumar, V.T.; Keshavarz, T.; Chandra, T.S.; Kyazze, G. The role of natural laccase redox mediators in simultaneous dye decolorization and power production in microbial fuel cells. *Energies* **2018**, *11*, 3455. [CrossRef]
27. Guo, Y.; Wang, Y.; Liu, Y.; Ma, Q.; Ji, C.; Zhao, L. Detoxification of the mycoestrogen zearalenone by *Bacillus licheniformis* spore CotA laccase and application of immobilized laccase in contaminated corn meal. *LWT* **2022**, *163*, 113548. [CrossRef]

28. Wang, X.; Qin, X.; Hao, Z.; Luo, H.; Yao, B.; Su, X. Degradation of four major mycotoxins by eight manganese peroxidases in presence of a dicarboxylic acid. *Toxins* **2019**, *11*, 566. [CrossRef] [PubMed]
29. Obleser, K.; Kalaus, H.; Seidl, B.; Kozich, M.; Stanetty, C.; Mihovilovic, M.D. An Organic Chemist's Guide to Mediated Laccase Oxidation. *ChemBioChem* **2022**, *23*, e202200411. [CrossRef] [PubMed]
30. Medina, F.; Aguila, S.; Baratto, M.C.; Martorana, A.; Basosi, R.; Alderete, J.B.; Vazquez-Duhalt, R. Prediction model based on decision tree analysis for laccase mediators. *Enzym. Microb.* **2013**, *52*, 68–76. [CrossRef]
31. Gu, Y.; Yuan, L.; Jia, L.; Xue, P.; Yao, H. Recent developments of a co-immobilized laccase–mediator system: A review. *RSC Adv.* **2021**, *11*, 29498–29506. [CrossRef]
32. Paciolla, C.; Fortunato, S.; Dipierro, N.; Paradiso, A.; De Leonardis, S.; Mastropasqua, L.; de Pinto, M.C. Vitamin C in Plants: From Functions to Biofortification. *Antioxidants* **2019**, *8*, 519. [CrossRef]
33. Tripathi, R.P.; Singh, B.; Bisht, S.S.; Pandey, J. L-Ascorbic Acid in Organic Synthesis: An Overview. *Curr. Org. Chem.* **2009**, *13*, 99–122. [CrossRef]
34. Matsui, T.; Kitagawa, Y.; Okumura, M.; Shigeta, Y. Accurate standard hydrogen electrode potential and applications to the redox potentials of vitamin C and NAD/NADH. *J. Phys. Chem. A* **2015**, *119*, 369–376. [CrossRef]
35. Paciolla, C.; Paradiso, A.; de Pinto, M.C. Cellular redox homeostasis as central modulator in plant stress. In *Redox State as a Central Regulator of Plant-Cell Stress Responses*; Gupta, D.K., Palma, J.M., Corpas, F.J., Eds.; Springer: Cham, Switzerland, 2016; pp. 1–23. [CrossRef]
36. Noctor, G.; Lelarge-Trouverie, C.; Mhamdi, A. The metabolomics of oxidative stress. *Phytochemistry* **2015**, *112*, 33–53. [CrossRef] [PubMed]
37. Vignault, A.; Gombau, J.; Jourdes, M.; Moine, V.; Canals, J.M.; Fermaud, M.; Roudet, J.; Zamora, F.; Teissedre, P.L. Oenological tannins to prevent *Botrytis cinerea* damage in grapes and musts: Kinetics and electrophoresis characterization of laccase. *Food Chem.* **2020**, *316*, 126334. [CrossRef] [PubMed]
38. Dewhirst, R.A.; Fry, S.C. The oxidation of dehydroascorbic acid and 2,3-diketogulonate by distinct reactive oxygen species. *Biochem. J.* **2018**, *475*, 3451–3470. [CrossRef] [PubMed]
39. Okwara, P.C.; Afolabi, I.S.; Ahuekwe, E.F. Application of laccase in aflatoxin B_1 degradation: A review. *IOP Conf. Ser. Mater. Eng.* **2021**, *1107*, 012178. [CrossRef]
40. Zhou, Z.; Li, R.; Ng, T.B.; Lai, Y.; Yang, J.; Ye, X. A New Laccase of Lac 2 from the White Rot Fungus *Cerrena unicolor* 6884 and Lac 2-Mediated Degradation of Aflatoxin B_1. *Toxins* **2020**, *12*, 476. [CrossRef]
41. Qin, X.; Xin, Y.; Zou, J.; Su, X.; Wang, X.; Wang, Y.; Zhang, J.; Tu, T.; Yao, B.; Luo, H.; et al. Efficient degradation of aflatoxin B_1 and zearalenone by laccase-like multicopper oxidase from *Streptomyces thermocarboxydus* in the presence of mediators. *Toxins* **2021**, *13*, 754. [CrossRef]
42. Johnson, W.W.; Harris, T.M.; Guengerich, F.P. Kinetics and mechanism of hydrolysis of aflatoxin B_1 exo-8, 9-epoxide and rearrangement of the dihydrodiol. *J. Am. Chem. Soc.* **1996**, *118*, 8213–8220. [CrossRef]
43. Mykkänen, H.; Zhu, H.; Salminen, E.; Juvonen, R.O.; Ling, W.; Ma, J.; Polychronaki, N.; Kemiläinen, H.; Mykkänen, O.; Salminen, S.; et al. Fecal and urinary excretion of aflatoxin B1 metabolites (AFQ1, AFM1 and AFB-N7-guanine) in young Chinese males. *Int. J. Cancer* **2005**, *115*, 879–884. [CrossRef]
44. Chen, L.; Wang, W.; Zhang, J.; Wang, W.; Ni, D.; Jiang, H. Dehydroascorbic acid affects the stability of catechins by forming conjunctions. *Molecules* **2020**, *25*, 4076. [CrossRef]
45. Chavan, S.N.; De Kesel, J.; Desmedt, W.; Degroote, E.; Singh, R.R.; Nguyen, G.T.; Demeestere, K.; De Meyer, T.; Kyndt, T. Dehydroascorbate induces plant resistance in rice against root-knot nematode *Meloidogyne graminicola*. *Mol. Plant Pathol.* **2022**, *23*, 1303–1319. [CrossRef]
46. Ślesak, I.; Libik, M.; Karpinska, B.; Karpinski, S.; Miszalski, Z. The role of hydrogen peroxide in regulation of plant metabolism and cellular signalling in response to environmental stresses. *Acta Biochim. Pol.* **2007**, *54*, 39–50. [CrossRef] [PubMed]
47. Smirnoff, N.; Arnaud, D. Hydrogen peroxide metabolism and functions in plants. *New Phytol.* **2019**, *221*, 1197–1214. [CrossRef] [PubMed]
48. Nardai, G.; Braun, L.; Csala, M.; Csermely, P.; Benedetti, A.; Mandl, J.; Banhegyi, G. Protein disulfide isomerase and protein thiol dependent dehydroascorbate reduction and ascorbate accumulation in the lumen of the endoplasmic reticulum. *J. Biol. Chem.* **2001**, *276*, 8825–8828. [CrossRef] [PubMed]
49. Fiorani, M.; De Sanctis, R.; Scarlatti, F.; Vallorani, L.; De Bellis, R.; Serafini, G.; Bianchi, M.; Stocchi, V. Dehydroascorbic acid irreversibly inhibits hexokinase activity. *Mol. Cell. Biochem.* **2000**, *209*, 145–153. [CrossRef]
50. Cárcamo, J.M.; Pedraza, A.; Bórquez-Ojeda, O.; Zhang, B.; Sanchez, R.; Golde, D.W. Vitamin C is a kinase inhibitor: Dehydroascorbic acid inhibits IkappaBalpha kinase beta. *Mol. Cell Biol.* **2004**, *24*, 6645–6652. [CrossRef] [PubMed]
51. Lyagin, I.; Efremenko, E. Enzymes for detoxification of various mycotoxins: Origins and mechanisms of catalytic action. *Molecules* **2019**, *24*, 2362. [CrossRef]
52. Kumar, V.; Bahuguna, A.; Ramalingam, S.; Dhakal, G.; Shim, J.J.; Kim, M. Recent technological advances in mechanism, toxicity, and food perspectives of enzyme-mediated aflatoxin degradation. *CritiCal Rev. Food Sci. Nutr.* **2022**, *62*, 5395–5412. [CrossRef]
53. Alberts, J.F.; Davids, I.; Moll, W.-D.; Schatzmayr, G.; Burger, H.-M.; Shephard, G.S.; Gelderblom, W.C.A. Enzymatic Detoxification of the Fumonisin Mycotoxins during Dry Milling of Maize. *Food Control* **2021**, *123*, 107726. [CrossRef]

54. Podgórska-Kryszczuk, I.; Solarska, E.; Kordowska-Wiater, M. Reduction of the *Fusarium* Mycotoxins: Deoxynivalenol, Nivalenol and Zearalenone by Selected Non-Conventional Yeast Strains in Wheat Grains and Bread. *Molecules* **2022**, *27*, 1578. [CrossRef]
55. Liu, X.; Zhao, F.; Chitrakar, B.; Wei, G.; Wang, X.; Sang, Y. Three recombinant peroxidases as a degradation agent of aflatoxin M_1 applied in milk and beer. *Food Res. Int.* **2022**, *166*, 112352. [CrossRef]
56. Yang, P.; Xiao, W.; Lu, S.; Jiang, S.; Zheng, Z.; Zhang, D.; Zhan, M.; Jiang, S.; Jiang, S. Recombinant Expression of *Trametes versicolor* Aflatoxin B_1-Degrading Enzyme (TV-AFB1D) in Engineering *Pichia pastoris* GS115 and Application in AFB_1 Degradation in AFB_1-Contaminated Peanuts. *Toxins* **2021**, *13*, 349. [CrossRef] [PubMed]
57. Dini, I.; Alborino, V.; Lanzuise, S.; Lombardi, N.; Marra, R.; Balestrieri, A.; Ritieni, A.; Woo, S.L.; Vinale, F. Trichoderma Enzymes for Degradation of Aflatoxin B1 and Ochratoxin A. *Molecules* **2022**, *27*, 3959. [CrossRef] [PubMed]
58. de Oliveira Garcia, S.; Sibaja, K.V.M.; Nogueira, W.V.; Feltrin, A.C.P.; Pinheiro, D.F.A.; Cerqueira, M.B.R.; Furlong, E.; Garda-Buffon, J. Peroxidase as a simultaneous degradation agent of ochratoxin A and zearalenone applied to model solution and beer. *Food Res. Int.* **2020**, *131*, 109039. [CrossRef] [PubMed]
59. Marimón Sibaja, K.V.; de Oliveira Garcia, S.; Feltrin, A.C.P.; Diaz Remedi, R.; Cerqueira, M.B.R.; Badiale-Furlong, E.; Garda-Buffon, J. Aflatoxin biotransformation by commercial peroxidase and its application in contaminated food. *J. Chem. Technol. Biotechnol.* **2019**, *94*, 1187–1194. [CrossRef]
60. Loi, M.; Quintieri, L.; Fanelli, F.; Caputo, L.; Mulè, G. Application of a recombinant laccase-chlorogenic acid system in protein crosslink and antioxidant properties of the curd. *Food Res. Int.* **2018**, *106*, 763–770. [CrossRef]
61. Loi, M.; Quintieri, L.; De Angelis, E.; Monaci, L.; Logrieco, A.F.; Caputo, L.; Mule, G. Yield improvement of the Italian fresh Giuncata cheese by laccase–induced protein crosslink. *Int. Dairy J.* **2020**, *100*, 104555. [CrossRef]
62. *AOAC 971.22-1988*; Association of Official Analytical Chemists. Standards for Aflatoxins. Thin-Layer Chromatographic Method. 2000. Available online: http://www.aoacofficialmethod.org/index.php?main_page=product_info&products_id=625 (accessed on 29 September 2022).
63. Ciasca, B.; Lanubile, A.; Marocco, A.; Pascale, M.; Logrieco, A.F.; Lattanzio, V.M.T. Application of an Integrated and Open Source Workflow for LC-HRMS Plant Metabolomics Studies. Case-Control Study: Metabolic Changes of Maize in Response to *Fusarium verticillioides* Infection. *Front. Plant Sci.* **2020**, *11*, 664. [CrossRef]
64. *AOAC 991.31*; Association of Official Analytical Chemists. Aflatoxins in Corn, Raw Peanuts, and Peanut Butter. Immunoaffinity Column (Aflatest). 2002. Available online: http://www.aoacofficialmethod.org/index.php?main_page=product_info&products_id=1723 (accessed on 29 September 2022).
65. Villani, A.; Tommasi, F.; Paciolla, C. The arbuscular mycorrhizal fungus *Glomus viscosum* improves the tolerance to verticillium wilt in artichoke by modulating the antioxidant defense systems. *Cells* **2021**, *10*, 1944. [CrossRef]
66. Lanubile, A.; Maschietto, V.; De Leonardis, S.; Battilani, P.; Paciolla, C.; Marocco, A. Defense responses to mycotoxin-producing fungi *Fusarium proliferatum*, *F. subglutinans*, and *Aspergillus flavus* in kernels of susceptible and resistant maize genotypes. *Mol. Plant-Microbe Interact.* **2015**, *28*, 546–557. [CrossRef]
67. Bradford, M.M. A Rapid and Sensitive Method for the Quantitaton of Microgram Quantities of Protein Utilizing the Principle of Protein-Dye Binding. *Anal. Biochem.* **1976**, *72*, 248–254. [CrossRef]

toxins

MDPI

Article

Epigallocatechin Gallate and Glutathione Attenuate Aflatoxin B_1-Induced Acute Liver Injury in Ducklings via Mitochondria-Mediated Apoptosis and the Nrf2 Signalling Pathway

Yanan Wang [1], Jiayu Wu [1], Lingfeng Wang [1], Ping Yang [1], Zuhong Liu [2], Shahid Ali Rajput [3], Mubashar Hassan [1] and Desheng Qi [1,*]

1 Department of Animal Nutrition and Feed Science, College of Animal Science and Technology, Huazhong Agricultural University, Wuhan 430070, China
2 Institute of Animal Husbandry and Veterinary Sciences, Wuhan Academy of Agricultural Sciences, Wuhan 430208, China
3 Department of Animal Feed and Production, Faculty of Veterinary and Animal Sciences, Muhammad Nawaz Shareef University of Agriculture, Multan 66000, Pakistan
* Correspondence: qds@mail.hzau.edu.cn

Abstract: Aflatoxin B_1 (AFB_1) exists widely in feed and food with severe hazards, posing a serious threat to human and animal health. Epigallocatechin gallate (EGCG) and glutathione (GSH) have been reported as having anti-oxidative and other functions. The present study aimed to investigate the detoxification effect of EGCG and GSH alone or in combination on AFB_1 exposure in ducklings. Fifty one-day-old male ducklings were randomly assigned into five experimental groups (n = 10): 1. Control (CTR); 2. 0.3 mg/kg BW AFB_1 (AFB_1); 3. 0.3 mg/kg BW AFB_1 + 100 mg/kg BW EGCG (AFB_1 + EGCG); 4. 0.3 mg/kg BW AFB_1 + 30 mg/kg BW GSH (AFB_1 + GSH); 5. 0.3 mg/kg BW AFB_1 + 100 mg/kg BW EGCG + 30 mg/kg BW GSH (AFB_1 + EGCG + GSH). The experiment lasted for seven days. Compared with the CTR group, AFB_1 reduced growth performance, total serum protein and albumin content, increased serum enzyme activity (alanine aminotransferase, aspartate aminotransferase, alkaline phosphatase, and γ-glutamyl transpeptidase), and caused pathological damage to the ducklings' livers. AFB_1 exposure increased malondialdehyde content and decreased superoxide dismutase, total antioxidant capacity, catalase, glutathione peroxidase activities, and glutathione content in the liver. EGCG and GSH alone or in combination mitigated these adverse effects. Meanwhile, EGCG and GSH attenuate apoptosis of hepatocytes, and regulated AFB_1-induced changes in the abundance of genes contained in the Keap1/Nrf2 signalling and apoptotic pathways. Collectively, these results suggest that EGCG and GSH alleviate the hepatocyte injury induced by AFB_1 by inhibiting oxidative stress and attenuating excessive mitochondria-mediated apoptosis.

Keywords: aflatoxin B_1; duckling; epigallocatechin gallate; glutathione; oxidative stress; Keap1/Nrf2 signalling; apoptosis

Key Contribution: EGCG and GSH alleviate AFB_1-induced liver injury in ducklings by increasing antioxidant capacity and antagonising apoptosis.

Citation: Wang, Y.; Wu, J.; Wang, L.; Yang, P.; Liu, Z.; Rajput, S.A.; Hassan, M.; Qi, D. Epigallocatechin Gallate and Glutathione Attenuate Aflatoxin B_1-Induced Acute Liver Injury in Ducklings via Mitochondria-Mediated Apoptosis and the Nrf2 Signalling Pathway. *Toxins* **2022**, *14*, 876. https://doi.org/10.3390/toxins14120876

Received: 3 November 2022
Accepted: 13 December 2022
Published: 15 December 2022

1. Introduction

Mycotoxins are harmful naturally occurring secondary metabolites produced by fungi [1]. Aflatoxins (AFs) are poisonous mycotoxins produced principally by *Aspergillus flavus* and *Aspergillus parasiticus*, of which Aflatoxin B_1 (AFB_1) is the most hepatotoxic [2]. Corn, wheat, and other grains have a high detection rate of AFB_1 [3], which seriously affects the health of poultry and humans throughout the food chain [4]. AFB_1 has been reported to cause diarrhoea, poor feather quality, weight loss, multifocal hepatic necrosis, bile duct hyperplasia, skeletal deformation, and altered muscle alignment in poultry [5,6]. Liver cancer, immunosuppression, and stunted children have all been linked with foods contaminated with AFB_1 [7]. Over 70% of the world's ducks are raised in China [8]. As ducklings are more sensitive to AFB_1 than turkeys, lower doses of AFB_1 can cause them bodily damage [9]. AFB_1 exhibits its toxic action by being metabolised to exo-8, 9-epoxide (AFBO) in the liver [10]. The Keap1 (Kelch-like ECH-associated protein 1)/Nrf2 (nuclear factor erythroid 2-related factor 2) pathway enables cells to adapt to oxidative stress caused by external stimuli, and some plant extracts may activate this pathway [11]. Apoptosis plays an essential role in maintaining stability in the internal environment [12]. AFB_1 can dysregulate the Keap1/Nrf2 pathway and cause excessive apoptosis in hepatocytes [13]. Hence, it will be an effective measure to improve the antioxidant capacity and inhibit excessive apoptosis caused by liver injury.

Epigallocatechin gallate (EGCG) is the main active ingredient in green tea. Other catechins include epicatechin-3-gallate, epigallocatechin and epicatechin, but EGCG is the most abundant [14]. It not only promotes animal growth performance and egg quality, ameliorates body fatty acid metabolism, and regulates intestinal health [15,16], but also contributes to cardioprotection, renoprotection, hepatoprotection, and neuroprotection in humans [17]. More importantly, EGCG is a scavenger of reactive oxygen/nitrogen and has potent antioxidant capacities. Moreover, it reduces the damage to cells caused by oxidative stress by capturing oxygen free radicals, restoring their redox status and mitochondrial function [18,19]. Previous studies have shown that EGCG can attenuate bleomycin-induced pulmonary fibrosis through the Keap1/Nrf2 signalling pathway [20]. Four hundred mg/kg of EGCG in the diet significantly alleviated heat stress in quail [21]. The preventive effect of EGCG against AFB_1-induced liver injury and the mechanisms involved are not clarified. Therefore, it is necessary to proceed with this work.

Glutathione (GSH) is a tripeptide comprising cysteine, glutamic acid, and glycine. It exists in two forms in animals, oxidised (GSSG) and reduced (GSH), both of which play a significant role in bodily redox status [22]. The leading absorption site of exogenous GSH is the small intestine. Oral GSH in animals and humans can increase the content of GSH in the body [23], improve antioxidant capacity (i.e., protect cells from oxidative damage), protect intestinal mucosa, and enhance the transport and absorption of nutrients [22]. Studies have shown that GSH can increase the resistance of carp to nitric oxide stress and lipopolysaccharide (LPS) stimulation [24]. Adding GSH to the diet alleviated the oxidative damage caused by ochratoxin A and significantly inhibited cell apoptosis in rats [25]. Therefore, GSH may exhibit a positively beneficial effect in mitigating the hazards of AFB_1.

The present study used EGCG and GSH to alleviate the damage caused by AFB_1 in ducklings. In particular, we investigated whether EGCG and GSH could alleviate liver damage through the Keap1/Nrf2 signalling pathway and the inhibition of apoptosis and whether there was a mutual effect between them. The present experiment hoped to prompt the individual and combined use of EGCG and GSH to ameliorate toxin damage in animals.

2. Results

2.1. EGCG and GSH Inhibit AFB_1-Induced Changes in Growth Performance and Liver Index of Ducklings

The effects of AFB_1 and EGCG or GSH on ducklings' growth are shown in Figure 1A,B. In this experiment, ducks were treated with gavage, and each group of ten ducks was individually numbered but housed in a combined pen, so only the mean values of feed intake were calculated. As can be seen from the Figure 1A,B, there was no significant difference in the initial body weight, but after acute attacks, the AFB_1 group had significantly reduced body weight ($p < 0.01$) compared with the CTR group, and the feed intake decreased by 13.1%. However, the EGCG and GSH alone and in combination significantly increased the ducklings' body weights compared with the AFB_1 group, while the feed intake increased by 7.7%, 2.5%, and 9.8%, respectively, showing that the combination of EGCG and GSH was more effective. As Figure 1C shows, AFB_1 significantly increased the relative weight of the ducklings' livers ($p < 0.01$), while EGCG and GSH significantly decreased as compared with the AFB_1 group. The results indicate that EGCG and GSH alleviated the damage caused by AFB_1, but a combination was more effective.

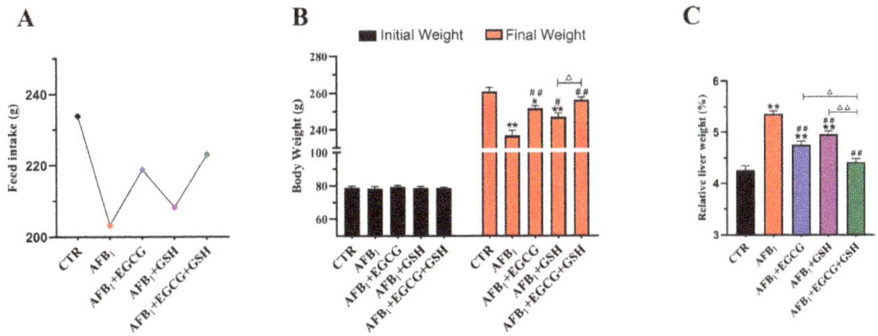

Figure 1. Effect of Epigallocatechin gallate (EGCG) and Glutathione (GSH) on Aflatoxin B_1 (AFB_1)-induced changes in growth performance and liver index of ducklings. (**A**) Average total feed intake per duckling during the experiment. (**B**) Body weight of each duckling at the beginning and end of the experiment. (**C**) Relative weight of the liver. Results are expressed as means ± SEM (n = 10). * $p < 0.05$, ** $p < 0.01$ vs. control (CTR) group; # $p < 0.05$, ## $p < 0.01$ vs. AFB_1 group; Δ $p < 0.05$, ΔΔ $p < 0.01$ significant difference between AFB_1 + EGCG + GSH and AFB_1 + EGCG or AFB_1 + GSH groups.

2.2. EGCG and GSH Protect against AFB_1-Induced Liver Damage in Ducklings

The effects of EGCG and GSH alone or in combination on the serum biochemistry of AFB_1-exposed ducklings are shown in Figure 2. Serum biochemistry was affected adversely by AFB_1 as the enzyme activities of alanine aminotransferase (ALT), aspartate aminotransferase (AST), alkaline phosphatase (ALP), and γ-glutamyl transpeptidase (γ-GT) were elevated ($p < 0.01$, Figure 2A–C,F). The use of EGCG or GSH mitigated these adverse effects. Compared with the AFB_1 group, the enzyme activities of ALT, AST, ALP, and γ-GT decreased by 19.4%, 41.2%, 23.0%, and 35.2%, respectively, in the combined detoxification group. The AFB_1-treated group reduced total serum protein (TP, 37.9%) and albumin (ALB, 49.2%) levels extremely significantly ($p < 0.01$; Figure 2D,E). Both EGCG or GSH increased the levels of TP and ALB compared with the AFB_1 group, with the combined group increasing by 45.0% and 47.1%, respectively. This was still lower than the control group. The results indicate that EGCG and GSH alleviated the negative effects caused by AFB_1 on the serum biochemistry of ducklings, but a combination was more effective.

2.3. EGCG and GSH Mitigate AFB_1-Induced Oxidative Stress in the Livers of Ducklings

To evaluate the damage caused by AFB_1 and the protective effect of EGCG and GSH, we examined the antioxidant capacity of the ducklings' livers. Compared with the CTR group, the AFB_1 group highly significantly elevated malondialdehyde (MDA) content ($p < 0.01$), while EGCG, GSH, and a combination of both decreased MDA content by 27.8%, 25.7%, and 37.7%, respectively (Figure 3A). Meanwhile, the levels of antioxidant enzymes and GSH were also negatively affected, with the enzymatic activities of superoxide dismutase (SOD), glutathione peroxide (GPX), total antioxidant capacity (T-AOC), and catalase (CAT) decreasing by 23.0%, 26.0%, 34.2%, and 38.4% (Figure 3B,D–F), respectively, compared with the CTR group, while the levels of GSH decreased by 40.6% (Figure 3C). Thus, EGCG and GSH effectively prevented their alteration, especially in MDA, T-AOC, and CAT, where there was a significant joint effect ($p < 0.05$). The results indicate that EGCG or GSH alleviated the oxidative damage caused by AFB_1, but a combination was more effective.

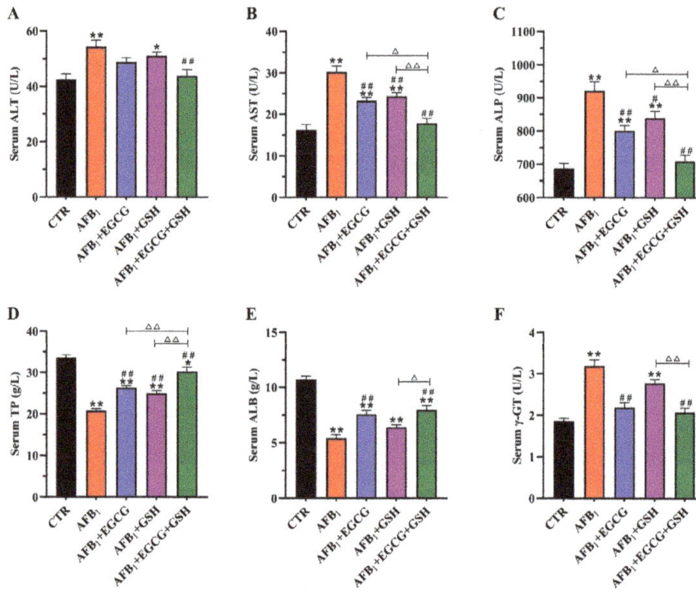

Figure 2. Effect of EGCG and GSH on AFB_1-induced changes in serum biochemical parameters. (**A**) ALT, alanine aminotransferase; (**B**) AST, aspartate aminotransferase; (**C**) ALP, alkaline phosphatase; (**D**) TP, total protein; (**E**) ALB, albumin; (**F**) γ-GT, γ-glutamyl transpeptidase. Results are expressed as means ± SEM ($n = 10$). * $p < 0.05$, ** $p < 0.01$ vs. CTR group; # $p < 0.05$, ## $p < 0.01$ vs. AFB_1 group; Δ $p < 0.05$, ΔΔ $p < 0.01$ significant difference between AFB_1 + EGCG + GSH and AFB_1 + EGCG or AFB_1 + GSH groups.

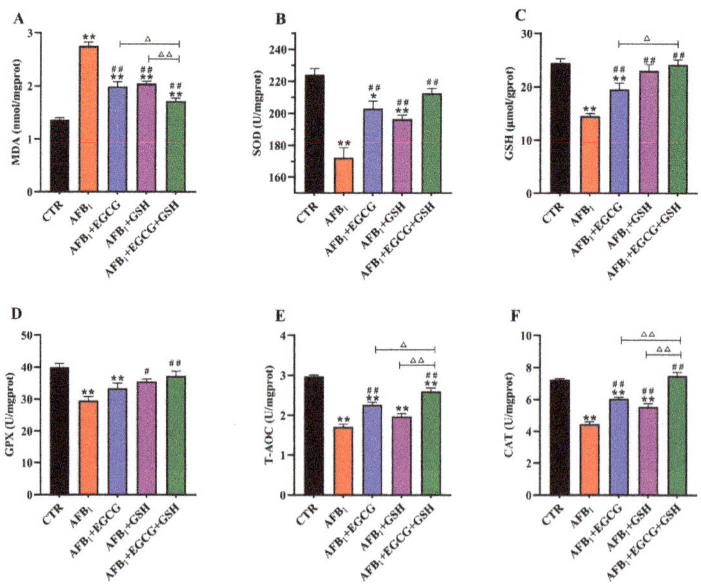

Figure 3. Effect of EGCG and GSH on AFB_1-induced oxidative stress in the livers of ducklings. (**A**) MDA, malondialdehyde; (**B**) SOD, superoxide dismutase; (**C**) GSH, glutathione; (**D**) GPX, glutathione peroxidase; (**E**) T-AOC, total antioxidant capacity; (**F**) CAT, catalase. Results are expressed as means ± SEM ($n = 10$). * $p < 0.05$, ** $p < 0.01$ vs. CTR group; # $p < 0.05$, ## $p < 0.01$ vs. AFB_1 group; Δ $p < 0.05$, ΔΔ $p < 0.01$ significant difference between AFB_1 + EGCG + GSH and AFB_1 + EGCG or AFB_1 + GSH groups.

2.4. EGCG and GSH Prevent AFB_1-Induced Alterations in the Microstructure and Ultrastructure of Duckling Livers

The liver is the target organ of AFB_1 action, so we observed its microscopic and ultrastructural structure. As Figure 4 shows, the liver tissue structure of the CTR group was normal, with an intact hepatocyte structure and no fatty degeneration, necrosis, or inflammatory cell infiltration. However, we observed that large areas of hepatocytes were ill-defined, some cells were swollen and necrotic, and disappeared nuclei or pyknosis was present in the AFB_1 group. Compared with the AFB_1 group, inflammatory cell infiltration and hepatocyte necrosis were reduced in the detoxification group alone or in combination, especially in the combined detoxification group. However, lipid droplets were still present in some hepatocytes.

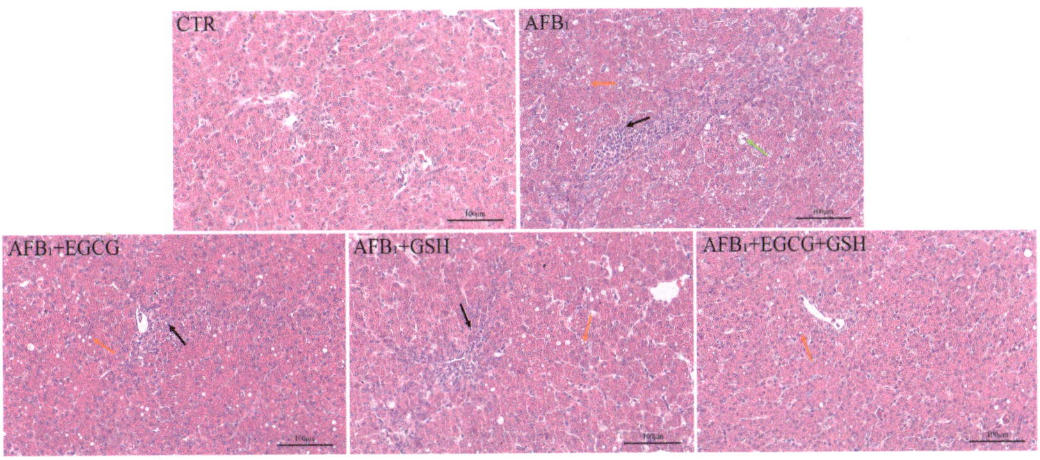

Figure 4. Effect of EGCG and GSH on the microscopic pathological structure of the livers of ducklings exposed to AFB_1. Magnification 200×, scale bars = 100 μm. Green arrows: cell swelling and necrosis, nuclear pyknosis; black arrows: inflammatory cell infiltration in the hepatic parenchyma; red arrows: a small number of lipid droplets can be seen in the cytoplasm.

To examine the internal structure of hepatocytes more closely, we performed transmission electron microscopy scans (Figure 5). In the CTR group, the nuclei and mitochondrial structures were normal, while in the AFB_1 group, the nuclei underwent significant wrinkling, and the mitochondrial structures were heavily abnormal, with swelling and disrupted mitochondrial ridges. Although the mitochondrial structure was also lesioned to varying degrees in the EGCG or GSH groups, it was largely improved relative to the AFB_1 group. The best results were seen in the combined group.

2.5. EGCG and GSH Alleviate the Interference of AFB_1 on the Keap1-Nrf2 Antioxidant Signalling Pathway

As Figure 6 shows, the abundance of related genes in the Nrf2 signalling pathway was significantly downregulated in the AFB_1 group compared with the CTR group ($p < 0.01$). However, the gene expression of Keap1, an Nrf2 repressor, was significantly elevated ($p < 0.01$). These changes were back-regulated to varying degrees in the EGCG and GSH groups and in the combined group. The combined group Nrf2, HO-1 and SOD1 gene expression reached significant levels ($p < 0.05$) compared to the group used alone. The results indicate that EGCG and GSH can alleviate the oxidative damage caused by AFB_1 and that they interact to some degree.

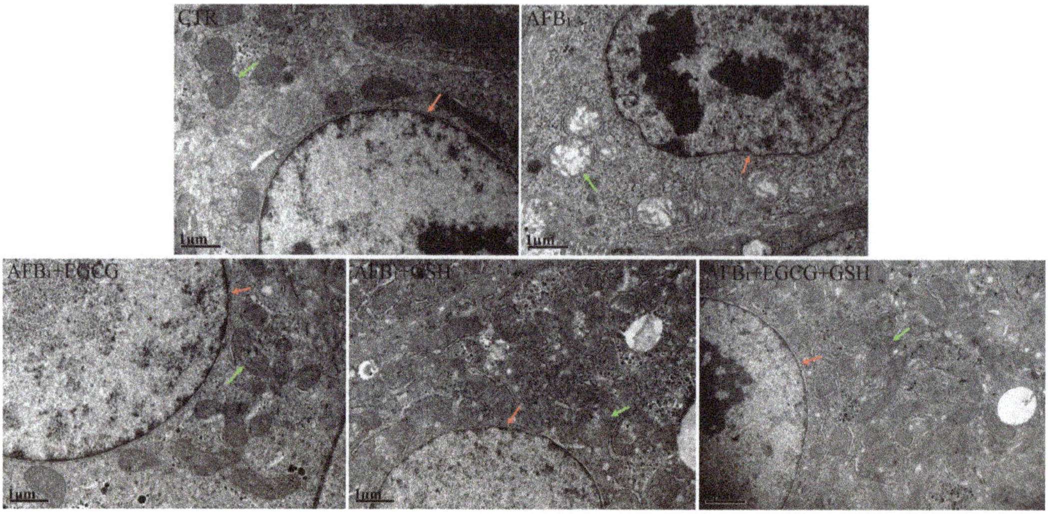

Figure 5. Effect of EGCG and GSH on the ultramicro-pathological structure of the livers of ducklings exposed to AFB_1. Magnification 20,000×, scale bars = 1 μm. Red arrows represent the nucleus of the cell, and green arrows represent mitochondria.

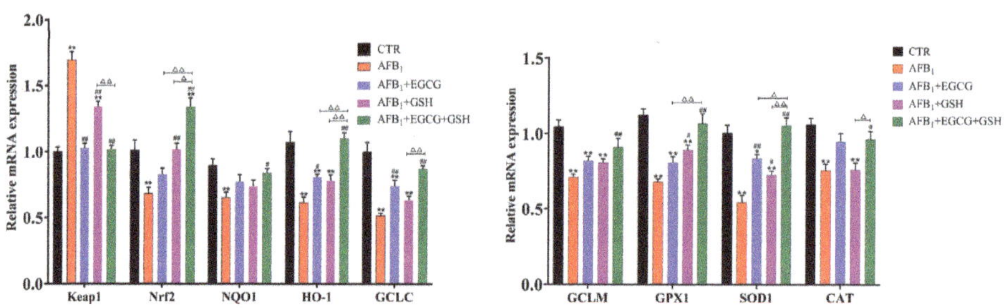

Figure 6. Effect of EGCG and GSH on the abundance of genes involved in the Keap1-Nrf2 signalling pathway in the livers of ducklings exposed to AFB_1. All results are expressed as means ± SEM (n = 6). * $p < 0.05$, ** $p < 0.01$ vs. CTR group; # $p < 0.05$, ## $p < 0.01$ vs. AFB_1 group; Δ $p < 0.05$, ΔΔ $p < 0.01$ significant difference between AFB_1 + EGCG + GSH and AFB_1 + EGCG or AFB_1 + GSH groups.

2.6. Protective Effects of EGCG and GSH on AFB_1-Induced Apoptosis of Duckling Hepatocytes

In order to evaluate the protective effect of EGCG and GSH alone or in combination against AFB_1-induced apoptosis, hepatocyte apoptosis and the expression of genes related to apoptosis mediated by mitochondria were examined by terminal deoxynucleotidyl transferase dUTP nick end-labelling (TUNEL) staining and RT-qPCR, respectively. Green fluorescence was significantly enhanced (as evidenced in the increased number of apoptotic cells) in all AFB_1-exposed groups (Figure 7A); the apoptosis rate (TUNEL positive rate) was elevated to 8.31% in the AFB_1 group, while the apoptosis rate decreased by 61.5%, 49.2%, and 74.0% in the EGCG, GSH, and combined groups, respectively, compared with the AFB_1 group (Figure 7B). Compared with the CTR group, the gene abundance of the pro-apoptotic gene Bax, as well as Cyt-c, Caspase-3, and p53 were significantly up-regulated ($p < 0.01$), and the anti-apoptotic gene Bcl-2 was significantly down-regulated ($p < 0.01$, Figure 7C) in the AFB_1 group; EGCG and GSH alone or in combination had a positive effect. It was concluded that EGCG and GSH attenuated AFB_1-induced apoptosis in hepatocytes.

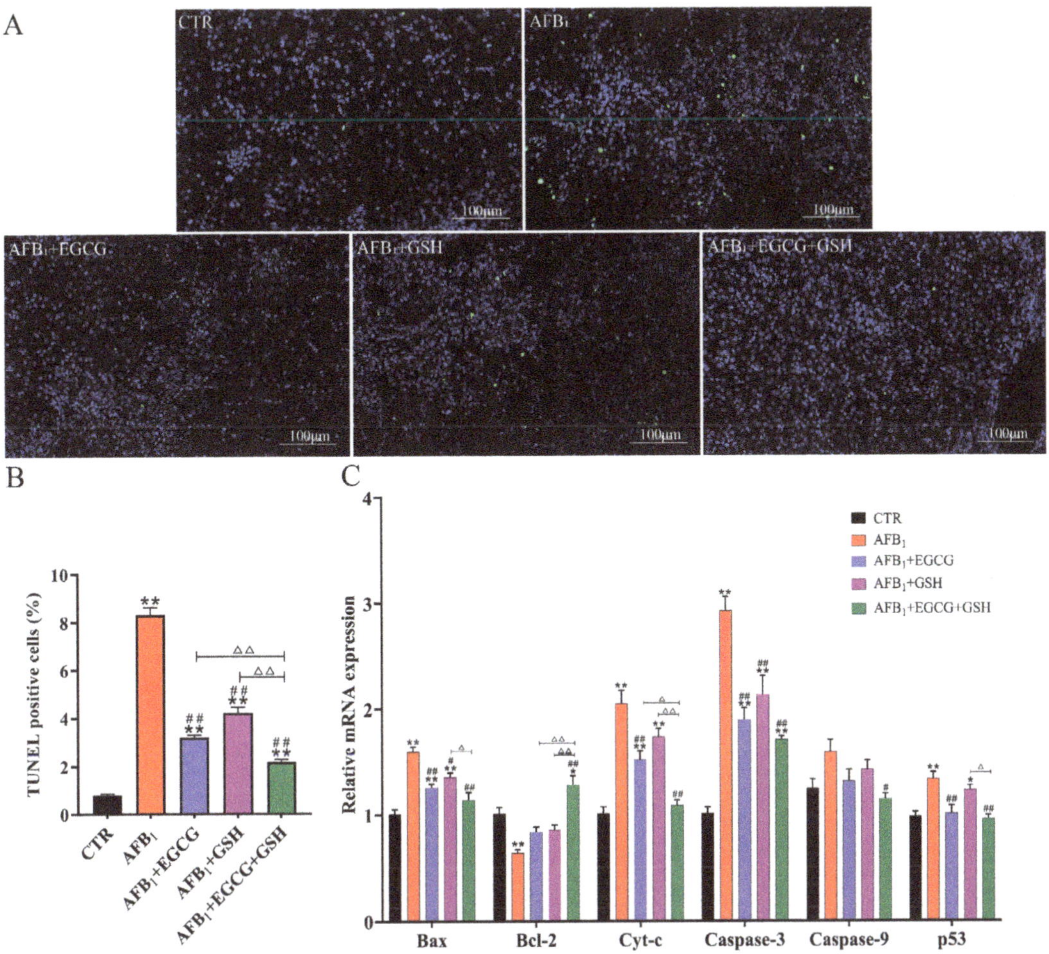

Figure 7. Effects of EGCG and GSH on AFB_1-induced apoptosis of ducklings' hepatocytes. (**A**) Terminal deoxynucleotidyl transferase dUTP nick end-labelling (TUNEL) staining to detect apoptosis. Magnification 200 ×, scale bars = 100 μm. Fluorescently labelled green indicates apoptotic cells, and blue indicates the nucleus. (**B**) TUNEL positive cells. (**C**) Expression of genes associated with mitochondria-mediated apoptosis, Cyt-c, Bax, Bcl-2, Caspase-3, Caspase-9, p53. All results are expressed as means ± SEM (n = 6). * $p < 0.05$, ** $p < 0.01$ vs. CTR group; # $p < 0.05$, ## $p < 0.01$ vs. AFB_1 group; Δ $p < 0.05$, ΔΔ $p < 0.01$ significant difference between AFB_1 + EGCG + GSH and AFB_1 + EGCG or AFB_1 + GSH groups.

3. Discussion

It is well-known that AFB_1 is commonly found in feed and causes severe damage to commercial animals, especially ducks [26,27]. Growth retardation and hepatic lesion are among the most important symptoms of AFB_1 poisoning. In the present study, we discovered that AFB_1 reduced the ducklings' feed intake and body weight, as well as caused liver damage. Our findings are consistent with previous research showing that AFB_1 causes a decrease in food intake, metabolic capacity, body weight, and significantly higher liver coefficients [28–30]. The results of the present study indicated that EGCG and GSH significantly increased body weight and decreased liver indices. Serum ALT, AST, and ALP are the most sensitive indicators for evaluating liver damage, and AFB_1 in the diet can increase the levels of these enzymes [31]. In one study, when ducklings were fed a diet of 0.1 mg/kg AFB_1, the levels of AST, ALT, and the ratio of AST/ALT increased [32]. Because AFB_1 inhibits protein biosynthesis, serum TB and ALB can be used to evaluate its impact [33]. Our results confirmed this: AFB_1 elevated the levels of ALT, AST, ALP, and γ-GT while decreasing the content of TB and ALB compared with the CTR group. The addition of EGCG and GSH slowed down the change.

Oxidative stress can promote the formation of reactive oxygen species (ROS) in animal target organs [34]. Numerous studies have shown that excessive ROS can damage macromolecules such as proteins and nucleic acids, thereby producing a large amount of MDA. Therefore, MDA, as the end product of lipid peroxidation, is an important indicator for detecting oxidative damage [35,36]. However, excessive ROS in the body can be scavenged by antioxidant enzymes, including SOD, GPX, CAT, and GSH. In the present study, exposure to AFB_1 increased the amount of MDA, while the content of GSH and the enzymatic activities of T-AOC, SOD, CAT, and GPX decreased. Apparently, AFB_1 induced oxidative stress in the ducklings' livers. The addition of EGCG and GSH also alleviated oxidative stress. Previous studies have shown that EGCG can attenuate carbon tetrachloride-induced oxidative stress in mouse livers and protect against H_2O_2-induced cellular oxidative damage [37,38]. Exogenous GSH has been found to have similar antioxidant effects in acute kidney injury in rats [39]. At the same time, we observed that AFB_1 induced pathological changes in the liver. These results indicated that AFB_1 caused oxidative damage to the liver, but EGCG and GSH could protect it by enhancing its antioxidant status. The antioxidant properties of EGCG and GSH themselves, as well as the fact that GSH can act as a substrate for GPX and GST, may explain the common effect exhibited by them.

Oxidative stress caused by AFB_1 can regulate the expression of a series of genes involved in the antioxidant system through the Keap1-Nrf2 signalling pathway. Nrf2 is the main regulator of cells that respond to environmental stress, inducing the expression of detoxification and antioxidant enzymes. The activity of Nrf2 is dependent on the regulation of the Keap1 adaptor protein, which is a negative regulator of the former [40,41]. Under non-stress conditions, Nrf2 binds to Keap1 in the cytoplasm to promote Nrf2 ubiquitination and proteasomal degradation; when stimulated, Nrf2 dissociates from Keap1 into the nucleus and combines with nuclear receptors to regulate the expression of downstream target genes (NQO1, HO-1, GCLC, GCLM, and so on) [42], thereby performing antioxidant or detoxification functions. It has been demonstrated that EGCG can strengthen cellular defences against chemical carcinogens as well as ultraviolet (UV) and oxidative stress through the Keap1-Nrf2 signalling pathway [43]. In one experiment, the EGCG treatment group normalised the expression of Keap1-Nrf2 and its downstream regulatory proteins in fluoride-treated rat kidneys [44]. We noted a significant upregulation of Keap1 mRNA expression in the AFB_1-treated group compared with the CTR group, indicating an enhanced negative regulation of Nrf2 by Keap1, while Nrf2 and its related genes (NQO1, HO-1, GCLC, GCLM, SOD1, GPX1, and CAT) were downregulated. Treatment with EGCG significantly reversed these effects (a finding that is consistent with previous studies). However, the effect of GSH on this pathway has not been investigated, so we speculated that it might regulate the expression of genes by balancing ROS production. This deserves more investigation. We concluded that EGCG and GSH contribute to the antioxidant capacity of the body through the Keap1-Nrf2 signalling pathway and that they are most effective in combination.

Apoptosis (i.e., programmed cell death) plays an essential role in controlling cell numbers and maintaining the homeostasis of multicellular organisms. Abnormal regulation of apoptosis has been associated with the development of a variety of diseases [45]. AFB_1 has been reported to induce apoptosis in hepatic, pulmonary, and bone marrow cells [46,47]. The present study found hepatocytes undergoing significant apoptosis in the AFB_1 group. It is widely known that mitochondria perform a central role in apoptosis initiated by many kinds of stimuli and that key events in apoptosis are associated with mitochondria [48]. Livers have been observed with severe mitochondrial lesions. In such cases, membrane permeability is altered, and Cyt-c enters the cytoplasm from the mitochondria, binding to the apoptosis protease activator Apaf-1 and caspase-9 and activating caspase-9, which in turn induces the activation of caspase-3 and subsequently triggers apoptosis mediated by the mitochondria [49]. However, mitochondria-mediated apoptosis is regulated by the pro-apoptotic protein Bax and the anti-apoptotic protein Bcl-2 [50]. The present study showed that AFB_1 reduced the mRNA expression of Bcl-2 and elevated the mRNA expression of Bax, while the expression of associated apoptotic genes (Cyt-c, caspase-3, caspase-9, and so on) was significantly elevated. However, EGCG and GSH inhibited the excessive apoptosis of hepatocytes caused by AFB_1 by regulating the expression of these genes. Studies have shown that EGCG protects against apoptosis in human umbilical vein endothelial cells by regulating the mitochondria-dependent apoptotic signalling pathway [51], while exogenous GSH defends IPEC-J2 cells from oxidative stress-induced apoptosis [52]. Apoptosis can be activated by oxidative stress [36]. Therefore, we speculate that EGCG and GSH alleviate apoptosis caused by AFB_1 in hepatocytes either directly or by inhibiting oxidative stress. However, the mechanism of interaction between EGCG and GSH needs to be further researched.

4. Conclusions

In the present study, AFB_1 caused serious damage to the ducklings. The results suggest that EGCG and GSH can alleviate acute liver injury by improving hepatic antioxidant capacity through the Keap1-Nrf2 signalling pathway and inhibiting the excessive apoptosis of hepatocytes mediated by mitochondria. This explains the protective mechanism of EGCG and GSH alone or in combination against AFB_1-induced liver injury. The present study also provides a theoretical basis for their application, and we suggest that EGCG and GSH could be used as promising duck feed additives to counteract AFB_1 damage.

5. Materials and Methods

5.1. Animals and Experimental Design

Age is an important factor affecting the bird's resistance to AFB_1, and male ducklings are more sensitive (male ducklings produce more AFBO than females), so we chose younger males to complete the experiment [53]. One-day-old male Cherry Valley ducks were purchased from Wuhan Yongsheng Duck Industry Co., Ltd. (Wuhan, China). The ducklings were kept in a controlled environment at a temperature of 30 ± 2 °C and 60 ± 5% humidity. The experiment (No. HZAUDU-2022-0002) was approved by the Animal Ethics Committee of Huazhong Agricultural University (Wuhan, China).

After three days of acclimatisation, 50 male ducklings were randomly divided into five groups (n = 10): 1. Control group (CTR); 2. Treated with AFB_1 (>99%, Pribolab, Qingdao, China) 0.3 mg/kg BW (AFB_1); 3. Treated with AFB_1 0.3 mg/kg BW + EGCG (98%, Shanghai Yuanye Biotech Co., Ltd., Shanghai, China) 100 mg/kg BW (AFB_1 + EGCG); 4. Treated with AFB_1 0.3 mg/kg BW + GSH (Reduced, 98%, Aladdin, Shanghai, China) 30 mg/kg BW (AFB_1 + GSH); 5. Treated with AFB_1 0.3 mg/kg BW + EGCG 100 mg/kg BW + GSH 30 mg/kg BW (AFB_1 + EGCG + GSH). Each group of ten ducklings was kept in a pen, marked and weighed individually. All ducklings were gavaged with the same concentration and 1 mL Volume/200 g BW. The acute liver injury experiment cycle lasted for 7 days. On Days 1–6, they were weighed daily and gavaged with distilled water, distilled water, EGCG, GSH, and EGCG + GSH, respectively. On Day 4, Groups 2 to 5 were treated with AFB_1 0.5 h after the first gavage, and the control group was given the corresponding solvent gavage (4% dimethyl sulfoxide). Slaughter sampling took place on Day 7. The composition and nutrient levels of the basal diet are shown in Appendix A, Table A1. The acute toxic dose of AFB_1 was determined based on previous reports [54,55] and preliminary experiments. Gavage doses of EGCG and GSH refer to preliminary experiment. The health status of the ducklings was strictly observed, and the body weight and feed intake were recorded during the experiment.

5.2. Sample Collection

After the ducklings fasted for 12 h, blood samples were collected using wing venipuncture into a tube. The blood samples were centrifuged at 3500 rpm for 10 min to obtain serum, which was divided and stored at −80 °C for biochemical analysis. The ducklings were immediately sacrificed and dissected to remove the liver, rinsed in cold saline, and weighed. A portion of liver tissue was cut and placed in paraformaldehyde fixative and 2.5% glutaraldehyde, respectively, for hematoxylin and eosin (H&E) staining or TUNEL detection and ultrastructural observation. The remaining part of each liver was stored at −80 °C in a refrigerator to detect antioxidant indexes, gene expression, and so on. The relative weight of the livers was calculated using the following formula:

$$\text{Relative weight} = \text{liver weight (g)/body weight (g)} \times 100\%$$

5.3. Determination of Serum Biochemical Indicators

The serum enzyme activities of ALT, AST, ALP, and γ-GT, as well as the content of ALB and TP, reflect the function of the liver. These indicators were measured using an automatic biochemical analyser according to the manufacturer's set procedure (Mindray, Shenzhen, China). The serum samples were placed in a cryogenic sample tray. ALT, AST, ALP, and γ-GT were expressed as U/L, while ALB and TP were expressed as g/L. All of these kits were purchased from the same manufacturer (Mindray, Shenzhen, China).

5.4. Detection of Antioxidant Capacity of the Liver

Liver tissue homogenates were prepared according to the requirements of the corresponding kits, and the protein concentration of the homogenate supernatant was determined by the BCA protein assay kit (Beyotime Biotechnology, Shanghai, China). SOD, GPX, MDA, T-AOC, CAT, and GSH kits were procured and operated according to the manufacturer's instructions (Nanjing Jiancheng Biotech, Nanjing, China). Absorbance was measured by a microplate reader (Multiskan MK3, Thermo Fisher Scientific, Waltham MA, USA) or visible light spectrophotometer (722E, Shanghai Spectrum Instruments Co., Ltd., Shanghai, China), and the enzyme activity or substance content was calculated and analysed.

5.5. Histopathological Analysis

Fresh liver tissue samples were placed in 4% paraformaldehyde and fixed for more than 24 h. The tissues were dehydrated with different concentrations of alcohol and embedded in wax. The wax blocks were placed in a microtome (Leica RM2016, Wetzlar, Germany) and cut into sections of 4 μm thickness. Staining with hematoxylin and eosin was performed for histopathological observation.

5.6. Ultrastructural Pathology Observation

Liver samples were cut to around 1 mm^3 in size and placed in 2.5% glutaraldehyde for 24 h. After 24 h, the samples were washed three times with 0.1 M PBS and fixed with 1% osmium acid for 2 h. The samples were rewashed with 0.1 M PBS, then dehydrated with gradient acetone and embedded in resin. The embedded samples were cut into 60 nm sections using an ultramicrotome (Leica UC5, Wetzlar, Germany), stained with uranyl acetate and lead citrate solution, and observed under a transmission electron microscope (Hitachi H-7650, Tokyo, Japan) for scanning and photographing [56].

5.7. Detection of Apoptosis by TUNEL Staining

Following the TUNEL kit manufacturer's instructions (Roche, Basel, Switzerland), the embedded liver sections were dewaxed, rehydrated, and then incubated with proteinase K at 37 °C for 30 min and washed three times with PBS. Fifty 50 μL of TUNEL reaction solution were added dropwise to the tissue, incubated at 37 °C for 2 h in the dark, washed again with PBS three times, and then incubated with 4,6-diamidino-2-phenylindole (DAPI) staining solution for 10 min while keeping it out of the light. After blocking, the images were observed and collected using an inverted fluorescence microscope (Olympus IX51, Tokyo, Japan). Fluorescence signals were analysed with Image-Pro Plus 6.0 (Media Cybernetics, Silver Spring, MD, USA), and apoptosis rates were calculated.

5.8. Quantitative Real-Time PCR

Total RNA was extracted from the ducklings' livers using Trizol reagent (TaKaRa, Dalian, China), and the quality (A260/A280) and concentration were evaluated using a NanoDrop 2000 spectrophotometer (Thermo Fisher Scientific, Waltham, MA, USA). Genomic DNA was removed, and RNA was reverse transcribed into cDNA using PrimeScript^{TM} RT reagent Kit with gDNA Eraser (TaKaRa, Dalian, China) according to the steps in the instructions, and expressed genes were evaluated by Real-Time PCR using TB Green® Premix Ex Taq^{TM} II (TaKaRa, Dalian, China). The primer sequences are shown in Appendix A Table A2. All primers were designed by Sangon Biotech (Shanghai, China) and synthesised by Tsingke Biotechnology Co., Ltd. (Beijing, China). The relative mRNA abundance was analysed following the $2^{-\Delta\Delta Ct}$ formula and normalised with the housekeeping gene GAPDH [57].

5.9. Statistical Analysis

The results were analysed using one-way ANOVA with SPSS Version 26 (SPSS Incorporated, Armonk, NY, USA) and Tukey's multiple comparisons as the post-hoc test. Outcomes were expressed as mean $\pm$ standard error (SEM). GraphPad Prism Version 9.0 (GraphPad Prism, San Diego, CA, USA) was used to visualise the data. In all statistical analyses, $p < 0.05$ was considered significant and $p < 0.01$ was considered highly significant.

Author Contributions: D.Q., conceptualisation, writing—review and editing, project administration and funding acquisition; Y.W., methodology, formal analysis, data curation and writing—original draft preparation; J.W., L.W. and P.Y., formal analysis, data curation and visualisation; Z.L., S.A.R. and M.H., writing—review and editing. All authors have read and agreed to the published version of the manuscript.

Funding: This project was supported by the National Key Research and Development Program of China (2016YFD0501207).

Institutional Review Board Statement: This experiment was approved by the Animal Ethics Committee of Huazhong Agricultural University (No. HZAUDU-2022-0002).

Informed Consent Statement: Not applicable.

Data Availability Statement: The data presented in this study are available on request from the corresponding author.

Conflicts of Interest: The authors declare no conflict of interest.

Appendix A

Table A1. Composition and nutrient level of basal diet.

Ingredients	Content (%)	Nutrition Component [2]	Content (%)
Corn	52.98	Crude protein	20.16
Soybean meal (44% CP)	30.36	ME (MJ/kg)	12.34
Fish meal	2.50	Calcium	0.91
Wheat bran	3.00	Available phosphorus	0.42
Rice bran	5.25	Methionine	0.45
Soybean oil	1.91	Lysine	1.12
Premix [1]	4.00	AFB_1 (μg/kg)	2.1
Total	100.00		

[1] Premix provided per kilogram of diet: 10,000 IU of vitamin A, 2500 IU of vitamin D_3, 35 IU of vitamin E, 2.5 mg of vitamin K_3, 2.5 mg of vitamin B_1, 9 mg of vitamin B_2, 0.02 mg of vitamin B_{12}, 15 mg of calcium pantothenate, 60 mg of niacin, 1.5 mg of folic acid, and 0.2 mg of biotin; 12 mg of Cu, 80 mg of Fe, 60 mg of Zn, 92 mg of Mn, 0.3 mg of Se, and 0.3 mg of I. [2] All nutrient levels were calculated.

Table A2. Primer sequences used in qRT-PCR.

Genes	Primer Sequences (5′ to 3′)	Product Lengths (bp)	Accession No.
Keap1	F: TCAAGACCTCACCCTCCATAAACCC R: AGTAGCCCAAGGACTGCCGATAG	110	KU048807.1
Nrf2	F: TGGCTGAGGTAAGCACAATCACAAG R: GGCTCTCAACAGTCTCCAGGAAATC	138	NM_001310777.1
NQO1	F: TGTCACCATCTCCGACCTCTATGC R: TCCTTCCACGCTTCTCCCATCTC	126	XM_027466610.2
HO-1	F: ATGAATGCCCTTGAGATGGACCTTG R: GTGACCGTTCTCCTGGCTCTTTG	132	XM_005015345.5
GCLC	F: TTCAGGTGACATTCCAGGCTTGC R: AGAACGGAGATGCAGCACTCAATG	108	XM_027455103.2
GCLM	F: TGTTGTGTGATGCCACCTGATCTC R: GTGCTTTGACGTTCTGGATGCTTTC	142	XM_027462629.2
SOD1	F: TCATCTCTCTGACTGGACCACACTG R: GTTAGCGTGCTCTCGTTGTCTCC	103	KU048808.1
GPX1	F: CCACCAGGAGAATGCCACCAAC R: CTTCCCGTTCACCTCGCACTTC	115	XM_027467953.2
CAT	F: ATGGACCAATGTGCGTGACTGAC R: CATGCGGCTCTCCTTCACAACAG	104	XM_027458335.2
Cyt-c	F: CCAGTGCCATACGGTTGAGAAGG R: TCTGTGTAGGAGAAGCCCTCAGC	105	XM_027447873.2
Bax	F: TCGTCGCCCTCTTCTACTTCGC R: CAGGAGACGATGGTGCGGAAAAG	88	KY788660.1
Bcl-2	F: CATGTGCGTGGAGAGCGTCAAC R: ACTGATCCAGCCTCCGTTGTCC	124	XM_027451679.2
Caspase-3	F: TGAGGCAGACAGTGGACCAGATG R: TCCTTCAGCACCCTACACAGAGAC	156	XM_021279218.3
Caspase-9	F: TGGATTGCGATTCACCCGAAGATG R: ATTACCCGAGGGAGCCTGGAAAG	83	XM_038166520.1
p53	F: AGGAGGAGAACTTCCGCAAGAGG R: GCAGGCAGAAGATCTCGTTGTCG	129	XM_038171818.1
GAPDH	F: GTCTCCTGCGACTTCAACGG R: CCTTGGATGCCATGTGGACC	160	XM_038180584.1

References

1. Reverberi, M.; Ricelli, A.; Zjalic, S.; Fabbri, A.A.; Fanelli, C. Natural functions of mycotoxins and control of their biosynthesis in fungi. *Appl. Microbiol. Biotechnol.* **2010**, *87*, 899–911. [CrossRef] [PubMed]
2. da Silva, J.B.; Dilkin, P.; Fonseca, H.; Correa, B. Production of aflatoxins by Aspergillus flavus and of fumonisins by Fusarium species isolated from Brazilian sorghum. *Braz. J. Microbiol.* **2004**, *35*, 182–186. [CrossRef]
3. Girolami, F.; Barbarossa, A.; Badino, P.; Ghadiri, S.; Cavallini, D.; Zaghini, A.; Nebbia, C. Effects of turmeric powder on aflatoxin M1 and aflatoxicol excretion in milk from dairy cows exposed to aflatoxin B1 at the EU maximum tolerable levels. *Toxins* **2022**, *14*, 430. [CrossRef] [PubMed]
4. Sana, S.; Anjum, A.A.; Yaqub, T.; Nasir, M.; Ali, M.A.; Abbas, M. Molecular approaches for characterisation of aflatoxin producing Aspergillus flavus isolates from poultry feed. *Pak. Vet. J.* **2019**, *39*, 169–174. [CrossRef] [PubMed]
5. Chen, K.; Fang, J.; Peng, X.; Cui, H.; Chen, J.; Wang, F.; Chen, Z.; Zuo, Z.; Deng, J.; Lai, W.; et al. Effect of selenium supplementation on aflatoxin B-1-induced histopathological lesions and apoptosis in bursa of Fabricius in broilers. *Food Chem. Toxicol.* **2014**, *74*, 91–97. [CrossRef]
6. Morrison, D.M.; Ledoux, D.R.; Chester, L.F.; Samuels, C.A.J.V.S. A limited survey of aflatoxins in poultry feed and feed ingredients in Guyana. *Vet. Sci.* **2017**, *4*, 60. [CrossRef]
7. Guo, Y.; Zhao, L.; Ma, Q.; Ji, C. Novel strategies for degradation of aflatoxins in food and feed: A review. *Food Res. Int.* **2021**, *140*, 109878. [CrossRef]

8. Deng, M.; Zhu, F.; Yang, Y.; Yang, F.; Hao, J.; Chen, S.; Hou, Z. Genome-wide association study reveals novel loci associated with body size and carcass yields in Pekin ducks. *BMC Genom.* **2019**, *20*, 1. [CrossRef]
9. Diaz, G.J.; Murcia, H.W. An unusually high production of hepatic aflatoxin B-1-dihydrodiol, the possible explanation for the high susceptibility of ducks to aflatoxin B-1. *Sci. Rep.* **2019**, *9*, 8010. [CrossRef]
10. Rushing, B.R.; Selim, M.I. Aflatoxin B1: A review on metabolism, toxicity, occurrence in food, occupational exposure, and detoxification methods. *Food Chem. Toxicol.* **2019**, *124*, 81–100. [CrossRef]
11. Stepkowski, T.M.; Kruszewski, M.K. Molecular cross-talk between the NRF2/KEAP1 signaling pathway, autophagy, and apoptosis. *Free Radic. Biol. Med.* **2011**, *50*, 1186–1195. [CrossRef] [PubMed]
12. Chen, Y.; Feng, X.; Hu, X.; Sha, J.; Li, B.; Zhang, H.; Fan, H. Dexmedetomidine ameliorates acute stress-induced kidney injury by attenuating oxidative stress and apoptosis through inhibition of the ROS/JNK signaling pathway. *Oxid. Med. Cell. Longev.* **2018**, *2018*, 4035310. [CrossRef] [PubMed]
13. Liu, Y.; Wang, W. Aflatoxin B1 impairs mitochondrial functions, activates ROS generation, induces apoptosis and involves Nrf2 signal pathway in primary broiler hepatocytes. *Anim. Sci. J.* **2016**, *87*, 1490–1500. [CrossRef]
14. Rains, T.M.; Agarwal, S.; Maki, K.C. Antiobesity effects of green tea catechins: A mechanistic review. *J. Nutr. Biochem.* **2011**, *22*, 1–7. [CrossRef] [PubMed]
15. Abd El-Hack, M.E.; Elnesr, S.S.; Alagawany, M.; Gado, A.; Noreldin, A.E.; Gabr, A.A. Impact of green tea (Camellia sinensis) and epigallocatechin gallate on poultry. *World's Poult. Sci. J.* **2020**, *76*, 49–63. [CrossRef]
16. Marin, L.; Miguelez, E.M.; Villar, C.J.; Lombo, F. Bioavailability of dietary polyphenols and gut microbiota metabolism: Antimicrobial properties. *BioMed Res. Int.* **2015**, *2015*, 905215. [CrossRef]
17. Chiu, H.; Venkatakrishnan, K.; Wang, C. The role of nutraceuticals as a complementary therapy against various neurodegenerative diseases: A mini-review. *J. Tradit. Complement. Med.* **2020**, *10*, 434–439. [CrossRef]
18. Aydogan, I.; Karsli, M.A.; Basalan, M.; Yildirim, E.; Cinar, M.; Sen, G.; Sumer, T. Effects of supplemental epigallocatechin gallate in the diet of broilers exposed to fluoride intoxication. *Biol. Trace Elem. Res.* **2018**, *186*, 258–266. [CrossRef]
19. Kim, H.-S.; Quon, M.J.; Kim, J.-a. New insights into the mechanisms of polyphenols beyond antioxidant properties; lessons from the green tea polyphenol, epigallocatechin 3-gallate. *Redox Biol.* **2014**, *2*, 187–195. [CrossRef]
20. Sriram, N.; Kalayarasan, S.; Sudhandiran, G. Epigallocatechin-3-gallate augments antioxidant activities and inhibits inflammation during bleomycin-induced experimental pulmonary fibrosis through Nrf2-Keap1 signaling. *Pulm. Pharmacol. Ther.* **2009**, *22*, 221–236. [CrossRef]
21. Orhan, C.; Tuzcu, M.; Gencoglu, H.; Sahin, N.; Hayirli, A.; Sahin, K. Epigallocatechin-3-gallate exerts protective effects against heat stress through modulating stress-responsive transcription factors in poultry. *Br. Poult. Sci.* **2013**, *54*, 447–453. [CrossRef] [PubMed]
22. Lui, S.M.; Eady, S.J. Glutathione: Its implications for animal health, meat quality, and health benefits of consumers. *Aust. J. Agric. Res.* **2005**, *56*, 775–780. [CrossRef]
23. Lomaestro, B.M.; Malone, M. Glutathione in health and disease: Pharmacotherapeutic issues. *Ann. Pharmacother.* **1995**, *29*, 1263–1273. [CrossRef]
24. Saeij, J.P.J.; van Muiswinkel, W.B.; van de Meent, M.; Amaral, C.; Wiegertjes, G.F. Different capacities of carp leukocytes to encounter nitric oxide-mediated stress: A role for the intracellular reduced glutathione pool. *Dev. Comp. Immunol.* **2003**, *27*, 555–568. [CrossRef]
25. Atef, H.A.; El-Shafei, H.M.; Mansour, M.K.; Al Kalamawy, N.M.; Hassan, N.; Lashin, A.J.I.J.C.M.A.S. Prevalence of ochratoxigenic fungi and ochratoxin A residues in animal feeds and modulation of their toxic effects by glutathione. *Int. J. Curr. Microbiol. Appl. Sci.* **2018**, *7*, 2559–2582. [CrossRef]
26. Fan, Y.; Liu, L.; Zhao, L.; Wang, X.; Wang, D.; Huang, C.; Zhang, J.; Ji, C.; Ma, Q. Influence of Bacillus subtilis ANSB060 on growth, digestive enzyme and aflatoxin residue in Yellow River carp fed diets contaminated with aflatoxin B-1. *Food Chem. Toxicol.* **2018**, *113*, 108–114. [CrossRef] [PubMed]
27. Limaye, A.; Yu, R.; Chou, C.; Liu, J.; Cheng, K. Protective and detoxifying effects conferred by dietary selenium and curcumin against AFB1-mediated toxicity in livestock: A Review. *Toxins* **2018**, *10*, 25. [CrossRef] [PubMed]
28. Mehta, R.; Campbell, J.S.; Laver, G.W.; Stapley, R.; Mueller, R. Acute hepatic response to aflatoxin B1 in rats fed a methyl-deficient, amino acid-defined diet. *Cancer Lett.* **1993**, *69*, 93–106. [CrossRef]
29. Ruggeberg, K.-G.; O'Sullivan, P.; Kovacs, T.J.; Dawson, K.; Capponi, V.J.; Chan, P.P.; Golobish, T.D.; Gruda, M.C. Hemoadsorption improves survival of rats exposed to an acutely lethal dose of aflatoxin B-1. *Sci. Rep.* **2020**, *10*, 799. [CrossRef] [PubMed]
30. Yan, J.; Chen, L.; Zhang, L.; Zhang, Z.; Zhao, Y.; Wang, Y.; Ou, J. New insights into the persistent effects of acute exposure to AFB(1) on rat liver. *Front Microbiol.* **2022**, *13*, 911757. [CrossRef]
31. Saei, M.M.; Sadeghi, A.A.; Ahmadvand, H. The effect of Myrtus communis oil extract on growth performance, serum biochemistry and humoral immune responses in broiler chicks fed diet containing aflatoxin B1. *Arch. Tierz.* **2013**, *56*, 842–850. [CrossRef]
32. Mendez-Albores, A.; Del Rio-Garcia, J.C.; Moreno-Martinez, E. Decontamination of aflatoxin duckling feed with aqueous citric acid treatment. *Anim. Feed Sci. Technol.* **2007**, *135*, 249–262. [CrossRef]
33. Mojgani, N.; Razmgah, N.; Torshizi, M.A.K.; Sanjabi, M.R. Effects of three Bacillus specious on hatchability, growth performance and serum biochemistry in Japanese quails fed diet contaminated with Aflatoxin B1. *Acta Sci. Anim. Sci.* **2020**, *42*, e50184. [CrossRef]

34. Yener, Z.; Celik, I.; Ilhan, F.; Bal, R. Effects of Urtica dioica L. seed on lipid peroxidation, antioxidants and liver pathology in aflatoxin-induced tissue injury in rats. *Food Chem. Toxicol.* **2009**, *47*, 418–424. [CrossRef] [PubMed]
35. Bai, K.; Huang, Q.; Zhang, J.; He, J.; Zhang, L.; Wang, T. Supplemental effects of probiotic Bacillus subtilis fmbJ on growth performance, antioxidant capacity, and meat quality of broiler chickens. *Poult. Sci.* **2017**, *96*, 74–82. [CrossRef] [PubMed]
36. Li, X.; Lv, Z.; Chen, J.; Nepovimova, E.; Long, M.; Wu, W.; Kuca, K. Bacillus amyloliquefaciens B10 can alleviate liver apoptosis and oxidative stress induced by aflatoxin B1. *Food Chem. Toxicol.* **2021**, *151*, 112124. [CrossRef] [PubMed]
37. Cia, D.; Vergnaud-Gauduchon, J.; Jacquemot, N.; Doly, M. Epigallocatechin gallate (EGCG) prevents H2O2-induced oxidative stress in primary rat retinal pigment epithelial cells. *Curr. Eye Res.* **2014**, *39*, 944–952. [CrossRef]
38. Tipoe, G.L.; Leung, T.M.; Liong, E.C.; Lau, T.Y.H.; Fung, M.L.; Nanji, A.A. Epigallocatechin-3-gallate (EGCG) reduces liver inflammation, oxidative stress and fibrosis in carbon tetrachloride (CCl4)-induced liver injury in mice. *Toxicology* **2010**, *273*, 45–52. [CrossRef]
39. Babaeenezhad, E.; Dezfoulian, O.; Hadipour Moradi, F.; Rahimi Monfared, S.; Fattahi, M.D.; Nasri, M.; Amini, A.; Ahmadvand, H.J.D.; Toxicology, C. Exogenous glutathione protects against gentamicin-induced acute kidney injury by inhibiting NF-κB pathway, oxidative stress, and apoptosis and regulating PCNA. *Drug Chem. Toxicol.* **2022**. [CrossRef]
40. Kovesi, B.; Pelyhe, C.; Zandoki, E.; Mezes, M.; Balogh, K. Changes of lipid peroxidation and glutathione redox system, and expression of glutathione peroxidase regulatory genes as effect of short-term aflatoxin B-1 exposure in common carp. *Toxicon* **2018**, *144*, 103–108. [CrossRef]
41. Suzuki, T.; Yamamoto, M. Molecular basis of the Keap1-Nrf2 system. *Free Radic. Biol. Med.* **2015**, *88*, 93–100. [CrossRef] [PubMed]
42. Baird, L.; Yamamoto, M. The molecular mechanisms regulating the KEAP1-NRF2 pathway. *Mol. Cell. Biol.* **2020**, *40*, e00099-20. [CrossRef] [PubMed]
43. Na, H.K.; Surh, Y.J. Modulation of Nrf2-mediated antioxidant and detoxifying enzyme induction by the green tea polyphenol EGCG. *Food Chem. Toxicol.* **2008**, *46*, 1271–1278. [CrossRef]
44. Thangapandiyan, S.; Miltonprabu, S. Epigallocatechin gallate supplementation protects against renal injury induced by fluoride intoxication in rats: Role of Nrf2/HO-1 signaling. *Toxicol. Rep.* **2014**, *1*, 12–30. [CrossRef] [PubMed]
45. Patel, B.K. Targeting apoptosis pathways for cancer therapy. In *Cancer Drug Discovery and Development: The Oncogenomics Handbook*; LaRochelle, W.J., Shimkets, R.A., Eds.; Humana Press: Totowa, NJ, USA, 2005; pp. 229–244, ISBN 978-1-59259-893-9.
46. Meki, A.R.M.; Abdel-Ghaffar, S.K.; El-Gibaly, I.J.N.L. Aflatoxin B1 induces apoptosis in rat liver: Protective effect of melatonin. *Neuroendocrinol. Lett.* **2001**, *22*, 417–426. [PubMed]
47. Raj, H.G.; Kohli, E.; Rohil, V.; Dwarakanath, B.S.; Parmar, V.S.; Malik, S.; Adhikari, J.S.; Tyagi, Y.K.; Goel, S.; Gupta, K.; et al. Acetoxy-4-methylcoumarins confer differential protection from aflatoxin B-1-induced micronuclei and apoptosis in lung and bone marrow cells. *Mutat. Res., Genet. Toxicol. Environ. Mutagen.* **2001**, *494*, 31–40. [CrossRef] [PubMed]
48. Bensassi, F.; Gallerne, C.; El Dein, O.S.; Lemaire, C.; Hajlaoui, M.R.; Bacha, H. Involvement of mitochondria-mediated apoptosis in deoxynivalenol cytotoxicity. *Food Chem. Toxicol.* **2012**, *50*, 1680–1689. [CrossRef]
49. Twiddy, D.; Brown, D.G.; Adrain, C.; Jukes, R.; Martin, S.J.; Cohen, G.M.; MacFarlane, M.; Cain, K. Pro-apoptotic proteins released from the mitochondria regulate the protein composition and caspase-processing activity of the native Apaf-1/caspase-9 apoptosome complex. *J. Biol. Chem.* **2004**, *279*, 19665–19682. [CrossRef]
50. Brunelle, J.K.; Letai, A. Control of mitochondrial apoptosis by the Bcl-2 family. *J. Cell Sci.* **2009**, *122*, 437–441. [CrossRef]
51. Liu, S.; Sun, Z.; Chu, P.; Li, H.; Ahsan, A.; Zhou, Z.; Zhang, Z.; Sun, B.; Wu, J.; Xi, Y.; et al. EGCG protects against homocysteine-induced human umbilical vein endothelial cells apoptosis by modulating mitochondrial-dependent apoptotic signaling and PI3K/Akt/eNOS signaling pathways. *Apoptosis* **2017**, *22*, 672–680. [CrossRef]
52. Chen, Q.; Yu, M.; Tian, Z.; Cui, Y.; Deng, D.; Rong, T.; Liu, Z.; Song, M.; Li, Z.; Ma, X.; et al. Exogenous glutathione protects IPEC-J2 cells against oxidative stress through a mitochondrial mechanism. *Molecules* **2022**, *27*, 2416. [CrossRef]
53. Dohnal, V.; Wu, Q.; Kuca, K. Metabolism of aflatoxins: Key enzymes and interindividual as well as interspecies differences. *Arch. Toxicol.* **2014**, *88*, 1635–1644. [CrossRef] [PubMed]
54. Monson, M.S.; Coulombe, R.A.; Reed, K.M.J.A. Aflatoxicosis: Lessons from toxicity and responses to aflatoxin B1 in poultry. *Agriculture* **2015**, *5*, 742–777. [CrossRef]
55. Yunus, A.W.; Razzazi-Fazeli, E.; Bohm, J. Aflatoxin B-1 in affecting broiler's performance, immunity, and gastrointestinal tract: A review of history and contemporary issues. *Toxins* **2011**, *3*, 566–590. [CrossRef]
56. Cao, J.; Song, Y.; Wu, H.; Qin, L.; Hu, L.; Hao, R. Ultrastructural studies on the natural leaf senescence of cinnamomum camphora. *Scanning* **2013**, *35*, 336–343. [CrossRef] [PubMed]
57. Zhong, G.; Wan, F.; Lan, J.; Jiang, X.; Wu, S.; Pan, J.; Tang, Z.; Hu, L. Arsenic exposure induces intestinal barrier damage and consequent activation of gut-liver axis leading to inflammation and pyroptosis of liver in ducks. *Sci. Total Environ.* **2021**, *788*, 147780. [CrossRef] [PubMed]

toxins

Communication

Inhibition of Essential Oils on Growth of *Aspergillus flavus* and Aflatoxin B1 Production in Broth and Poultry Feed

Bing Han [1,*], Guang-Wu Fu [2] and Jin-Quan Wang [1,*]

[1] Key Laboratory of Feed Biotechnology, Ministry of Agriculture and Rural Affairs, Institute of Feed Research, Chinese Academy of Agricultural Sciences, No. 12 Zhong Guan Cun South Street, Haidian District, Beijing 100081, China
[2] China Animal Husbandry Industry Corporation, Ltd., No. 156 Beiqing Road, Haidian District, Beijing 100095, China
* Correspondence: hanbing02@caas.cn (B.H.); wangjinquan@caas.cn (J.-Q.W.)

Abstract: Aflatoxin B1 (AFB1), a common contaminant in food and feed during storage, does great harm to human and animal health. Five essential oils (thymol, carvacrol, cinnamaldehyde, eugenol, and citral) were tested for their inhibition effect against *Aspergillus flavus* (*A. flavus*) in broth and feed. Cinnamaldehyde and citral were proven to be most effective against *A. flavus* compared to others and have a synergistic effect when used simultaneously. The broth supplemented with cinnamaldehyde and citral was inoculated with *A. flavus* (10^6 CFU/mL) by using the checkerboard method, and mold counts and AFB1 production were tested on days 0, 1, 3, and 5. Similarly, 100 g poultry feed supplemented with the mixture of cinnamaldehyde and citral at the ratio 1:1 was also inoculated with *A. flavus*, and the same parameters were tested on days 0, 7, 14, and 21. In poultry feed, cinnamaldehyde and citral significantly reduced mold counts and AFB1 concentrations ($p < 0.05$). Results showed that cinnamaldehyde and citral have a positive synergy effect and could both inhibit at least 90% the fungal growth and aflatoxin B1 production at 40 μg/mL in broth and poultry feed, and could be an alternative to control aflatoxin contamination in food and feed in future.

Keywords: aflatoxin B1; cinnamaldehyde; citral; inhibition; synergy

Key Contribution: Cinnamaldehyde and citral were proven to have a synergistic effect when used simultaneously and to be most effective against *A. flavus* and the production of AFB1 in broth and feed in the research.

Citation: Han, B.; Fu, G.-W.; Wang, J.-Q. Inhibition of Essential Oils on Growth of *Aspergillus flavus* and Aflatoxin B1 Production in Broth and Poultry Feed. *Toxins* **2022**, *14*, 655. https://doi.org/10.3390/toxins14100655

Received: 26 August 2022
Accepted: 20 September 2022
Published: 22 September 2022

1. Introduction

Aflatoxins (AF) constitute secondary metabolites produced by *Aspergillus flavus* and *Aspergillus parasitic* which contaminate a variety of feed ingredients, including peanuts, corn, and cottonseed [1]. Aflatoxin B1 (AFB1) is one of the most toxic members of the aflatoxin family [2]. Previous studies showed that young chicks were especially vulnerable to AF, which may depress feed conversion efficiency and body weight gain, and ultimately cause significant economic losses [3,4]. Because of the carcinogenicity of AF, AF residues in chicken may pose risks to human health [5,6]. AF severely influences the health and growth performance of animals. Therefore, it is necessary to develop a method by which to control the production of AF in feed. Essential oils derived from flavorants often have different capacities to inhibit the growth and toxicity of *A. flavus*.

Essential oils (EO) are complex mixtures of secondary plant metabolites. Moreover, because of their antimicrobial effects, essential oils have been widely used as food preservatives. Over the past several decades, some studies have discovered that various essential oils could also resist fungal growth [7,8]. Kumar [9] reported that the eugenol could inhibit the growth of *A. flavus*. Cinnamaldehyde, thymol, carvacrol et al. also have a strong inhibition to the growth of *A. flavus* [10–12]. Therefore, for essential oils derived from plants, it is a safe substitute for antibiotics. The objective of this study was to determine the inhibition of five different essential oils to *A. flavus* and the inhibition of AFB1 production in broth and poultry feed; in addition, a low concentration of essential oils was studied, which was not referred to in the previous studies. Low concentrations of essential oil could not only reduce the harm to animals but reduce the cost of feed industry.

2. Results

2.1. The Inhibition of Different Concentration of Essential Oils against A. flavus

The fungal growth inhibition was assessed by testing absorbance at 600 nm. In Figure 1, cinnamaldehyde inhibited 94.4% of the four fungal growth types, while other tested essential oils inhibited 93.7%, 86.9%, 80.1%, 78%, respectively. Consequently, it was concluded that cinnamaldehyde and citral had the most significant inhibitory effects on the growth of *A. flavus* CGMCC 3.2890 at the concentration of 40 μg/mL, and inhibition rate of cinnamaldehyde was higher compared to carvacrol ($p < 0.05$) but no significance compared to other essential oils ($p > 0.05$).

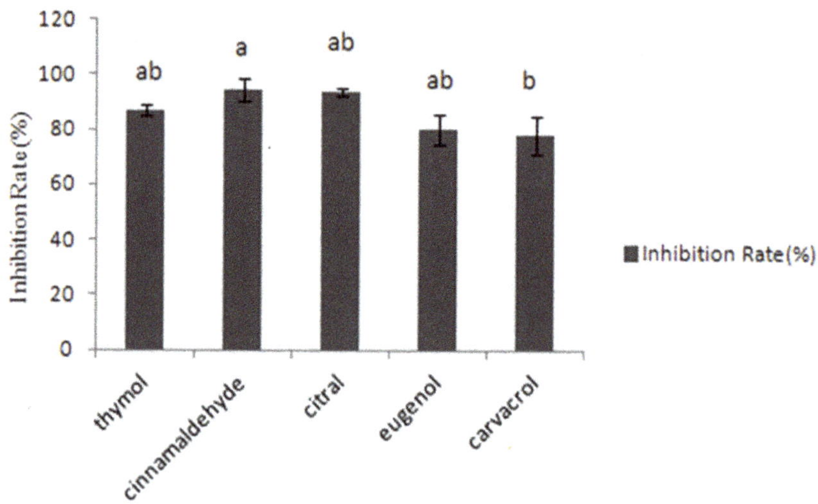

Figure 1. Inhibition of different essential oils at the concentration of 40 μg/mL. Different small letters in the same row (a, b) denote a significant difference ($p < 0.05$) among values, based on Tukey's test.

Consequently, cinnamaldehyde and citral were screened for further research with regard to their inhibitory effects on the growth of *A. flavus* in feed. Representative strain *A. flavus* CGMCC 3.2890 was considered for further research.

2.2. MIC Tests and Synergy Effects of the Best Effect of Essential Oils on Fungal Growth in Broth by Using Checkerboard

The synergy effect of cinnamaldehyde and citral against *A. flavus* CGMCC 3.2890 was tested by using the checkerboard method. The modal minimum inhibition concentration (MIC) results are presented in Table 1. The two essential oils synergistically affected *A. flavus* CGMCC 3.2890 as shown in Table 2.

Table 1. Modal MIC results by broth microdilution.

Essential Oils	MICs
Cinnamaldehyde	80 ppm
Citral	80 ppm

Table 2. Synergy method results.

Citral \ Cinnamaldehyde	0	1/4 × MIC	1/2 × MIC	1 × MIC	2 × MIC
0	0.439 ± 0.012	0.428 ± 0.022	0.427 ± 0.038	0.377 ± 0.034	0.376 ± 0.027
1/4 × MIC	0.419 ± 0.009	0.401 ± 0.028	0.369 ± 0.029	0.371 ± 0.011	0.371 ± 0.051
1/2 × MIC	0.418 ± 0.025	0.383 ± 0.003	0.425 ± 0.082	0.372 ± 0.028	0.374 ± 0.044
1 × MIC	0.419 ± 0.018	0.392 ± 0.008	0.372 ± 0.022	0.368 ± 0.036	0.375 ± 0.051
2 × MIC	0.406 ± 0.003	0.392 ± 0.053	0.366 ± 0.004	0.377 ± 0.017	0.376 ± 0.032

From the results of Table 2 and the FIC value (FIC = 0.5), it could be concluded that cinnamaldehyde and citral had the synergy effects when used simultaneously.

2.3. Effects of Essential Oils on Fungal Growth and AFB1 Production in Poultry Feed

Cinnamaldehyde and citral were screened for further research with regard to their inhibitory effects on the growth of *A. flavus* in poultry feed on days 0, 7, 14, and 21, respectively. Representative strain *A. flavus* CGMCC 3.2890 was considered for further research in feed. The effects of combination of two essential oils on the growth of *A. flavus* CGMCC 3.2890 in feed were shown in Figures 2 and 3, respectively.

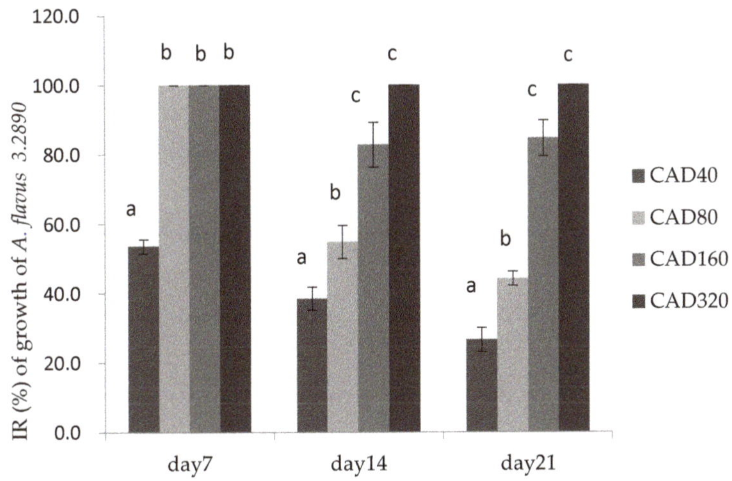

Figure 2. Inhibition rate (IR) of cinnamaldehyde and citral on the growth of *A. flavus* 3.2890 in feed at 40 μg/mL (CAD40), 80 μg/mL (CAD80), 160 μg/mL (CAD160), and 320 μg/mL (CAD320). Different small letters in the same row (a, b, c) denote a significant difference ($p < 0.05$) among values, based on Tukey's test.

Figure 2 showed that on day 7 there was no significance among the CAD80, CAD160, and CAD320 ($p > 0.05$) with IR almost 100%, but on days 14 and 21, the growth of molds was significantly decreased at 160 μg/mL (CAD160) and 320 μg/mL (CAD320) compared to CAD40 treatment, respectively, at the end of the storage period. Therefore, the addition of cinnamaldehyde and citral could suppress the molds sprouting in feed during the first week.

Figure 3 showed the inhibitory effect of cinnamaldehyde and citral on AFB1 production by *A. flavus* 3.2890 in feed, where it could be seen that CAD160 and CAD320 treatments could completely inhibit the productions of AFB1 on day 14 and day 21 ($p < 0.05$) compared to other groups, and CAD40 and CAD80 could reduce the production of AFB1 to some extent ($p < 0.05$). However, with the time going, a high concentration could still totally inhibit fungus growth and AFB1 production on day 21, while the IR of low centration decreased, but compared to CT treatment, the low concentration could still play an important inhibition role to some extent.

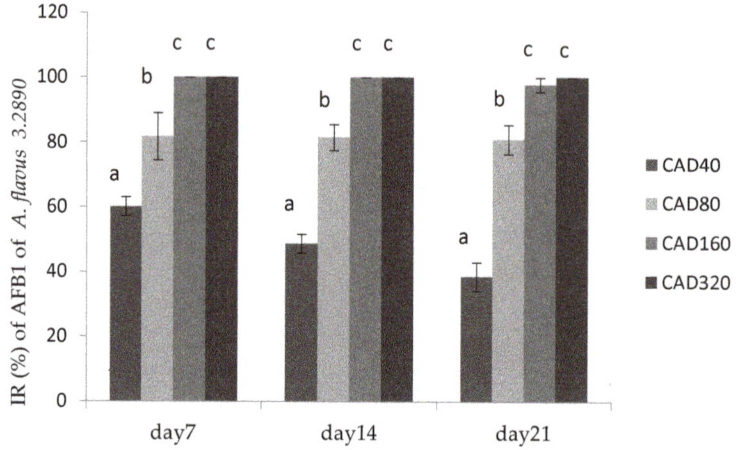

Figure 3. Inhibition rate (IR) of cinnamaldehyde on AFB1 production by *A. flavus* CGMCC 3.2890 in feed at 40 μg/mL (CAD40), 80 μg/mL (CAD80), and 160 μg/mL (CAD160). Different small letters in the same row (a, b, c) denote a significant difference ($p < 0.05$) among values based on Tukey's test.

3. Discussion

In this research, a high-concentration treatment could totally inhibit the growth of fungus and AFB1 production even on day 21, perhaps because the essential oils have killed the fungus, and the low concentration of essential oils probably just inhibited the fungus. In line with Mahnoud [13] and López–Malo et al. [14], the results of the present study showed that the growth of *A. flavus* CGMCC 3.2890 was inhibited by the five EOs, which was in accordance with Kumar et al. [9]. However, in the former studies, the concentrations of inhibition were all high, rarely studying the inhibition capability of low concentrations. Our research proved that low concentration also has inhibition ability against *A. flavus*. Our results showed that the cinnamaldehyde and citral suppressed the growth of *A. flavus* at low concentrations. Sun et al. [15] also reported that the germination of *A. flavus* was delayed by cinnamaldehyde in PDA medium when administered at 79.29 mg/L, whereas our concentration was only 40 μg/mL. The reason why cinnamaldehyde had the best inhibition ability was perhaps related to its special structure, such as aldehyde and phenol, which could attack the cell membrane or cell wall. Nogueira et al. [16] indicated that essential oil of *Ageratum conyzoides* changed the ultra-structure of *A. flavus*, which was more evident in the endomembrane system, such as mitochondria, thus inhibiting the growth of *A. flavus*. Sun et al. [15] reported that the diameter of the spore size linearly decreased with the increase of concentration of essential oils. The probable mechanism of cinnamaldehyde and citral needs further study.

The checkerboard method is a common method by which to evaluate the synergy effect of different drugs. The results showed that cinnamaldehyde and citral could synergistically affect *A. flavus*, which was rarely reported before.

Cinnamaldehyde and citral dose-dependently inhibited AFB1 production in the liquid medium, which was in accordance with our results in the feed. The mechanism of decreased AFB1 production by cinnamaldehyde and citral may be related to the downregulation of the expression of key genes for AFB biosynthesis, such as *aflC*, *nor1* and *norA* [16].

In the feed industry, essential oil is usually used as an odorant to increase feed intake, and acidifier product is often used to inhibit fungus, but the dosage of acidifier in feed is high, which may cause the negative effect on the performance of animals. Consequently, essential oil could be a good replacement for acidifier. Essential oils could not only inhibit the fungus, but have many other positive effects on animal production. Thymol was proven to increase the polyunsaturated fatty acid in egg yolk [17], perhaps for its antibacterial and antioxidant properties, which was also proven in other studies [18]. Moreover, essential oils could be improved by increasing the oleic acid content [19], and thiobarbituric acid reactive substances (TBARS) values could be lowered and the color parameters could be increased during storage when using essential oils [20]. Now only rosemary extract and oregano essential oils are permitted to be used as feed additives, but in the future, more essential oils would probably be applied as feed additives.

4. Conclusions

AFB1 do great harm to the health of human and animals. The research showed cinnamaldehyde and citral could be an alternative to control aflatoxin contamination in food and feed in future.

5. Material and Methods

5.1. Microbial Cultures

The following microbial strain were selected for their relevance in the feed industry: *A. flavus* CGMCC 3.2890 was obtained from China General Microbiological Culture Collection Center (CGMCC). The fungi strain was subcultured in potato dextrose agar (PDA) at 28 °C for five days. A spore suspension (approx. 10^6 CFU/mL) was prepared with potato dextrose broth (PDB).

5.2. Preparation of Essential Oils

Thymol (Sinopharm Chemical Reagent Co., Ltd., Beijing, China, ≥99.0%), cinnamaldehyde (Sinopharm Chemical Reagent Co., Ltd., Beijing, China, ≥99.0%), citral (Sinopharm Chemical Reagent Co., Ltd., Beijing, China, ≥97.0%), eugenol (Sinopharm Chemical Reagent Co., Ltd., Beijing, China, ≥98.5%), and carvacrol (J & KCHEMICA, Beijing, China, ≥98.0%) were mixed with potato dextrose broth (PDB) containing ethanol (5%; *v*/*v*) and tween 80 (0.5%; *v*/*v*) at 1000 μg/mL, 200 μg/mL, and 40 μg/mL, respectively. PDB solution was prepared according to the above method (devoid of essential oils) served as the control.

5.3. Screening of Best Effect of Essential Oils on Mold Growth in Broth

The modified micro-plate assay used in this study has already been described in detail by Gorran et al. [21]. Briefly, EOs at different concentrations (0, 40, 200 and 1000 μg/mL) were screened for inhibiting *A. flavus* growth. In 96-well micro-plates (Costar®, 3599, Corning, NY, USA), 160 μL PDB and 20 μL of different concentrations of EOs were mixed with 20 μL of four different strains of prepared *A. flavus* spores (at the concentration of 10^6 CFU per well), and shaken overnight at 28 °C. The fungal growth was determined by measuring the absorbance at 600 nm of fungal culture in 96-well micro-plates by using a micro-plate reader (model 680, BIO-RAD Laboratories, Inc., Hercules, CA, USA) for 24 h and 48 h [22]. All assays were performed in triplicates. The essential oils of best inhibition effect were chosen for the next trials.

5.4. MIC Tests and Synergy Effects of the Best Effect Essential Oils on Fungal Growth in Broth

The MICs of two essential oils' best inhibition effects were separately determined by broth microdilution method by using 96 kits. The MICs were tested in replicates of six. The MIC was defined as the lowest concentration of completely inhibiting the growth of *A. flavus*. The inoculums were approximately 1 × 10^6 CFU/mL in each well.

The synergy effects of the two essentials were determined by using the checkerboard method. The concentration range of each essential oil in combination ranged from 1/4 MIC to 2 MIC. Dilutions of two essential oils were made with a twofold diluter [14]. The initial inoculum was approximately 1 × 10^6 CFU/mL. The fungal growth was determined by measuring the absorbance at 600 nm of fungal culture in 96-well micro-plates by using a micro-plate reader (model 680, BIO-RAD Laboratories, Inc., Hercules, CA, USA) for 24 h.

To evaluate the effects of combinations, the fractional inhibition concentration (FIC) was calculated for each essential oil in each combination [22]. The following formulas were used to calculate the FIC index. The results were expressed as four situations, including synergy (FIC $\leq$ 0.5), additive (0.5 < FIC $\leq$ 1), indifference (1 < FIC $\leq$ 4) or antagonism (FIC > 4). A and B separately stands for the two tested essential oils:

$$FIC = (MIC_{A+B})/(MIC_A) + (MIC_{A+B})/(MIC_B). \quad (1)$$

5.5. Effects of the Best Effect Essential Oil on Fungal Growth and AFB1 Production in Poultry Feed

Prior to mixing with cinnamaldehyde, broiler feed free of any toxin binder was sterilized at 121 °C for 20 min followed and the moisture of the feed was adjusted to 17% (on dry basis) with sterile water. Then the feed was inoculated with each mold separately by using the method described by Yin et al. [23], wherein *A. flavus* CGMCC 3.2890 was added to 200-g portions of feed to obtain 5.5 log CFU/g feed, and mixed well. After inoculation, cinnamaldehyde and citral were added at 0, 40, 80, 160, and 320 mg/kg (CT, CAD40, CAD80, CAD160, CAD320) feed totally followed by incubation at 28 °C in 500 mL Erlenmeyer flasks, sealed with rubber closures. A 20-g portion of the feed was sampled on days 0, 7, 14, and 21, of which 10 g was used for mold enumeration and 5 g for AFB1 detection, respectively, to calculate the inhibition rate (IR) of *A. flavus* CGMCC 3.2890 and AFB1. Each treatment was repeated three times.

The inhibition rate (IR) of *A. flavus* CGMCC 3.2890 was calculated according to the following formula:

$$IR\ (\%) = (\Delta OD_c - \Delta ODx)/\Delta OD_c \times 100. \quad (2)$$

IR means the inhibition rate of *A. flavus*, ΔOD_c means the difference value of the OD of CT treatment of day (7, 14, 21) and day0, ΔOD_x means the difference value of OD of the treatments (CAD40, CAD80, CAD160, CAD320) on day x (day 7, 14, and 21) and day 0 respectively;

IR of AFB1 was calculated according to the following formula:

$$IR\ (\%) = (\Delta AFB_c - \Delta AFBx)/\Delta AFB_c \times 100. \quad (3)$$

IR means the inhibition rate of AFB1, ΔAFB_c means the difference value of the concentration of AFB1 of CT treatment on day x (day 7, 14, 21) and day0, $\Delta AFBx$ means the difference value of the concentration of AFB1 of of the treatments (CAD40, CAD80, CAD160, CAD320) on day x (day 7, 14, 21) and day 0, respectively.

5.6. Determination of Mold Counts

To enumerate *A. flavus* in the control and treated feed, 10 g portions of feed samples were added to 100 mL of PBS in sterile glass flasks, and blended in a shaker for 30 min. The feed homogenate was serially diluted (1:10) in PBS, and 0.1 mL aliquots from appropriate dilutions were surface-plated on duplicate PDA plates, and incubated as previously described.

5.7. Determination of AFB1 in Feed

The concentrations of AFB1 were determined by using a commercial ELISA Kit (HEM 00496, Huaan Magnech Bio-Tech Co., Ltd., Beijing, China). All the procedures were performed on the basis of manufacturer's instructions and the absorbance was determined by using a micro-plate reader. The AFB1 kit is an indirect competitive enzyme-labeled immunoassay. The AFB1 antigen is pre-coated on the wells. The pre-coated antigen competes with the AFB1 antibody (antibody solution) with AFB1 in the sample, anti-AFB1 antibody binds to the AFB1-HRP enzyme conjugate. The substrate solution was pipetted into the wells to convert the color. The color of unknown samples is compared to the color of the standards and the AFB1 concentrations of the samples were derived.

Samples were prepared by weighing out a 5.0-g comminuted sample into a 100-mL triangular flask with a stopper. A total of 25 mL of 60% methanol solution was added and blended vigorously for 10 min on a vertex. The sample was transferred to a centrifuge tube and centrifuged for 5 min at 4000 r/min. A total of 1.0 mL of the top-layer liquid was transferred to a new tube, and 4.0 mL of deionized water was added and blended for 5 s. A total of 50 μL of the solution was taken for assay.

5.8. Statistical Analysis

Data from this study was analyzed with one-way ANOVA followed by Tukey's multiple range test; data were expressed as the mean ± SE by Tukey's multiple range test. Data were expressed as significant if p was less than 0.05. All statistical analyses were performed by SPSS 25.0 (SPSS Inc., Chicago, IL, USA).

Author Contributions: Conceptualization, B.H. and G.-W.F.; methodology, B.H. and G.-W.F.; software, B.H.; validation, B.H., G.-W.F. and J.-Q.W.; formal analysis, B.H.; investigation, B.H.; resources, B.H.; data curation, B.H.; writing—original draft preparation, B.H.; writing—review and editing, B.H.; visualization, B.H.; supervision, B.H.; project administration, B.H. All authors have read and agreed to the published version of the manuscript.

Funding: This research received no external funding.

Institutional Review Board Statement: Not applicable.

Informed Consent Statement: Not applicable.

Data Availability Statement: Not applicable.

Conflicts of Interest: The authors declare no conflict of interest.

References

1. Oguz, H.; Kececi, T.; Birdane, Y.O.; Onder, F. Effect of clinoptilolite on serum biochemical and haematological characters of broiler chickens during aflatoxicosis. *Res. Vet. Sci.* **2000**, *69*, 89–93. [CrossRef]
2. Yunus, A.W.; Razzazi-Fazeli, E.; Bohm, J. Aflatoxin B(1) in affecting broiler's performance, immunity, and gastrointestinal tract: A review of history and contemporary issues. *Toxins* **2011**, *3*, 566–590. [CrossRef]
3. Oguz, H.; Kurtoglu, F.; Kurtoglu, V.; Birdane, Y.O. Evaluation of biochemical characters of broiler chickens during dietary aflatoxin (50 and 100 ppb) and clinoptilolite exposure. *Res. Vet. Sci.* **2002**, *73*, 101–103. [CrossRef]
4. Tessari, E.N.; Oliveira, C.A.; Cardoso, A.L.; Ledoux, D.R. Effects of aflatoxin B1 and fumonisin B1 on body weight, antibody titres and histology of broiler chicks. *Brit. Poult. Sci.* **2006**, *47*, 357–364. [CrossRef] [PubMed]
5. Berry, C.L. The pathology of mycotoxins. *J. Pathol.* **1988**, *154*, 301–311. [CrossRef] [PubMed]
6. Chu, F.S. Mycotoxins: Food contamination, mechanism, carcinogenic potential and preventive measures. *Mutat. Res.* **1991**, *259*, 291–306. [CrossRef]
7. Burt, S. Essential oils: Their antibacterial properties and potential applications in foods–a review. *Int. J. Food Microbiol.* **2004**, *94*, 223–253. [CrossRef] [PubMed]
8. Kohiyama, C.Y.; Yamamoto Ribeiro, M.M.; Mossini, S.A.; Bando, E. Antifungal properties and inhibitory effects upon aflatoxin production of Thymus vulgaris L. by *Aspergillus flavus* Link. *Food Chem.* **2015**, *173*, 1006–1010. [CrossRef]
9. Kumar, A.; Shukla, R.; Singh, P.; Dubey, N.K. Chemical composition, antifungal and antiaflatoxigenic activities of Ocimum sanctum L. essential oil and its safety assessment as plant based antimicrobial. *Food Chem. Toxicol.* **2010**, *48*, 539–543. [CrossRef] [PubMed]
10. López, P.; Sánchez, C.B.R. Vapor-Phase Activities of Cinnamon, Thyme, and Oregano Essential Oils and Key Constituents against Foodbore Microorganisms. *J. Agric. Food Chem.* **2007**, *55*, 4348–4356. [CrossRef]
11. Chinnathambi, S.; Sivaranjani, R.; Rona, V. Mechanism of antioxidant and antifungal properties of *Pimenta dioica* (L.) leaf essential oil on Aspergillus flavus. *J. Food Sci. Technol.* **2021**, *58*, 2497–2506.
12. Shukla, R.; Kumar, A.; Singh, P.; Dubey, N.K. Efficacy of *Lippia alba* (Mill.) N.E. Brown essential oil and its monoterpene aldehyde constituents against fungi isolated from some edible legume seeds and aflatoxin B1 production. *Int. J. Food Microbiol.* **2009**, *135*, 165–170. [CrossRef] [PubMed]
13. Mahmoud, A.L.E. Antifungal action and antiaflatoxigenic properties of some essential oil constituents. *Lett. Appl. Microbiol.* **1994**, *19*, 110–113. [CrossRef] [PubMed]
14. Lopez-Malo, A.; Maris Alzamora, S.; Palou, E. *Aspergillus flavus* growth in the presence of chemical preservatives and naturally occurring antimicrobial compounds. *Int. J. Food Microbiol.* **2005**, *99*, 119–128. [CrossRef]

15. Sun, Q.; Wang, L.; Lu, Z.; Liu, Y. In vitro anti-aflatoxigenic effect and mode of action of cinnamaldehyde against aflatoxin B1. *Int. Biodeterior. Biodegrad.* **2015**, *104*, 419–425. [CrossRef]
16. Nogueira, J.H.; Goncalez, E.; Galleti, S.R.; Facanali, R. Ageratum conyzoides essential oil as aflatoxin suppressor of Aspergillus flavus. *Int. J. Food Microbial.* **2010**, *137*, 55–60. [CrossRef]
17. Fernandez, M.E.; Kembro, J.M.; Ballesteros, M.L.; Caliva, J.M.; Marin, R.H.; Labaque, M.C. Dynamics of thymol dietary supplementation in quail (*Cotrrnix japonica*): Linking bioavailability, effects on egg yolk total fatty acids and performance traits. *PLoS ONE* **2019**, *4*, 623–629.
18. Abo Ghanima, M.M.; Elsadek, M.F.; Taha, A.E.; Abd El-Hack, M.E.; Alagawany, M.; Ahmed, B.M.; Elshafie, M.M.; El-Sabrout, K. Effect of Housing System and Rosemary and Cinnamon Essential Oils on Layers Performance, Egg Qualiyt, Haematological Traits, Blood Chemistry, Immunity, and Antioxidant. *Animals* **2020**, *10*, 245. [CrossRef]
19. Kürekci, C.; Özsoy, B.; Hassan, E.; Özkan, H.; Gundoğdu, A.; Özsoy, Ş.Y.; Yakan, A. Effect of essential oil supplementation to diet on meat quality, fatty acid composition, performance parameters and intestinal microbiota of Japanese quails. *J.Anim. Physiol. Anim. Nutr.* **2021**, *105*, 927–937. [CrossRef]
20. Alessandro, S.; Nunziatina, R.; Antonino, M.; Carmine, S.; Antonio, N.; Carmine, S.; Antonio, N.; Cinzia, C.; Cinzia, L.R.; Massimiliano, L. Effect of essential oil supplementation to diet on meat quality, fatty acid composition, performance parameters and intestinal microbiota of Japanese quails. *Animals* **2020**, *10*, 640.
21. Gorran, A.; Farzaneh, M.; Shivazad, M.; Rezaeian, M. Aflatoxin B1-reduction of *Aspergillus flavus* by three medicinal plants (Lamiaceae). *Food Control* **2013**, *31*, 218–223. [CrossRef]
22. Karaca, N.; Sener, G.; Demirci, B.; Demirci, F. Synergistic antibacterial combination of *Lavandula latifolia* Medik. essential oil with camphor. *Z. Naturforsch. C J. Biosci.* **2021**, *76*, 169–173. [CrossRef] [PubMed]
23. Yin, H.B.; Chen, C.H.; Kollanoor-Johny, A.; Darre, M.J. Controlling *Aspergillus flavus* and *Aspergillus parasiticus* growth and aflatoxin production in poultry feed using carvacrol and trans-cinnamaldehyde. *Poult. Sci.* **2015**, *94*, 2183–2190. [CrossRef] [PubMed]

toxins

MDPI

Article

Whole-Transcriptome Analysis of Non-Coding RNA Alteration in Porcine Alveolar Macrophage Exposed to Aflatoxin B1

Huhe Chao [1,2,†], Haohai Ma [1,†], Jiadong Sun [3], Shuai Yuan [4], Peiyu Dong [1], Aihong Zhao [5], Lan Li [3], Wei Shen [3] and Xifeng Zhang [1,*]

1 College of Veterinary Medicine, Qingdao Agricultural University, Qingdao 266109, China; chaohuhe@126.com (H.C.); 20202113033@stu.qau.edu.cn (H.M.); 20212213004@stu.qau.edu.cn (P.D.)
2 Central Laboratory, Beijing Obstetrics and Gynecology Hospital, Capital Medical University, Beijing 100023, China
3 Key Laboratory of Animal Reproduction and Biotechnology in Universities of Shandong, College of Life Sciences, Qingdao Agricultural University, Qingdao 266109, China; sun-jiadong@163.com (J.S.); lli@qau.edu.cn (L.L.); wshen@qau.edu.cn (W.S.)
4 School of Medicine, Henan Polytechnic University, Jiaozuo 454000, China; wsys727@sina.com
5 Qingdao Academy of Agricultural Science, Qingdao 266100, China; zhaoaihong0@163.com
* Correspondence: zhangxf106@qau.edu.cn
† These authors contributed equally to this work.

Abstract: Aflatoxin B1 (AFB1) is a type of mycotoxin produced by the fungi Aspergillus flavus and Aspergillus parasiticus and is commonly found in cereals, oils and foodstuffs. In order to understand the toxic effects of AFB1 exposure on Porcine alveolar macrophages (3D4/2 cell), the 3D4/2 cells were exposed to 40 μg/mL AFB1 for 24 h in vitro, and several methods were used for analysis. Edu and TUNEL analysis showed that the proliferation of 3D4/2 cells was significantly inhibited and the apoptosis of 3D4/2 cells was significantly induced after AFB1 exposure compared with that of the control group. Whole-transcriptome analysis was performed to reveal the non-coding RNA alteration in 3D4/2 cells after AFB1 exposure. It was found that the expression of cell-cycle-related and apoptosis-related genes was altered after AFB1 exposure, and lncRNAs and miRNAs were also significantly different among the experimental groups. In particular, AFB1 exposure affected the expression of lncRNAs associated with cellular senescence signaling pathways, such as MSTRG.24315 and MSTRG.80767, as well as related genes, Cxcl8 and Gadd45g. In addition, AFB1 exposure affected the expression of miRNAs associated with immune-related genes, such as miR-181a, miR-331-3p and miR-342, as well as immune-related genes Nfkb1 and Rras2. Moreover, the regulation networks between mRNA-miRNAs and mRNA-lncRNAs were confirmed by the results of RT-qPCR and immunofluorescence. In conclusion, our results here demonstrate that AFB1 exposure impaired proliferation of 3D4/2 cells via the non-coding RNA-mediated pathway.

Keywords: aflatoxin B1; porcine; porcine alveolar macrophages; apoptosis; cell cycle

Key Contribution: AFB1 exposure impaired proliferation of 3D4/2 cells via non-coding RNA-mediated pathway.

Citation: Chao, H.; Ma, H.; Sun, J.; Yuan, S.; Dong, P.; Zhao, A.; Li, L.; Shen, W.; Zhang, X. Whole-Transcriptome Analysis of Non-Coding RNA Alteration in Porcine Alveolar Macrophage Exposed to Aflatoxin B1. *Toxins* **2022**, *14*, 373. https://doi.org/10.3390/toxins14060373

Received: 26 April 2022
Accepted: 25 May 2022
Published: 27 May 2022

1. Introduction

Mycotoxin is the toxic metabolite of mold, which is limited to some strains of a few toxigenic molds. Different molds can produce the same mycotoxin, while one strain can produce several mycotoxins [1]. At present, about 200 kinds of mycotoxins have been found, and a few of them can cause poisoning in animals and humans under natural conditions [2]. The most important mycotoxins are aflatoxin B1, ochratoxin A, zearalenone, fumin and deoxynivalenol. Mycotoxins can pollute all types of food and feed and can threaten human and animal health through food chain accumulation, producing hepatotoxicity, nephrotoxicity, neurotoxicity, hematopoietic tissue toxicity, etc. [3–5]. Some mycotoxins are mutagenic and carcinogenic [6,7].

Aflatoxin B1 is the most toxic mycotoxin; it was listed as a Group 1 carcinogen by the International Agency for Research on Cancer (IARC) in 1996 [8]. AFB1 mainly targets the liver of humans and animals, where it is metabolized by cytochrome 450 into carcinogenic AFB1-8,9-exo-epoxide (AFBO). AFBO combines with phase II enzymes such as glutathione-S-transferase (GST) to form afb1-thiol acid (AFB1-NAC), which is excreted with urine. AFBO can also combine with DNA to form AFB1-N7-Gua, causing DNA mutation [9,10]. AFB1 reduces steroid production by competitively binding StAR protein of rats, affects the secretion of estradiol-17β and progesterone in animal serum, inhibits the growth of oocytes and leads to the decrease of ovarian size and weight [11,12]. In male mice, AFB1 exposure is related to histological changes of testis, reduction of sperm number and differences in sperm motility and litter size [13,14]. In primary broiler hepatocytes, AFB1 results in an increase in mitochondrial ROS production, a decrease in mitochondrial membrane potential and an induction of apoptosis. This is related to the upregulation of *Nrf2* gene expression and downregulation of NAD(P)H: quinine oxidoreductase 1, SOD and HO-1 [15]. AFB1, as a potential endocrine disruptor, can affect the expression of aromatase enzymes (P450s or CYPs enzymes) [16].

Epigenetic modification includes DNA methylation, ncRNA (miRNA, lncRNA and circular RNA) and post-translational modification (PTM) (glycosylation, methylation, acetylation, phosphorylation and ubiquitination) [17,18]. ncRNA participates in various biological processes, and abnormal expression of ncRNA always destroys the balance in vivo and leads to diseases [19,20]. At present, most studies involving ncRNAs focus on miRNA, circRNA and lncRNA. LncRNA is involved in X chromosome silencing, genome imprinting, chromatin modification, transcription activation, transcription interference, nuclear transport and other important regulatory processes, such as apoptosis. LncRNA is often used to study toxicological mechanisms [21,22].

This study was designed to determine the mechanism of lncRNA and microRNA targeting regulatory genes in Porcine alveolar macrophage cells in response to the toxic effects of AFB1. We generated differential expression profiles of lncRNA and miRNA in Porcine alveolar macrophage cells with and without AFB1 exposure. The findings of this research provide the molecular mechanisms involved in the development of AFB1-induced hepatotoxicity and enrich the valuable resources for lncRNA and miRNA in toxicological research.

2. Results

2.1. AFB1 Inhibited Cell Proliferation and Induced Cell Apoptosis

EdU assay is a commonly used method for detecting cell proliferation. In order to deeply understand the molecular mechanism of porcine 3D4/2 cell cytotoxicity induced by AFB1 exposure, porcine 3D4/2 cells were treated with 40 μm AFB1 in vitro for 24 h. The whole experimental design is shown in Figure 1A. Compared with the untreated group, cells treated with 40 μm AFB1 had significant morphological differences (Figure 1B). The proliferation ability of porcine 3D4/2 cells was checked with an EdU kit, and the results showed that the number of EdU-positive cells in the AFB1 treatment groups decreased significantly compared with the control group (Figure 1C). After AFB1 treatment, the number of TUNEL-positive cells was significantly increased (23.5%) compared with the control group (2.2%) (Figure 1D).

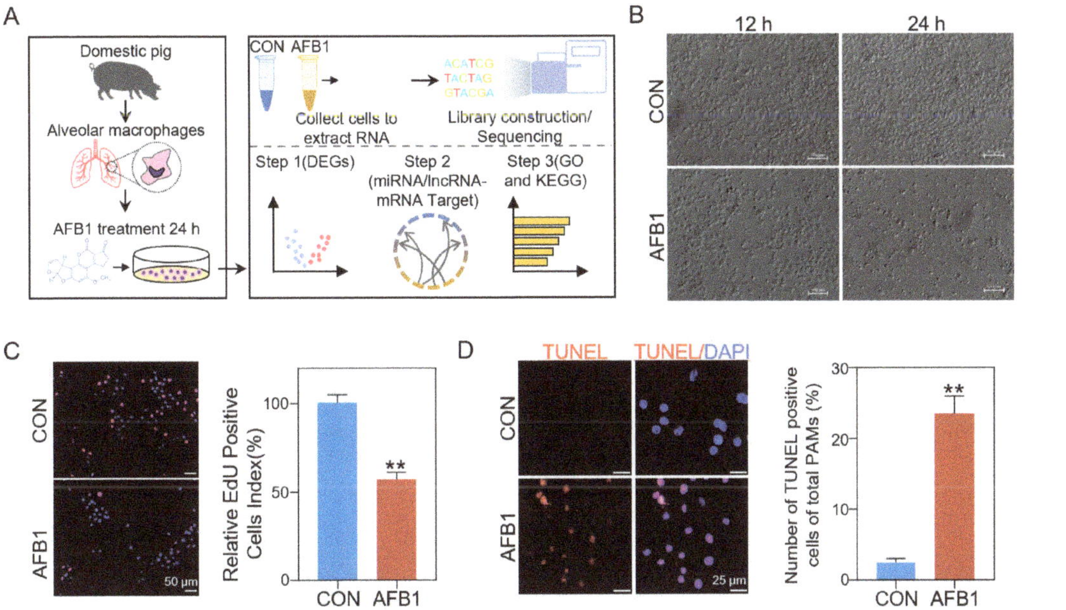

Figure 1. Sequence and data preprocessing of porcine 3D4/2 cell. (**A**) The Schematic diagram of sample treatment and RNA sequencing procedure. (**B**) The morphological changes of porcine 3D4/2 cells exposed to AFB1 for 12 and 24 h in vitro. Scale bar, 100 mm. (**C**) Representative immunofluorescent images of EdU positive cells (red) and the cell nuclei (blue) in the control and AFB1-exposed cells after 24 h (**left**), and the percentages of EdU positive cells (**right**). ** $p < 0.01$. (**D**) Percentages of TUNEL-positive cells treated for 24 h in different groups (**left**) and number of TUNEL-positive cells of the total cells (%, (**right**)). ** $p < 0.01$. All experiments were repeated 3 times.

2.2. AFB1 Exposure Affected lncRNA and mRNA Expression of Porcine 3D4/2 Cells

Ribonucleic acid sequencing (RNA-seq) was utilized to explore the effect of AFB1 on the expression of lncRNAs and mRNAs in porcine 3D4/2 cells. Figure 2A shows the number of lncRNA and mRNA transcripts in the control and AFB1-treated groups. Based on the principal component analysis (PCA), the various lncRNA and mRNA datasets with the same treated methods were highly similar, respectively (Figure 2B). According to the volcano plots of differentially expressed genes (DEmRNAs) and lncRNAs (DElncRNAs) of porcine 3D4/2 cells (Figure 2C), there were 4589 and 1308 downregulated mRNA and lncRNA in the control versus the AFB1-treated group, respectively, and the upregulated mRNAs and lncRNAs were 7069 and 2195, respectively (Figure 2D). The distribution and expression of each lncRNA and mRNA on each chromosome are displayed by the chord diagram (Figure 2E). Gene Ontology (GO) and Kyoto Encyclopedia of Genes and Genomes (KEGG) enrichment analyses were performed based on DEmRNAs. The enriched GO terms included cell adhesion, biological adhesion, vesicle mediated transport, response to endogenous stimulus, negative regulation of signal transduction, and others (Figure 2F). KEGG analysis was used to examine pathway enrichment (Figure 2G). The top 15 enriched KEGG pathways involved the MAPK signaling pathway, P13K-Akt signaling pathway, Hippo signaling pathway, Camp signaling pathway, mTOR signaling pathway, TNF signaling pathway and Focal adhesion (Figure 2G).

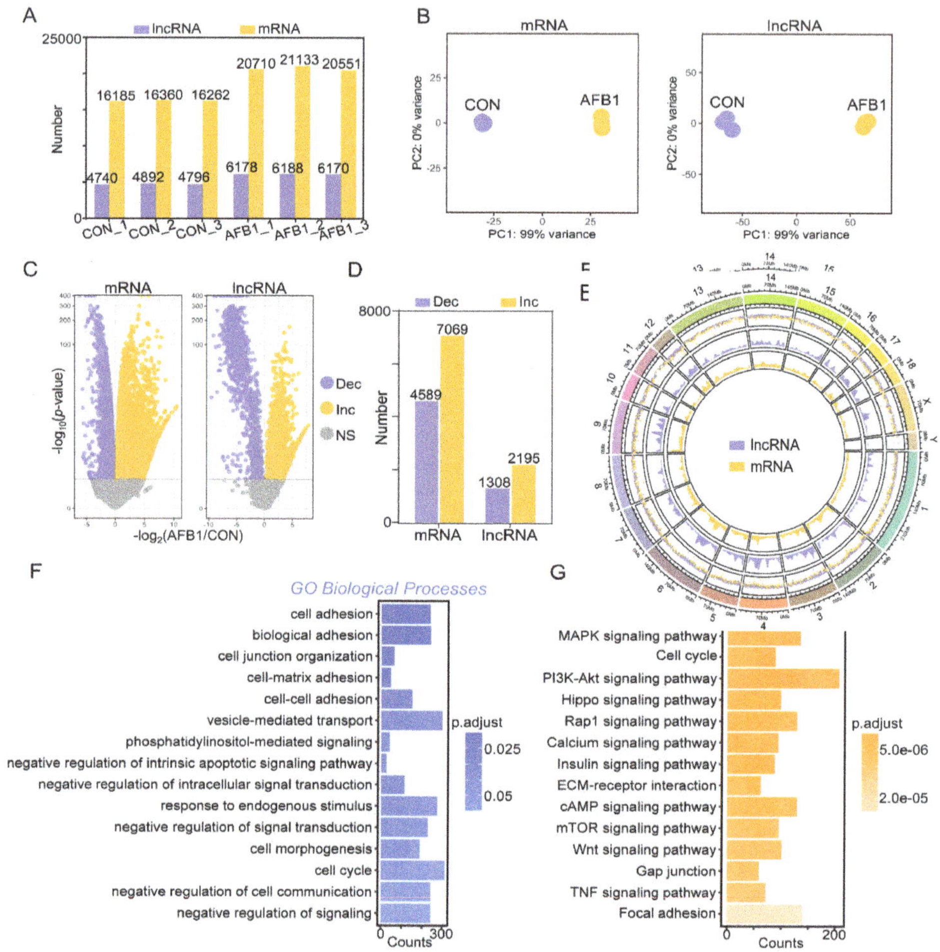

Figure 2. Divergent expression patterns of lncRNA and mRNA of porcine 3D4/2 cells. (**A**) The number of lncRNA and mRNA transcripts in control and AFB1-treated groups. (**B**) Principal component analysis (PCA) based on lncRNA and mRNA. (**C**) The volcano plots of differentially expressed genes (DEGs) and lncRNAs (DELs) of porcine 3D4/2 cells. (**D**) The number of upregulated and downregulated DEGs and DELs of control vs. AFB1-treated group. (**E**) The chord diagram showing the distribution and expression of differentially expressed lncRNAs and mRNAs in the chromosome. (**F**) GO and (**G**) KEGG enrichment of DEGs.

2.3. Co-Expression Analysis of DELs and DEGs in Porcine 3D4/2 Cells

To accurately identify the regulatory mechanisms of lncRNAs and mRNAs, we performed co-expression analysis based on Differentially expressed genes (DEGs) and Differentially expressed lncRNAs(DELs). After filtering according to *p*-value ($p < 0.01$) and Pearson correlation coefficient, 3479 mRNAs and 248 lncRNAs were obtained (Figure 3A,B). The heatmap was plotted according to the expression of mRNA and lncRNA (Figure 3C), which were all related to component organization biogenesis, process regulation metabolic, cycle mitotic cell and localization transport establishment (Figure 3D). KEGG pathway analysis was performed to elucidate the function of co-expressed genes (Figure 3E). We obtained similar enrichment results as above, including cellular senescence, cell cycle, mitogen-activated protein kinase (MAPK) signaling pathway, Tumor necrosis factor (TNF) signaling pathway, p53 signaling pathway and phosphatidylinositol 3 kinase-protein kinase B (PI3K-Akt) signaling pathway, which were all related to apoptosis, indicating that AFB1 exposure affected mRNA and lncRNA expression of 3D4/2 cells and led to cell apoptosis (Figure 3E). For trend analysis of RNA data sets, we also performed GSEA analysis. The results showed that the gene expression related to the focal adhesion pathway was upregulated (Figure 3F). The heatmap shows the gene expression in cellular senescence signaling pathway (Figure 3G).

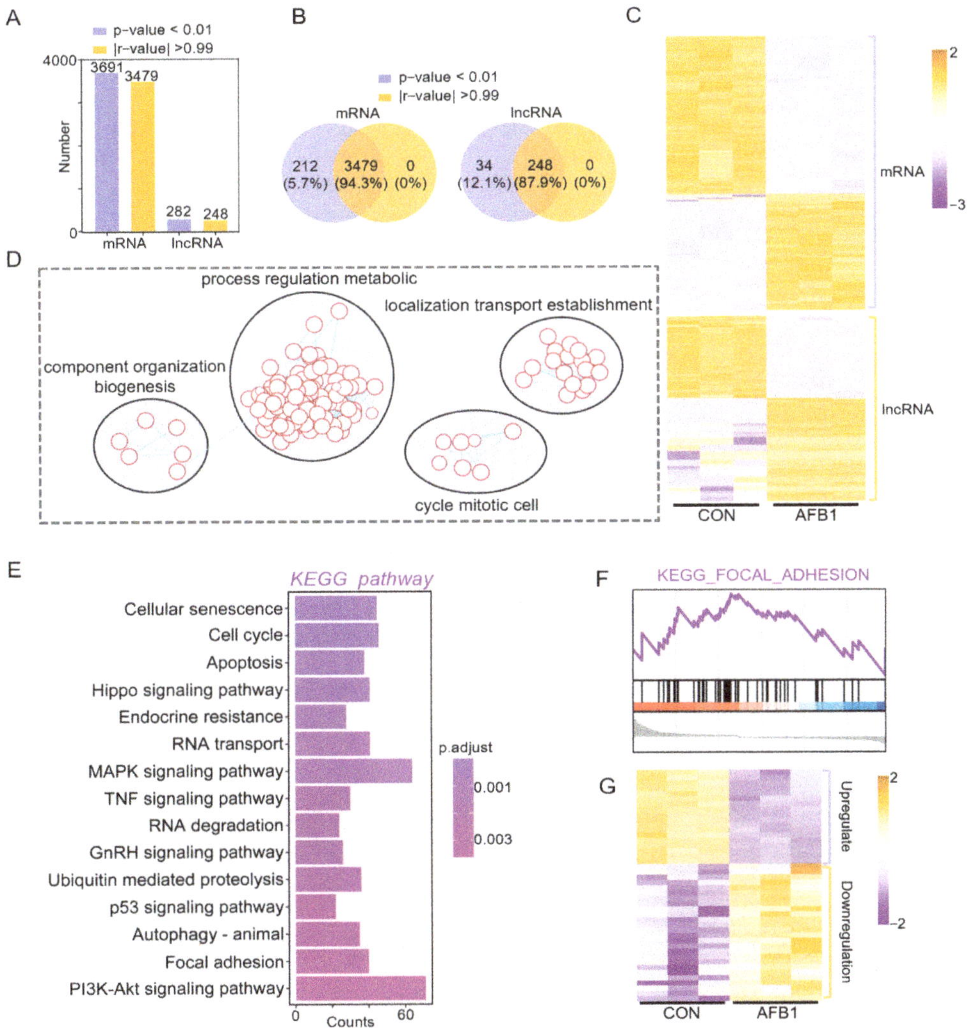

Figure 3. Co-expression analysis of DELs and DEGs in porcine 3D4/2 cells. (**A**) Identification of pairwise lncRNA-mRNA in co-expression analysis; the histogram showing the number of mRNA and lncRNA with p-value < 0.01 and |r-value| > 0.99. (**B**) The Venn diagram showing the number of mRNA and lncRNA with p-value < 0.01 and |r-value| > 0.99. (**C**) The heatmap showing the expression of mRNA and lncRNA. (**D**) Functional enrichment analysis of co-expressed genes. (**E**) KEGG enrichment results of co-expressed genes. (**F**) GSEA enrichment results of co-expressed genes. (**G**) The heatmap showing gene expression in the cellular senescence signaling pathway.

2.4. Cis-Regulation of mRNA and lncRNAs with Target Genes

Based on the Venn plots of unique lncRNAs or mRNA in cis-regulation with co-expressed lncRNAs or mRNA, 196 key lncRNAs and 1704 key mRNAs were found (Figure 4A). Subtype statistics of key cis-regulatory mRNA were assayed. According to the illustration, "Genic" includes the subtypes of "overlapping", "containing", and "nested"; "Intergenic" consists of "same strand", "convergent", and "divergent" subtypes (Figure 4B). Genome location statistics showed that the key cis-regulatory mRNAs were located upstream (8.99%), intronic (50.83%), exonic (34.59%) and downstream (5.59%), respectively (Figure 4C). To understand the function of DElncRNAs target genes, we explored the function of these target genes using KEGG analysis (Figure 4D). KEGG analysis showed that there are 15 significantly enriched signal pathways with DElncRNAs, including cellular senescence, cell cycle, MAPK signaling pathway, IL-17 signaling pathway, TNF signaling pathway, autophagy, etc. (Figure 4D). Next, based on co-expression and co-localization analysis, we found that DElncRNAs regulated key genes in cellular senescence signaling pathways, and we display them through chord diagrams (Figure 4E). The relative expression of these genes was determined by Immunofluorescence Staining (Figure 5). Compared to the control group, the expression levels of CXCL8 and GADD45G were significantly up-regulated (Figure 5), which was consistent with the whole transcriptome sequence results.

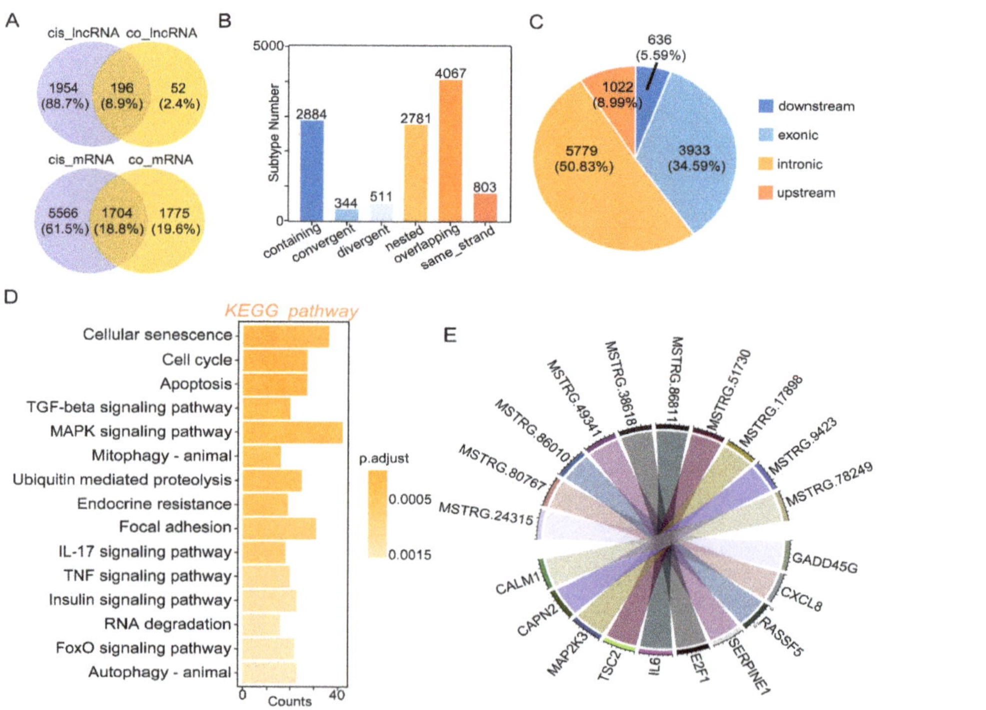

Figure 4. Cis-regulation of lncRNAs with target genes. (**A**) The Venn plot of unique lncRNAs or mRNAs in cis-regulation with co-expressed lncRNAs or mRNA. (**B**) Subtype statistics of key cis-regulatory mRNA. According to the illustration, "Genic" includes the subtypes of "overlapping", "containing", and "nested"; "Intergenic" consists of "same strand", "convergent", and "divergent" subtypes. (**C**) Genome location statistics of key cis-regulatory mRNA. (**D**) KEGG enrichment analysis of target genes regulated by lncRNA homeopathy. (**E**) The chord graph showing the targeting relationship between key genes in the cellular senescence signaling pathway and lncRNA.

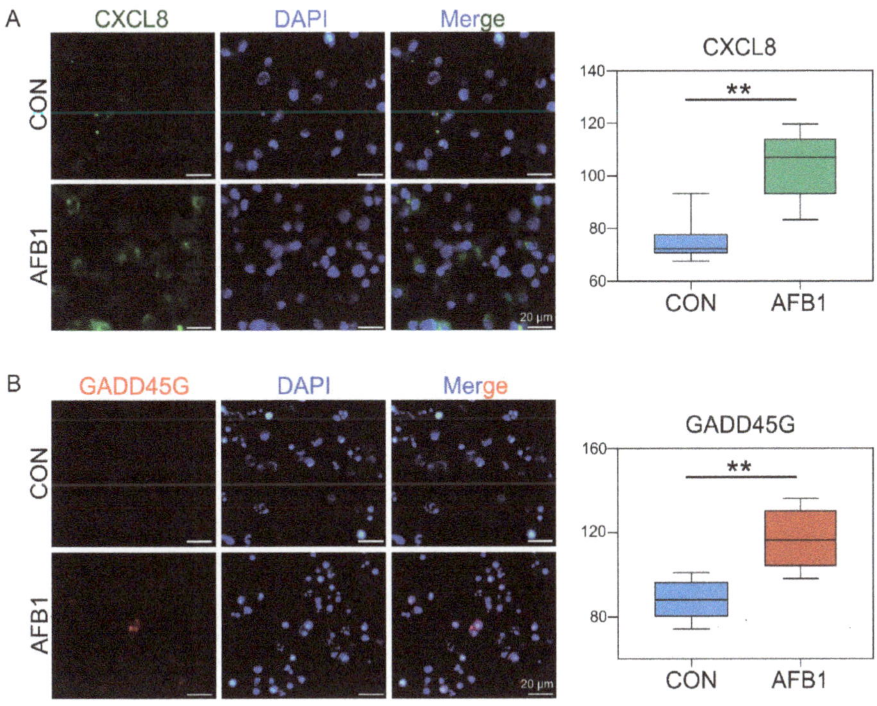

Figure 5. Cell immunofluorescence assay of the expression of CXCL8 and GADD45G proteins. (**A**) The fluorescence intensity and positive percentages of CXCL8. Nuclear staining (blue) and CXCL8-positive cells (green). (**B**) The fluorescence intensity and positive percentages of GADD45G. Nuclear staining (blue) and GADD45G-positive cells (red). ** Indicates extremely significant differences ($p < 0.01$). All experiments were repeated 3 times.

2.5. AFB1 Exposure Altered miRNA Expression

To investigate the impact of AFB1 on miRNA in porcine 3D4/2 cells, we performed differential expression analysis of miRNA between three control groups and three AFB1-treated groups (Figure 6A). The volcano plot was used to show the distribution of differentially expressed miRNAs (Figure 6B). Compared with the control group, a total of 5 DEmiRNAs were upregulated and 6 DEmiRNAs were downregulated in the AFB1-treated group. The change of DEmiRNA expression in the different groups is shown in the heatmap (Figure 6C). Moreover, TargetScan, miRanda and RNAhybrid software (http://www.targetscan.org/, http://www.microrna.org/microrna/home.do, and http://bibiserv.techfak.uni-bielefeld.de/rnahybrid/, accessed on 25 April 2022) was used for predicting the DEmiRNA-related genes (Figure 6D,E), and finally 205 genes were predicted (Figure 6E). The genes were found to be enriched in GO terms associated with the negative regulation of biological processes, execution phase of apoptosis, system development, translation, regulation of cell communication, regulation of growth and regulation of cellular response to growth factor stimulus (Figure 6F). The KEGG analysis showed that the target genes were enriched in the Ras signaling pathway, MAPK cell signaling pathway, Rap1 signaling pathway, P13k-Akt signaling pathway, cAMP signaling pathway, calcium signaling pathway, etc. (Figure 6G). We show the targeting relationship between key genes and miRNAs in the Ras signaling pathway (Figure 6H).

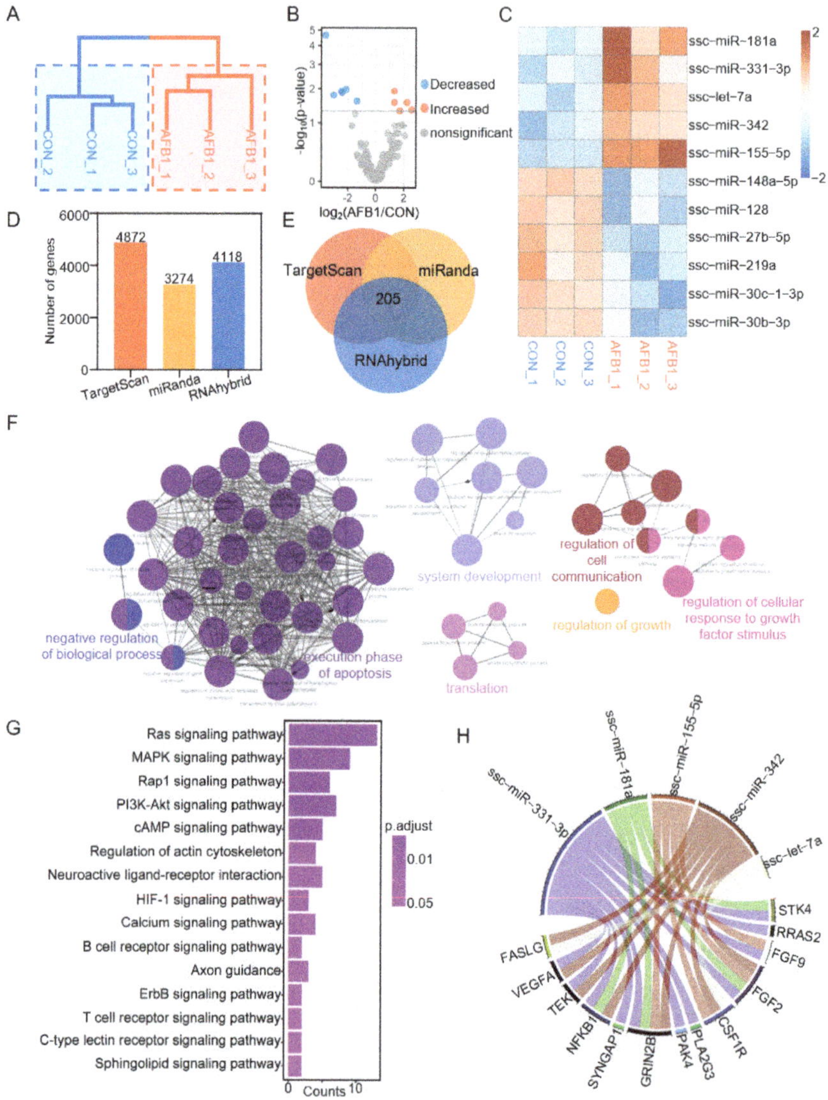

Figure 6. AFB1 exposure alters the miRNA expression levels of porcine 3D4/2 cells. (**A**) The cluster dendrogram between AFB1-treated groups and the control groups based on miRNA. (**B**) The volcano diagram showing the distribution of DEmiRNAs between the AFB1-treated group and the control group. (**C**) The heatmap demonstrating the expression level of DEmiRNAs in six samples. (**D**) The histogram showing the number of miRNAs and the number of target genes. (**E**) The Venn diagram showing the number of target genes shared by DEmiRNAs in TargetScan, miRanda and RNAhybrid. (**F**) GO enrichment analysis of DEmiRNA target genes. (**G**) KEGG enrichment analysis of DEmiRNA target genes. (**H**) The circos diagram indicating the targeting relationship of the miRNA-mRNA network in porcine 3D4/2 cells.

We selected genes shared among the DEmRNA and DEmiRNA target genes (Figure S2A). We use the heatmap to show the expression of key genes in different groups (Figure S2B). Subsequently, we performed KEGG enrichment analysis on the 74 target genes (Figure S2C).

The results indicated that the Ras signaling pathway was mainly regulated by miRNAs. Using genetic interactions and co-expression networks, we found that NFKB1 and RRAS2 play key roles in the Ras signaling pathway (Figure S2D). The targeting relationships between miRNAs and key genes are shown in Figure S2E. The relative expression of DEmiRNAs was determined by RT-qPCR. Compared with the control group, the expression levels of miR-181a, miR-331-3p and miR-342 in the AFB1 treatment group were significantly upregulated (Figure 7A). To verify the expression of the previous related genes *Nfkb1* and *Rras2* after AFB1 exposure, we further analyzed their expression using immunofluorescence. The fluorescent intensity of NFKB1 and RRAS2 genes was significantly decreased in the AFB1-treated group compared with that of the control group (Figure 7B,C). These results were consistent with the data of the whole-transcriptome sequence.

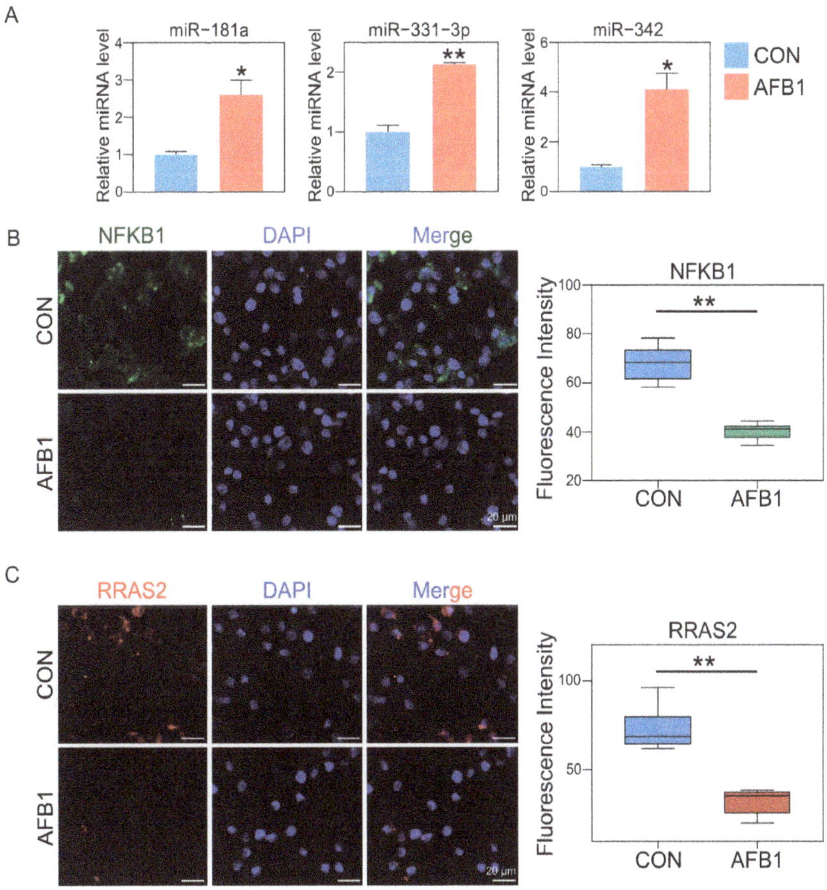

Figure 7. Validation of miRNA-seq data availability with RT-qPCR and examination of the expression of NFKB1 and RRAS2 proteins with cell immunofluorescence assay. (**A**) Expression of miR-181a, miR-331-3p and miR-342 in 3D4/2 cells after 24 h AFB1 exposure. * Indicates extremely significant differences ($p < 0.05$), ** Indicates extremely significant differences ($p < 0.01$). All experiments were repeated 3 times. (**B**) The fluorescence intensity and positive percentages of NFKB1. Nuclear staining (blue) and NFKB1 positive cells (green). (**C**) The fluorescence intensity and positive percentages of RRAS2. Nuclear staining (blue) and NFKB1 positive cells (red). ** Indicates extremely significant differences ($p < 0.01$). All experiments were repeated 3 times.

3. Discussion

Aflatoxin can induce mutation, inhibit immunity and cause cancer. The liver tissue is the main target organ of aflatoxin, which can lead to liver cancer and even death in severe cases [23,24]. Acute poisoning of animals can lead to serious damage to blood vessels and the central nervous system, and animals may die within several hours to several days after poisoning. Chronic poisoning is characterized by poor appetite, weight loss, decreased production performance, decreased carcass and eggshell quality, liver injury, inhibition of animal immune function and carcinogenesis. Aflatoxin has immunosuppressive properties [25]. Intake of contaminated feed will increase the susceptibility to infection and reduce the immunity of vaccines. AFB1 mainly affects cellular immunity. It can reduce the total number of lymphocytes, especially the total number of circulating activated lymphocytes, inhibit the production of lymphocytes and damage the delayed hypersensitivity and graft-versus-host reaction of skin [26]. AFB1 can also reduce the lysis of natural killer cells and the function of macrophages, such as thiophene swallowing activity, intracellular killing or production of oxidative free radicals [27]. In vitro analysis of mouse peritoneal macrophages exposed to AFB1 showed that the expression of IL-lα and IL-6 or TNF-α increased [28]. Blood lymphocytes of pigs fed with food containing AFB1 feed contaminant were catalyzed by mitogens, and the expression of IL-lP decreased while the expression of IL-10 increased [29]. In addition, studies have shown that AFB1 affects swine growth performance, apparent total tract digestibility and intestinal health, seriously impairing the development of the swine industry [30].

LncRNA plays an important role in many life activities, such as dose compensation effect, epigenetic regulation, cell cycle regulation and cell differentiation regulation [31]. Similarly, microRNA (miRNA) plays a variety of important regulatory roles in cells. Like transcription factors, miRNA regulates gene expression and plays a great role in cell differentiation, biological development and disease occurrence and development, which has attracted more and more attention from researchers [32]. RNA-seq data analysis of mRNA, microRNA and lncRNA provides new clues for gene expression profile and transcriptional regulation in animal cells in response to mycotoxin exposure, and helps to detect biomarkers and drug targets for predicting and controlling mycotoxin contamination [33]. The expressions of lncRNA and miRNA are analyzed by lncRNA microarray, which proves that Zearalenone (Zea) and imprinted lncRNAs are closely related to reproduction and development [34]. Zhang et al. showed that Zearalenone (ZEA) can activate the JAK2–STAT3 signaling pathway through the two lncRNAs MSTRG.22680 and MSTRG.23882 to induce cell apoptosis [35]. ZEN causes toxicological effects by regulating the expression of miRNA and miRNA target genes [36]. Some reports have focused on the role of ncRNAs (miRNA and lncRNA) in AFB1-induced toxicity, especially the relationship between AFB1 and HCC [37]. ADAMTS4, the targeted mir-1268a gene, is affected by pre miRNA polymorphism to reveal the risk of AFB1 related hepatocellular carcinoma(HCC) [38]. In the toxicological study of AFB1, it was also found that AFB1 can affect the expression of miRNAs and lncRNA in the liver, result in liver fat deposition and hepatocyte apoptosis, and induce hepatotoxicity [21]. Our results revealed that AFB1 exposure affected the expression of miRNAs such as ssc-miR-181a, ssc-miR-331-3p and ssc-miR-342 and affected the expression of lncRNAs such as MSTRG.24315 and MSTRG.80767.

AFB1 can affect the expression of many genes. Chemokines are small proteins that control a variety of tissue functions, including cell recruitment and activation under homeostatic and inflammatory conditions. CXCL8 (Interleukin-8) is a member of the chemokine family and acts on CXCR1 and CXCR2 receptors. CXCL8 and its receptors help eliminate pathogens but may also contribute significantly to disease-related processes, including tissue damage, fibrosis, angiogenesis and tumorigenesis [39]. IL-8 is related to a variety of inflammation and chemotaxis and participates in the occurrence of many diseases. The main biological function of IL-8 is to contribute to the chemotaxis of neutrophils, T lymphocytes and basophil (Basophils) during inflammation, and its chemotaxis are different in different cells [40]. GADD45G protein is a stress protein that responds to the environment.

As a stress-sensitive factor, it plays an important role in response to toxic and non-toxic stress responses. It also plays an important regulatory role in many cell functions such as DNA repair, cell cycle regulation and senescence, toxic stress response of genes, inducing cell cycle arrest and apoptosis [41,42]. After AFB1 treatment, as one of the major mediators of the inflammatory response, CXCL8 was upregulated with the highest fold change [43]. In addition, AFB1 exposure induced the expression of Cxcl8 and Gadd45g genes in 3D4/2 cells (Figure 5) in our study, leading to cell inflammation, DNA repair, cell cycle arrest and apoptosis, which is consistent with the results of Figures 2F and 3E. Similarly, the mRNA level of Il6 in the liver of broilers exposed to AFB1 was significantly higher than that of the control group [44].

The *Nfkb1* gene is considered to be anti-apoptotic. In liver cells, increased expression of Nfkb1 has been shown to upregulate other inflammatory genes, such as Tnfa and Il6 [45,46]. However, improper activation of NF-κB is associated with a variety of inflammatory diseases, while persistent inhibition of NF-κB leads to improper development of immune cells or delayed cell growth [47]. AFB1 exposure affected the development of macrophages by inhibiting NF-κB (Figure 7). RRAS2 is necessary for the proliferation of human CLL cells. Rras2 encodes a protein that binds to the plasma membrane and plays an important role in activating the signal transduction pathway that controls cell proliferation. RRAS2 is associated with the BCR in leukemic cells and is required for human CLL cell proliferation [48]. The treatment of AFB1 decreased the expression of RRAS2, thus inhibiting the proliferation of cells (Figures 1C and 7). It is worth noting that studies have shown that curcumin successfully alleviated AFB1-induced oxidative stress, inflammation and apoptosis in broiler liver by regulating the expression of lncRNA [49]. This suggests that our study may provide a therapeutic target for the swine industry to control AFB1 toxicity.

4. Materials and Methods

4.1. In Vitro AFB1 Treatment of Porcine Alveolar Macrophages

AFB1 (AFB1, A832707, Macklin, Shanghai, China) was dissolved in Dimethyl sulfoxide (DMSO) and stored at −20 °C until use. Porcine alveolar macrophages (3D4/2, ATCC: CRL-2845) were cultured in 96-well plates or 6 cm culture dishes (Corning, 430166, New York, NY, USA) for AFB1 treatment at the concentration of 40 μg/mL.

4.2. EdU Staining for Proliferation

An EdU Assay/EdU Staining Proliferation Kit (Beyotime, C00755, Shanghai, China) was used to detect and quantify cell proliferation in porcine alveolar macrophage cells using flow cytometry. Proliferating cells were stained for incorporated EdU against total DNA content using Hoechst.

4.3. TUNEL Staining

Cells were collected after 24 h of AFB1 treatment and then fixated by 4% Paraformaldehyde. The One Step TUNEL Apoptosis Assay Kit (Beyotime, C1086, Shanghai, China) was used to examine apoptosis cells according to the instructions. After sealing with anti-fluorescence quenching sealing solution, cells were observed under a fluorescence microscope (Olympus, BX51, Tokyo, Japan). TUNEL-positive cell rates were counted and analyzed using IPWIN software (Meedia Cybernetics, Rockville, MD, USA).

4.4. RNA Extraction and Sequencing of Whole Transcriptome RNA

The total RNA from cells was extracted using RNAprep Microkit pure (Aidlab, RN07, Shanghai, China) according to the manufacturer's instructions. The Illumina TruSeq™ RNA preparation kit (Illumina, San Diego, CA, USA) was used to prepare samples, and the Novogene (Beijing, China) HiSeq 4000 platform was used for sequences.

4.5. Pipeline of Data Processing

FastQC software (http://www.bioinformatics.bbsrc.ac.uk/projects/fastqc/, accessed on 25 April 2022) and Fastp software (https://github.com/OpenGene/fastp/, accessed on 25 April 2022) were used to analyze the quality control of sequencing data and eliminate the low-quality reads from the raw data. The clean reads of samples were mapped using STAR software for mRNA. SAMtools was also used to remove reads not mapping in a proper mate-pair, and the featureCounts software was used to assign sequence reads to genomic features (Figure S1A).

4.6. Screening for Candidate lncRNAs

Preliminary filtering was done based on "class_code" type. We then used Coding Potential Calculator (CPC) [50], Coding–Non-coding Index (CNCI) [51] and PfamScan software (http://xfam.org/, accessed on 25 April 2022) to identify lncRNAs [52] (Figure S1B).

4.7. Discovering Differentially Expressed Genes (DEGs) and RNA Target Prediction

Differentially expressed lncRNAs (DElncRNAs) and miRNAs (DEmiRNAs) were assessed using the R Bioconductor/DESeq2 package (https://support.bioconductor.org/, accessed on 25 April 2022). The targeting relationship of DElncRNAs and DEmRNAs (Differentially expressed mRNA) was predicted using the R package Hmisc (https://github.com/harrelfe/Hmisc, accessed on 25 April 2022). We then used FEELnc (v 0.1.1) (https://github.com/tderrien/FEELnc, accessed on 25 April 2022) to find mRNAs cis-regulated by lncRNAs [53]. The target genes of the miRNAs were predicted using TargetScan, miRanda and RNAhybrid [54,55].

4.8. GO Classification and KEGG Enrichment Analysis

Gene Ontology (GO) enrichment analysis was performed using the "org.Ss.eg.db" database (http://bioconductor.org/packages/release/data/annotation/html/org.Ss.eg.db.html, accessed on 25 April 2022) to convert the gene SYMOL to ENTREZID. The Kyoto Encyclopedia of Genes and Genomes (KEGG) pathways was analyzed using the R Bioconductor/Pathview package (http://bioconductor.org/developers/how-to/buildingPackagesForBioc/, accessed on 25 April 2022).

4.9. RT-qPCR

The whole transcriptome RNA from cells was extracted with the EASYspin Plus cellular RNA rapid extraction kit (Aidlab, RN2802, Beijing, China), and reverse transcription was conducted using the MiRcute Plus miRNA First-Strand cDNA Kit (Tiangen, KR211-01, Beijing, China). The MiRcute Plus miRNA qPCR Kit (SYBR Green) (Tiangen, FP411-01, Beijing, China) was employed (primers are shown in Table S1). Relative quantitative PCR data analysis was performed using the difference multiple = $2^{-\Delta\Delta ct}$ method.

4.10. Immunofluorescence Staining

Cells were treated with AFB1 for 24 h in order to perform immunofluorescence staining. The following primary antibodies were used: CXCL8 (IL8, DF6998, Affinity Biosciences, Cincinnati, OH, USA), GADD45G (GADD45G, DF2376, Affinity Biosciences, Cincinnati, OH, USA) and NFKB1 (NFKB1, BF0466, Affinity Biosciences, Cincinnati, OH, USA), RRAS2 (RRAS2, DF9840, Affinity Biosciences, Cincinnati, OH, USA). Goat anti-Rabbit IgG (H + I) (Beyotime, A0521, Nantong, China) was used as the second antibody. The methods of immunofluorescence staining followed the published methods [56].

4.11. Statistical Method

The differences between mean values were statistically tested using Student's t test or one-way ANOVA followed by the Tukey test for multiple comparisons. Comparisons were considered significant at $p < 0.05$ (*) and $p < 0.01$ (**).

Supplementary Materials: The following supporting information can be downloaded at: https://www.mdpi.com/article/10.3390/toxins14060373/s1, Figure S1: Data preprocessing and Candidate lncRNA identification; Figure S2: Target genes of DEmRNAs, DEmiRNAs, and miRNA-mRNA network.; Table S1: The sequences of primers.

Author Contributions: Conceptualization, X.Z.; methodology, S.Y.; software, H.C. and H.M.; investigation, H.C., H.M., J.S. and P.D.; resources, A.Z.; data curation, L.L.; writing—original draft preparation, H.C. and H.M.; writing—review and editing, X.Z.; project administration, W.S. and X.Z.; funding acquisition, X.Z. All authors have read and agreed to the published version of the manuscript.

Funding: This work was supported by High level talents research fund project of Qingdao Agricultural University in China (1120043) to Zhang XF, Science & Technology Fund Planning Projects of Qingdao City (21-1-4-ny-7-nsh), the Natural Science Foundation of Shandong Province of China (ZR2021MC191), the Taishan Scholar Construction Foundation of Shandong province of China (ts20190946), Gene expression analysis of hormone receptor-negative breast cancer with low HER-2 expression and its potential influence on neoadjuvant chemotherapy (212102310658), the Cultivating Fund of Capital Medical University (Grant No. PYZ2017002).

Institutional Review Board Statement: Not applicable.

Informed Consent Statement: Not applicable.

Data Availability Statement: Not applicable.

Conflicts of Interest: The authors declare no conflict of interest.

References

1. Tkaczyk, A.; Jedziniak, P. Mycotoxin Biomarkers in Pigs-Current State of Knowledge and Analytics. *Toxins* **2021**, *13*, 586. [CrossRef] [PubMed]
2. Liew, W.P.; Mohd-Redzwan, S. Mycotoxin: Its Impact on Gut Health and Microbiota. *Front. Cell. Infect. Microbiol.* **2018**, *8*, 60. [CrossRef] [PubMed]
3. Bennett, J.W.; Klich, M. Mycotoxins. *Clin. Microbiol. Rev.* **2003**, *16*, 497–516. [CrossRef] [PubMed]
4. Pleadin, J.; Frece, J.; Markov, K. Mycotoxins in Food and Feed. *Adv. Food Nutr. Res.* **2019**, *89*, 297–345.
5. Alshannaq, A.; Yu, J.H. Occurrence, Toxicity, and Analysis of Major Mycotoxins in Food. *Int. J. Environ. Res. Public Health* **2017**, *14*, 632. [CrossRef]
6. Benkerroum, N. Chronic and Acute Toxicities of Aflatoxins: Mechanisms of Action. *Int. J. Environ. Res. Public Health* **2020**, *17*, 423. [CrossRef]
7. Pócsi, I.; Király, G.; Bánfalvi, G. Antineoplastic potential of mycotoxins. *Acta Microbiol. Immunol. Hung.* **2018**, *65*, 267–307. [CrossRef]
8. Dai, Y.; Huang, K.; Zhang, B.; Zhu, L.; Xu, W. Aflatoxin B1-induced epigenetic alterations: An overview. *Food Chem Toxicol.* **2017**, *109*, 683–689. [CrossRef]
9. Guengerich, F.P.; Johnson, W.W.; Shimada, T.; Ueng, Y.F.; Yamazaki, H.; Langouët, S. Activation and Detoxication of Aflatoxin B1. *Mutat. Res. Fundam. Mol. Mech. Mutagenesis* **1998**, *402*, 121–128. [CrossRef]
10. Gouas, D.; Shi, H.; Hainaut, P. The Aflatoxin-Induced TP53 Mutation at Codon 249 (R249S): Biomarker of Exposure, Early Detection and Target for Therapy. *Cancer Lett.* **2009**, *286*, 29–37. [CrossRef]
11. Supriya, C.H.; Akhila, B.; Pratap, R.K.; Girish, B.P.; Sreenivasula, R.P. Effects of Maternal Exposure to Aflatoxin B1 during Pregnancy on Fertility Output of Dams and Developmental, Behavioral and Reproductive Consequences in Female Offspring Using a Rat Model. *Toxicol. Mech. Methods* **2016**, *26*, 202–210. [CrossRef] [PubMed]
12. Kourousekos, G.D.; Theodosiadou, E.K.; Lymberopoulos, A.G.; Belibasaki, S.; Boscos, C. Effect of Aflatoxin B1 on Blood Serum Oestradiol-17β and Progesterone Concentrations during the Luteal Phase and the Synchronized Oestrus of Goats. *Anim. Reprod.* **2018**, *15*, 75–83. [CrossRef] [PubMed]
13. Huang, W.; Cao, Z.; Zhang, J.; Ji, Q.; Li, Y. Aflatoxin B1 Promotes Autophagy Associated with Oxidative Stress-Related PI3K/AKT/mTOR Signaling Pathway in Mice Testis. *Environ. Pollut.* **2019**, *255*, 113317. [CrossRef]
14. Cao, Z.; Shao, B.; Xu, F.; Liu, Y.; Li, Y.; Zhu, Y. Protective Effect of Selenium on Aflatoxin B1-Induced Testicular Toxicity in Mice. *Biol. Trac. Elem. Res.* **2017**, *180*, 233–238. [CrossRef] [PubMed]
15. Liu, Y.; Wang, W. Aflatoxin B1 Impairs Mitochondrial Functions, Activates Ros Generation, Induces Apoptosis and Involves Nrf2 Signal Pathway in Primary Broiler Hepatocytes. *Anim. Sci. J.* **2016**, *87*, 1490–1500. [CrossRef]
16. Storvik, M.; Huuskonen, P.; Kyllönen, T.; Lehtonen, S.; El-Nezami, H.; Auriola, S.; Pasanen, M. Aflatoxin B1—A Potential Endocrine Disruptor–up-Regulates CYP19A1 in JEG-3 cells. *Toxicol. Lett.* **2011**, *202*, 161–167. [CrossRef]

17. Skinner, M.K.; Ben, M.M.; Sadler-Riggleman, I.; Beck, D.; Nilsson, E.; McBirney, M.; Klukovich, R.; Xie, Y.M.; Tang, C.; Yan, W. Alterations in Sperm DNA Methylation, Non-Coding RNA and Histone Retention Associate with DDT-Induced Epigenetic Transgenerational Inheritance of Disease. *Epigenet. Chromatin* **2018**, *11*, 8. [CrossRef]
18. Beck, D.; Ben, M.M.; Skinner, M.K. Integration of Sperm ncRNA-Directed DNA Methylation and DNA Methylation-Directed Histone Retention in Epigenetic Transgenerational Inheritance. *Epigenet. Chromatin* **2021**, *14*, 6. [CrossRef]
19. Mathieu, E.L.; Belhocine, M.; Dao, L.T.; Puthier, D.; Spicuglia, S. Functions of lncRNA in Development and Diseases. *Med. Sci.* **2014**, *30*, 790–796.
20. Ma, Y.; Zhang, J.; Wen, L.; Lin, A. Membrane-Lipid Associated lncRNA: New Regulator in Cancer Signaling. *Cancer Lett.* **2018**, *419*, 27–29. [CrossRef]
21. Liu, X.; Kumar, M.S.; Wang, T.; Xu, Z.; Zhao, X.; Wang, Y.; Yin, H.; Fan, X.; Zeng, B.; Yang, M.; et al. AFB1 Induced Transcriptional Regulation Related to Apoptosis and Lipid Metabolism in Liver of Chicken. *Toxins* **2020**, *12*, 290. [CrossRef] [PubMed]
22. Merrick, B.A.; Chang, J.S.; Phadke, D.P.; Bostrom, M.A.; Shah, R.R.; Wang, X.; Gordon, O.; Wright, G.M. HAfTs Are Novel lncRNA Transcripts from Aflatoxin Exposure. *PLoS ONE* **2018**, *13*, e0190992. [CrossRef] [PubMed]
23. Meki, A.R.M.; Abdel-Ghaffar, S.K.; El-Gibaly, I. Aflatoxin B1 Induces Apoptosis in Rat Liver: Protective Effect of Melatonin. *Neuroendocrinol. Lett.* **2001**, *22*, 417–426. [PubMed]
24. Jackson, P.E.; Groopman, J.D. Aflatoxin and Liver Cancer. *Best Pract. Res. Clin. Gastroenterol.* **1999**, *13*, 545–555. [CrossRef] [PubMed]
25. Wang, J.; Ogata, M.; Hirai, H.; Kawagishi, H. Detoxification of Aflatoxin B1 by Manganese Peroxidase from the White-Rot Fungus Phanerochaete Sordida YK-624. *FEMS Microbiol. Lett.* **2011**, *314*, 164–169. [CrossRef] [PubMed]
26. Afshar, P.; Roozbeh, N.L.; Shokrzadeh, M.; Ghorbani, H.A.; Naghizadeh, R.S. Bio-Protective Effects of *Lactobacillus plantarum* subsp. *plantarum* against Aflatoxin b1 Genotoxicity on Human Blood Lymphocytes: A Native Probiotic Strain Isolated from Iranian Camel Milk. *Curr. Med. Mycol.* **2020**, *6*, 54–61.
27. Ma, J.; Liu, Y.; Guo, Y.; Ma, Q.; Ji, C.; Zhao, L. Transcriptional Profiling of Aflatoxin B1-Induced Oxidative Stress and Inflammatory Response in Macrophages. *Toxins* **2021**, *13*, 401. [CrossRef]
28. Dugyala, R.R.; Sharma, R.P. The Effect of Aflatoxin B1 on Cytokine mRNA and Corresponding Protein Levels in Peritoneal macrophages and Splenic Lymphocytes. *Int. J. Immunopharmacol.* **1996**, *18*, 599–608. [CrossRef]
29. Marin, D.E.; Taranu, I.; Bunaciu, R.P.; Pascale, F.; Tudor, D.S.; Avram, N.; Sarca, M.; Cureu, I.; Criste, R.D.; Suta, V.; et al. Changes in Performance, Blood Parameters, Humoral and Cellular Immune Responses in Weanling Piglets Exposed To Low Doses of Aflatoxin. *J. Anim. Sci.* **2002**, *80*, 1250–1257. [CrossRef]
30. Pu, J.; Yuan, Q.; Yan, H.; Tian, G.; Chen, D.; He, J.; Zeng, P.; Yu, J.; Mao, X.; Huang, Z.; et al. Effects of Chronic Exposure to Low Levels of Dietary Aflatoxin B1 on Growth Performance, Apparent Total Tract Digestibility and Intestinal Health in Pigs. *Animals* **2021**, *11*, 336. [CrossRef]
31. Liu, N.; Wang, Z.Z.; Zhao, M.; Zhang, Y.; Chen, N.H. Role of Non-Coding RNA in the Pathogenesis of Depression. *Gene* **2020**, *735*, 144276. [CrossRef]
32. Zhang, B.; Wang, Q.; Pan, X. MicroRNAs and Their Regulatory Roles in Animals and Plants. *J. Cell Physiol.* **2007**, *210*, 279–289. [CrossRef] [PubMed]
33. Wang, H.; Jin, J.; Wu, J.; Qu, H.; Wu, S.; Bao, W. Transcriptome and Chromatin Accessibility in Porcine Intestinal Epithelial Cells upon Zearalenone Exposure. *Sci. Data* **2019**, *6*, 298. [CrossRef] [PubMed]
34. Xie, H.; Wang, H.; Li, X.; Ji, H.; Xu, Y. ZEA Exerts Toxic Effects on Reproduction and Development by Mediating Dio3os in Mouse Endometrial Stromal Cells. *J. Biochem. Mol. Toxicol.* **2019**, *33*, e22310. [CrossRef] [PubMed]
35. Zhang, F.L.; Li, N.; Wang, H.; Ma, J.M.; Shen, W.; Li, L. Zearalenone Exposure Induces the Apoptosis of Porcine Granulosa Cells and Changes Long Noncoding RNA Expression to Promote Antiapoptosis by Activating the JAK2–STAT3 pathway. *J. Agric. Food Chem.* **2019**, *67*, 12117–12128. [CrossRef]
36. Wang, M.; Wu, W.; Li, L.; He, J.; Huang, S.; Chen, S.; Chen, J.; Long, M.; Yang, S.; Li, P. Analysis of the miRNA Expression Profiles in the Zearalenone-Exposed TM3 Leydig Cell Line. *Int. J. Mol. Sci.* **2019**, *20*, 635. [CrossRef]
37. Zhu, L.Y.; Yuhan, J.Y.; Huang, K.L.; He, X.Y.; Liang, Z.H.; Xu, W.T. Multidimensional Analysis of the Epigenetic Alterations in Toxicities induced by Mycotoxins. *Food Chem. Toxicol.* **2021**, *153*, 112251. [CrossRef]
38. Long, X.D.; Huang, X.Y.; Ya, O.J.G.; Liao, P.H.; Tang, Y.J.; Ma, Y.; Xia, Q. Polymorphisms in the Precursor microRNAs and Aflatoxin B1-Related Hepatocellular Carcinoma. *Mol. Carcinog.* **2016**, *55*, 1060–1072. [CrossRef]
39. Russo, R.C.; Garcia, C.C.; Teixeira, M.M.; Amaral, F.A. The CXCL8/IL-8 Chemokine Family and Its Receptors in Inflammatory Diseases. *Expert Rev. Clin. Immu.* **2014**, *10*, 593–619. [CrossRef]
40. Krieger, M.; Brunner, T.; Bischoff, S.C.; von Tscharner, V.; Walz, A.; Moser, B.; Baggiolini, M.; Dahinden, C.A. Activation of Human Basophils through the IL-8 Receptor. *J. Immunol.* **1992**, *149*, 2662–2667.
41. Zhang, L.; Yang, Z.; Liu, Y. GADD45 Proteins: Roles in Cellular Senescence and Tumor Development. *Exp. Biol. Med.* **2014**, *239*, 773–778. [CrossRef] [PubMed]
42. Rosemary, S.A.; Richardson, D.R. Growth Arrest and DNA Damage-45 Alpha (GADD45 Alpha). *Int. J. Biochem. Cell B* **2009**, *41*, 986–989. [CrossRef] [PubMed]
43. Pauletto, M.; Tolosi, R.; Giantin, M.; Guerra, G.; Barbarossa, A.; Zaghini, A.; Dacasto, M. Insights into Aflatoxin B1 Toxicity in Cattle: An in vitro Whole-Transcriptomic Approach. *Toxins* **2020**, *12*, 429. [CrossRef] [PubMed]

44. Ma, Q.; Li, Y.; Fan, Y.; Zhao, L.; Wei, H.; Ji, C.; Zhang, J. Molecular Mechanisms of Lipoic Acid Protection against Aflatoxin b1-Induced Liver Oxidative Damage and Inflammatory Responses in Broilers. *Toxins* **2015**, *7*, 5435–5447. [CrossRef] [PubMed]
45. Baker, R.G.; Hayden, M.S.; Ghosh, S. NF-κB, Inflammation, and Metabolic Disease. *Cell Metab.* **2011**, *13*, 11–22. [CrossRef]
46. Shi, J.; He, J.; Lin, J.; Sun, X.; Sun, C.; Ou, C.; Jiang, C. Distinct Response of the Hepatic Transcriptome to Aflatoxin B1 Induced Hepatocellular Carcinogenesis and Resistance in Rats. *Sci. Rep.* **2016**, *6*, 31898. [CrossRef]
47. Somma, D.; Kok, F.O.; Kerrigan, D.; Wells, C.A.; Carmody, R.J. Defining the Role of Nuclear Factor (NF)-κB p105 Subunit in Human Macrophage by Transcriptomic Analysis of NFKB1 Knockout THP1 Cells. *Front Immunol.* **2021**, *12*, 669906. [CrossRef]
48. Hortal, A.M.; Oeste, C.L.; Cifuentes, C.; Alcoceba, M.; Fernández-Pisonero, I.; Clavaín, L.; Tercero, R.; Mendoza, P.; Domínguez, V.; García-Flores, M.; et al. Overexpression of Wild Type RRAS2, without Oncogenic Mutations, Drives Chronic Lymphocytic Leukemia. *Mol. Cancer* **2022**, *21*, 35. [CrossRef]
49. Li, S.; Liu, R.; Wei, G.; Guo, G.; Yu, H.; Zhang, Y.; Ishfaq, M.; Fazilani, S.A.; Zhang, X. Curcumin Protects against Aflatoxin B1-Induced Liver Injury in Broilers via the Modulation of Long Non-Coding RNA Expression. *Ecotoxicol. Environ. Saf.* **2021**, *208*, 111725. [CrossRef]
50. Kang, Y.J.; Yang, D.C.; Kong, L.; Hou, M.; Meng, Y.Q.; Wei, L.; Gao, G. CPC2: A Fast and Accurate Coding Potential Calculator Based on Sequence Intrinsic Features. *Nucleic Acids Res.* **2017**, *45*, 12–16. [CrossRef]
51. Sun, L.; Luo, H.; Bu, D.; Zhao, G.; Yu, K.; Zhang, C.; Liu, Y.; Chen, R.; Zhao, Y. Utilizing Sequence Intrinsic Composition to Classify Protein-Coding and Long Non-Coding Transcripts. *Nucleic Acids Res.* **2013**, *41*, e166. [CrossRef]
52. Finn, R.D.; Coggill, P.; Eberhardt, R.Y.; Eddy, S.R.; Mistry, J.; Mitchell, A.L.; Potter, S.C.; Punta, M.; Qureshi, M.; Sangrador-Vegas, A.; et al. The Pfam Protein Families Database: Towards a More Sustainable Future. *Nucleic Acids Res.* **2016**, *44*, D279–D285. [CrossRef] [PubMed]
53. Wucher, V.; Legeai, F.; Hedan, B.; Rizk, G.; Lagoutte, L.; Leeb, T.; Jagannathan, V.; Cadieu, E.; David, A.; Lohi, H.; et al. FEELnc: A Tool for Long Non-Coding RNA Annotation and Its Application to the Dog Transcriptome. *Nucleic Acids Res.* **2017**, *45*, e57. [CrossRef] [PubMed]
54. Agarwal, V.; Bell, G.W.; Nam, J.W.; Bartel, D.P. Predicting Effective microRNA Target Sites in Mammalian mRNAs. *elife* **2015**, *4*, e05005. [CrossRef] [PubMed]
55. Kruger, J.; Rehmsmeier, M. RNAhybrid: microRNA Target Prediction Easy, Fast and Flexible. *Nucleic Acids Res.* **2006**, *34*, 451–454. [CrossRef] [PubMed]
56. Tian, Y.; Zhang, M.Y.; Li, N.; Wang, J.J.; Ge, W.; Tan, S.J.; Shen, W.; Li, L. Zearalenone Exposure Triggered Porcine Granulosa Cells Apoptosis via microRNAs-Mediated Focal Adhesion Pathway. *Toxicol. Lett.* **2020**, *330*, 80–89. [CrossRef] [PubMed]

toxins

MDPI

Article

Isolation, Purification, and Characterization of a Laccase-Degrading Aflatoxin B1 from *Bacillus amyloliquefaciens* B10

Dongwei Xiong †, Jun Wen †, Gen Lu †, Tianxi Li and Miao Long *

Key Laboratory of Livestock Infectious Diseases, Ministry of Education, College of Animal Science & Veterinary Medicine, Shenyang Agricultural University, Shenyang 110866, China; 2019220557@stu.syau.edu.cn (D.X.); 2020240593@stu.syau.edu.cn (J.W.); 2019220539@stu.syau.edu.cn (G.L.); 2020240619@stu.syau.edu.cn (T.L.)
* Correspondence: longmiao@syau.edu.cn
† These authors contributed equally to this work.

Abstract: Aflatoxins, widely found in feed and foodstuffs, are potentially harmful to human and animal health because of their high toxicity. In this study, a strain of *Bacillus amyloliquefaciens* B10 with a strong ability to degrade aflatoxin B1 (AFB1) was screened; it could degrade 2.5 μg/mL of AFB1 within 96 h. The active substances of *Bacillus amyloliquefaciens* B10 for the degradation of AFB1 mainly existed in the culture supernatant. A new laccase with AFB1-degrading activity was separated by ammonium sulfate precipitation, diethylaminoethyl (DEAE) and gel filtration chromatography. The results of molecular docking showed that B10 laccase and aflatoxin had a high docking score. The coding sequence of the laccase was successfully amplified from cDNA by PCR and cloned into *E. coli*. The purified laccase could degrade 79.3% of AFB1 within 36 h. The optimum temperature for AFB1 degradation was 40 °C, and the optimum pH was 6.0–8.0. Notably, Mg^{2+} and dimethyl sulfoxide (DMSO) could enhance the AFB1-degrading activity of B10 laccase. Mutation of the three key metal combined sites of B10 laccase resulted in the loss of AFB1-degrading activity, indicating that these three metal combined sites of B10 laccase play an essential role in the catalytic degradation of AFB1.

Keywords: aflatoxin; *Bacillus amyloliquefaciens*; laccase; degradation; molecular docking; mutagenesis

Key Contribution: A novel laccase was isolated and purified, and some key sites for the catalytic degradation of aflatoxin were identified by targeted mutagenesis.

Citation: Xiong, D.; Wen, J.; Lu, G.; Li, T.; Long, M. Isolation, Purification, and Characterization of a Laccase-Degrading Aflatoxin B1 from *Bacillus amyloliquefaciens* B10. *Toxins* **2022**, *14*, 250. https://doi.org/10.3390/toxins14040250

Received: 17 January 2022
Accepted: 29 March 2022
Published: 31 March 2022

1. Introduction

Aflatoxins are difuranocoumarin derivatives produced mainly by strains of *Aspergillus flavus* and *Aspergillus parasiticus*. They are produced during the growth and storage of crops and are chemically and thermally stable. Aflatoxin is highly hepatotoxic, nephrotoxic, acutely toxic, and immunotoxic, and belongs to a class of teratogenic, carcinogenic, and mutagenic compounds. In decreasing order of toxicity, the various metabolites are aflatoxin B1 (AFB1), AFM1, AFG1, AFB2, AFM2, and AFG2 [1–4]. Aflatoxins pose a significant risk to human health through the food chain [5].

It is essential to avoid AFB1 contamination and develop safe and effective detoxification methods to improve food safety. Traditional methods of AFB1 degradation include physical, chemical, and microbiological techniques [6]. Physical methods possess the disadvantage of being time consuming and less efficient in removing aflatoxin, while chemical methods lead to the loss of nutrients in food or feed [7]. In previous research focused on microbial degradation, AFB1 was degraded into nontoxic or less toxic metabolites by microorganisms or enzymes. The furan and lactone rings are the two key sites influencing the toxicity of AFB1 [8]. The existing studies on the microbial degradation of AFB1 also revolve around these two key sites. Some microorganisms are not probiotics, and their safety should be evaluated. Meanwhile, some microorganisms may inadvertently disrupt the nutritional properties of the product or even introduce other toxic substances that are harmful to the organism [9]. Enzymes are a promising choice, being ecofriendly and endowed with high substrate specificity and catalytic efficiency [10].

Enzymes identified to degrade aflatoxins include laccase, oxidoreductase, peroxidase, and manganese peroxidase [11]. Laccase is widely found in bacteria, fungi, insects, and higher plants [12]. Previous studies indicated that laccase can catalyse the oxidation of various compounds such as phenol, aniline, aromatic amines, ascorbic acid, and certain inorganic compounds, and can be coupled to the four-electron reduction of dioxygen to water [6,13–15]. The fungus laccase with high redox potential, isolated from the white-rot fungus *Cerrena unicolor* 6884, could efficiently degrade AFB1 to AFQ1 [16]. The CotA protein, as the best-known bacterial laccase, is predominantly located in the outer endospore layer of *Bacillus subtilis* and other *Bacillus* species and has a larger substrate binding pocket than other laccases [17,18]. Guo et al. cloned and expressed a novel CotA laccase from *Bacillus licheniformis* that converts AFB1 to AFQ1 and epi-AFQ1 [19]. Interestingly, Loi et al. also found that peroxidase converted AFB1 to AFQ1 [20]. Furthermore, many studies have verified that the toxicity of AFQ1 is one order of magnitude lower than that of AFB1 [21,22]. Typically, bacterial laccase exhibits higher thermal and alkaline stability than fungus laccase and other laccase [23]. Therefore, bacterial laccase is a promising biocatalyst to degrade AFB1 in feed and food.

In this study, a strain of *Bacillus amyloliquefaciens* B10 with efficient AFB1-degrading activity was isolated from 74 strains, and the active substance responsible for AFB1 degradation by this strain was localised. A new laccase was found to efficiently degrade AFB1. The laccase was expressed in *E. coli* and the degradation characteristics of recombinant laccase were determined. This study provides a theoretical basis for AFB1 degradation by laccase and promotes the development of the enzymatic degradation of AFB1 in feed and food.

2. Results

2.1. Isolation and Identification of AFB1-Degrading Bacteria

The initial screening of the 74 strains kept in the laboratory revealed that the strain labelled B10 grew relatively well in the medium containing different concentrations of AFB1. Strain B10 was white on LB (Luria-Bertani) medium, slightly elevated, with a rough surface and folded colonies; single colonies were round and 3–4 mm in diameter (Figure 1A). Gram staining was positive with bluntly rounded ends, and rod-shaped cells were seen (Figure 1B). The complete 16S rDNA gene of length 1462 bp and the complete *gyrB* gene of size 1186 bp were obtained through amplification and sequencing with 16S rDNA and *gyrB* universal primers. Phylogenetic analysis of the 16S rDNA gene showed that the homology of B10 with the *Bacillus amyloliquefaciens* strain GD4a was 98.8%. Phylogenetic analysis of the *gyrB* gene implied that B10 was 99.8% homologous to the *Bacillus amyloliquefaciens* strain JX014631.1 on the same branch (Figure 1E,F). This strain was identified as *Bacillus amyloliquefaciens* based on the biochemical response characteristics (Table 1), combining morphological and phylogenetic characteristics. The OD_{600} value of strain B10 was also measured using a UV spectrophotometer at 6 h intervals. The results indicated that the strain entered the log phase at 12–18 h of incubation (Figure 1C). In further AFB1 degradation tests, the efficiency of AFB1 degradation increased significantly with the increase in co-culture time; almost complete degradation of 2.5 μg/mL AFB1 was found after 96 h of co-culture (Figure 1D).

2.2. Localisation of Degradation Active Substances

The high-performance liquid chromatography (HPLC) results showed that the extracellular fluid of *B. amyloliquefaciens* B10 had the best degradation ability on AFB1, with a degradation rate of 72.9%. The other components of strain B10 caused a minor AFB1 degradation at a level far inferior to the extracellular fluid effect. Bacteria inactivated by high temperature could degrade 15.2% of AFB1, probably due to the physical adsorption of the cell membrane of the B10 strain (Figure 2A).

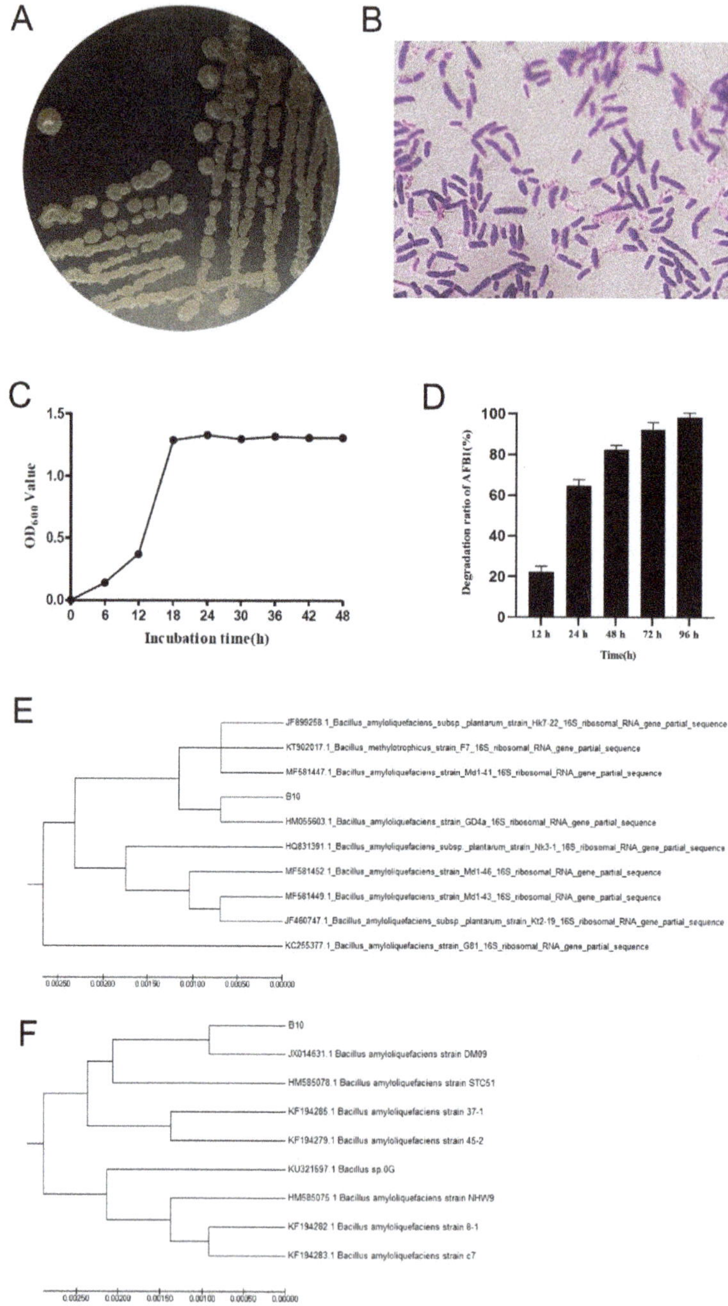

Figure 1. Characteristics of strain B10 and efficiency of AFB1 degradation: (**A**) Morphology of strain B10 cultured on LB medium for 24 h. (**B**) Gram stain of strain B10. (**C**) Growth curve of strain B10. (**D**) Effect of co-culture time on the degradation efficiency of AFB1 (2.5 μg/mL). (**E**) Phylogenetic tree of the 16S rDNA gene sequence of strain B10 using the neighbour-joining method. (**F**) Phylogenetic tree of the *gyrB* gene sequence of strain B10 using the neighbour-joining method.

Table 1. Physiological and biochemical characteristics of strain B10.

Indicator	Result
Methyl Red assay	+
Anaerobic	−
Oxidase	+
Contact enzyme assay	+
V-P test	+
Indole test	−
Starch hydrolysis	+
Glucose fermentation	+
Glucose gas production	−
Gelatine liquefaction	+
Citrate	+
Xylose	+
Nitrate	+
Propionate	−
7% NaCl growth	+

V-P Test: Voges–Proskauer test.

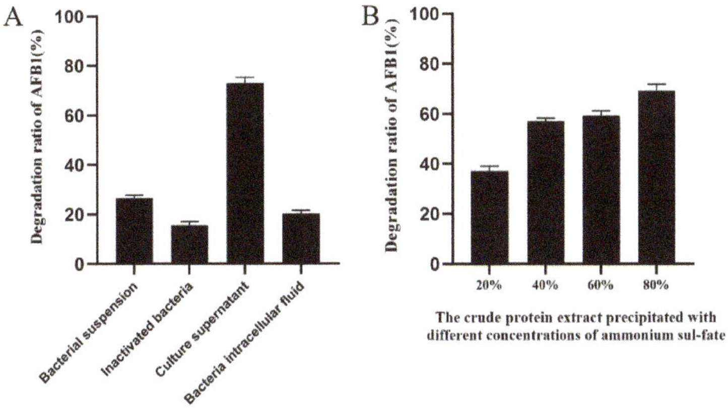

Figure 2. Localization of active substances for aflatoxin B1 degradation by strain B10: (**A**) Evaluation on the degradation effect of AFB1 (2.5 μg/mL) by each component of strain B10. (**B**) Evaluation on the degradation effect of AFB1 (2.5 μg/mL) by the crude protein extract of culture supernatant with different concentrations of ammonium sulfate.

The crude protein of the supernatant from strain B10 was extracted with different concentrations of ammonium sulfate and dialysed for desalination. The degradation efficiency of AFB1 was 69.13% when the ammonium sulfate crude protein concentration was 80%. (Figure 2B). Therefore, we further determined that the AFB1-degrading substance might be a type of enzyme.

2.3. Isolation of AFB1-Degrading Proteins

The crude protein from the extracellular fluid was concentrated and initially purified using a DEAE ion-exchange chromatography column to obtain an elution curve with four component peaks (Figure 3D). The degradation of AFB1 by each component showed that components 3 and 4 exhibited vigorous AFB1 degradation activity after 36 h incubation, and component 4 had the highest degradation activity with regard to AFB1 with a rate of 73.4% (Figure 3A). Therefore, component 4 was further purified by gel filtration chromatography and seven component peaks were obtained (Figure 3E). Among the seven protein components, components 4-3 demonstrated the highest degradation activity to AFB1 with a rate of 87.3% (Figure 3B). After SDS-PAGE and Coomassie brilliant blue R-250 staining of components 4-3, an obvious protein band of 35 kDa was observed. Components 4-3 were identified by protein mass spectrometry and matched to 60 proteins. The laccase component was contained in these 60 proteins by protein blast in NCBI. Therefore, we speculated that the proteins with AFB1 degradation activity in strain B10 probably contained laccase. However, the protein profile did not include a single protein, and whether all other proteins have AFB1-degrading activity warrants further research.

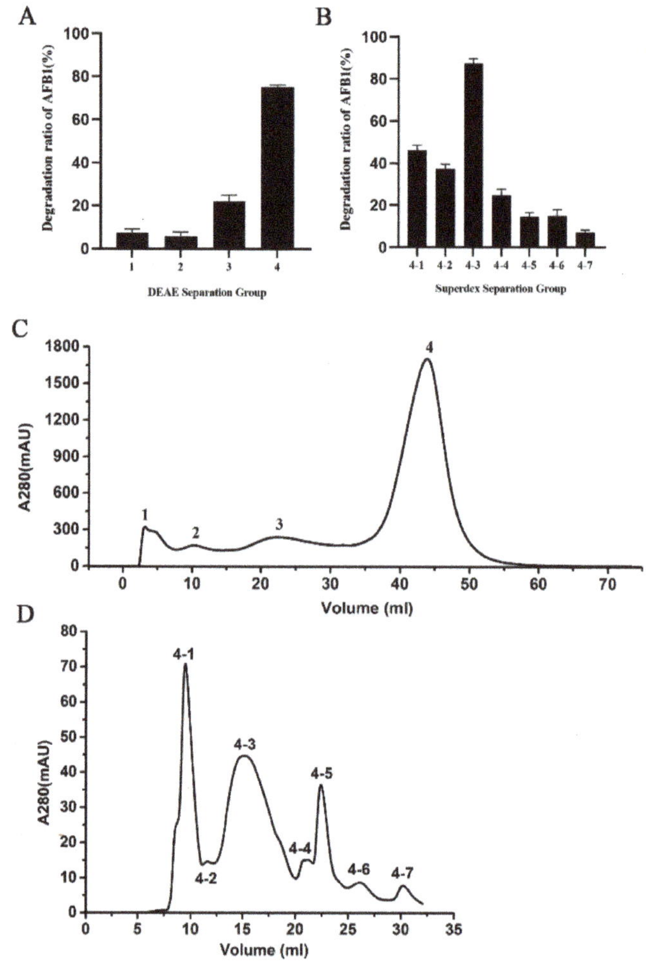

Figure 3. Isolation of AFB1 degrading enzyme from strain B10: (**A**) Degradation efficiency of AFB1 by each protein component precipitated by DEAE. (**B**) The efficiency of degradation of AFB1 by protein component precipitated by Superdex separation. (**C**) DEAE separation chromatogram (peaks 1 to 4 correspond to different protein fractions). (**D**) Superdex separation chromatogram (peaks 4-1 to 4-7 correspond to different protein fractions).

2.4. Molecular Docking for Function Prediction

Due to the excellent results achieved by AlphaFold2 in protein structure prediction [24], the predicted structure of a laccase from strain B10 was further refined and evaluated by Rasch plot, and 97% of the residues fell within the permissible interval (Figure 4A). This finding indicated that our predicted structure of B10 laccase could be used for molecular docking studies.

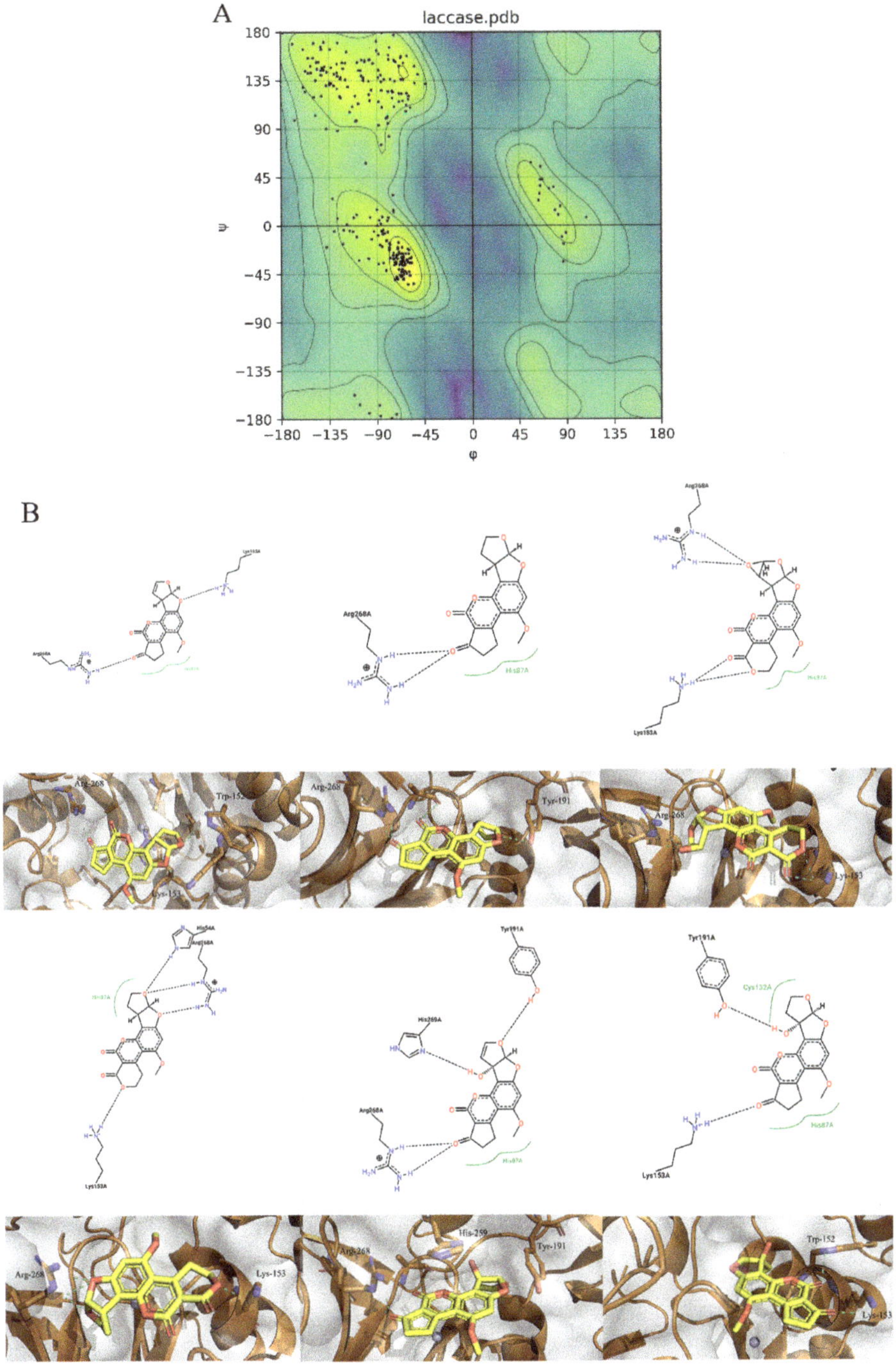

Figure 4. Molecular docking of aflatoxins with B10 laccase: (**A**) Ramachandran diagram of B10 laccase protein structure. (**B**) Models of the interaction of AFB1, AFB2, AFG1, AFG2, AFM1, and AFM2 with B10 laccase (grey, Zn^{2+}; yellow, ligand).

The molecular docking of the six aflatoxins (AFB1, AFB2, AFG1, AFG2, AFM1, and AFM2) was performed using the B10 laccase prediction model, and each aflatoxin had a high docking score with B10 laccase (Table 2, Figure 4B). Therefore, it is important to further verify the aflatoxin degradation ability of B10 laccase. The differences in docking scores with B10 laccase were also minor as the chemical structures of the six aflatoxins were similar. The docking score results indicated that AFB2 had the most vital binding capacity for laccase. Both Lys-153 and Arg-268 in B10 laccase produced hydrogen-bonding interactions with each aflatoxin, and these two residues were most likely to be the key amino acids for binding the toxins. His-87 made hydrophobic interactions with each aflatoxin, while His-87 was one of the amino acids with a coordination bond with Zn^{2+}. The O-1 site on the terminal furan ring of AFB1 that acted as an acceptor with O-5 on the five-membered ring of AFB2 formed two hydrogen bonds as an acceptor with the side chain of Arg-268 and with Tyr-191. AFM1 formed hydrogen bonds with Arg-268, His-259, and Tyr-191. AFM2 formed hydrogen bonds with Trp-152, Lys-153, and Tyr-191 (Table 2, Figure 4B). The results of the molecular docking analysis showed that B10 laccase had a stable binding mode with aflatoxin, which was predicted to play a role in the degradation of aflatoxin. Therefore, we continued to clone the B10 laccase gene and investigated its protein expression in order to verify the ability of B10 laccase to degrade aflatoxin.

Table 2. Molecular docking results.

Compound Name	Structure	Docking Score (kcal/mol)	Hydrogen Bond			Hydrophobic Effect
			Enzyme Residues	Distance D-A (Å)	Acceptor Atom	
AFB1	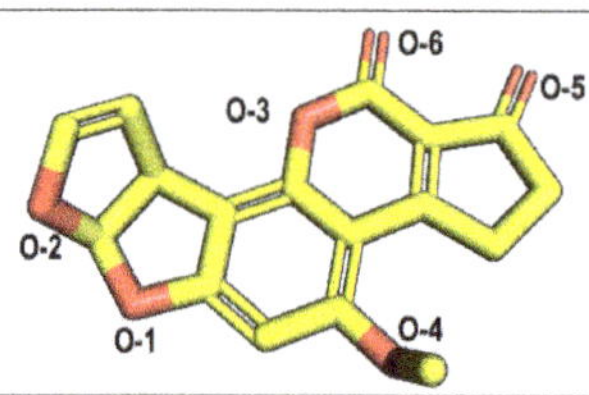	−5.60	Trp-152	3.17	O-2	His-87
			Lys-153	2.82	O-1	
			Arg-268	2.89	O-5	
AFB2	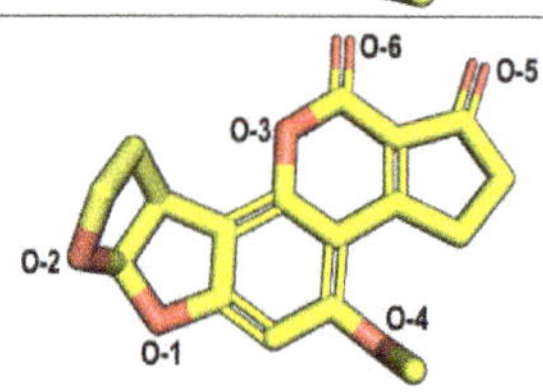	−6.82	Tyr-191	3.50	O-2	His-87
			Arg-268	2.63	O-5	
				2.62	O-5	
AFG1	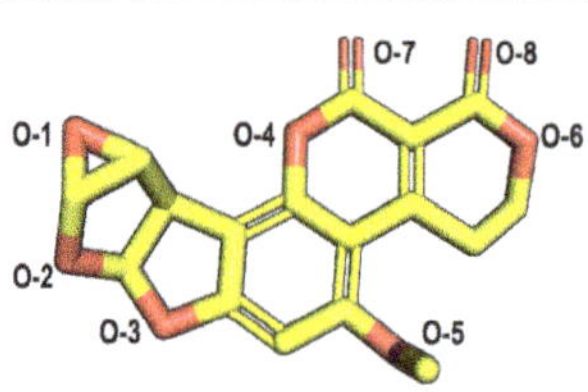	−6.58	Lys-153	2.84	O-6	His-87
				3.15	O-8	
			Arg-268	2.79	O-1	
				2.77	O-1	
AFG2	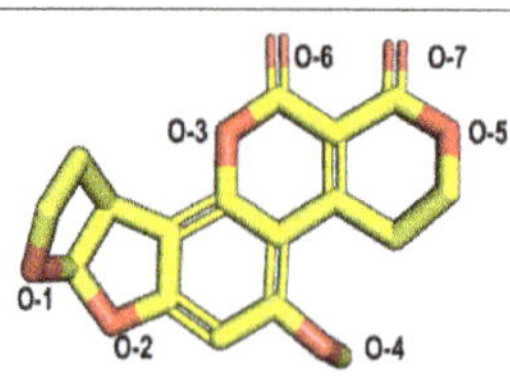	−5.31	Lys-153	2.77	O-5	His-87
			Arg-268	2.65	O-2	
				3.09	O-1	

Table 2. *Cont.*

Compound Name	Structure	Docking Score (kcal/mol)	Hydrogen Bond			Hydrophobic Effect
			Enzyme Residues	Distance D-A (Å)	Acceptor Atom	
AFM1	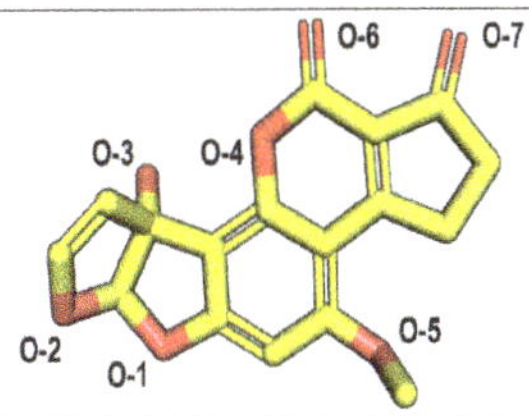	−6.30	Tyr-191	3.16	O-2	His-87
			His-259	2.87	O-3	
			Arg-268	2.84	O-7	
				2.82	O-7	
AFM2	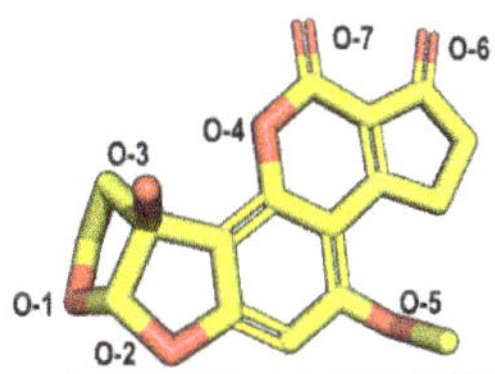	−6.13	Trp-152	2.78	O-7	His-87 Cys-132
			Lys-153	2.76	O-6	
			Tyr-191	2.65	O-3	

2.5. Cloning, Expression, and Purification of Laccase from Strain B10

The amino acid sequence of B10 laccase obtained in Section 2.3 was matched by NCBI BLAST and had the highest similarity with laccase No. ASB53002.1. The cloning primers were designed according to the base sequence of the CDs region of ASB53002.1 laccase, and the B10 laccase gene was cloned from strain B10 (Figure 4A). The open reading frame of the B10 laccase gene was 837 bp (Figure 5A), encoded 278 amino acids, and was predicted to have a molecular weight of 30.9 kDa and an isoelectric point 5.84. The amino acid homology of laccase cloned from the B10 strain to *Bacillus amyloliquefaciens* 629 laccase was 99.64% (GenBank: KNX34508.1) [25], to *Bacillus velezensis* RC218 laccase was 99.28% (GenBank: KUP42711.1) [26], and to *Bacillus* 916 laccase was 99.22% (GenBank: AIW29756.1) (Figure 5D) [27].

The laccase gene was transformed into *E. coli* DH5α and verified by double digestion (Figure 5B) and expressed in *E. coli* BL21(DE3). The purified recombinant laccase was then purified by Ni Sepharose 6 Fast Flow affinity chromatographic packing, and the purified recombinant laccase demonstrated a more distinct, but single, band by SDS-PAGE gel electrophoresis. The apparent molecular weight of the band was approximately 35 kDa, which was close to the predicted molecular weight (Figure 5C).

2.6. Efficiency of AFB1 Degradation by the Recombinant Laccase and the Effects of Different Conditions on AFB1 Degradation

A series of experiments proved that the recombinant laccase of strain B10 could efficiently degrade AFB1. The rate of degradation of AFB1 gradually increased with the increase in co-culture time, reaching 79.3% after 36 h. Meanwhile, the degradation rate of AFB1 reached a plateau at 48 h (Figure 6A,F). The results with regard to the effects of temperature on the degradation of AFB1 by the recombinant laccase showed that the degradation rate of AFB1 was gradually increased with the increase in temperature from 20 to 40 °C at pH 7.2; the degradation rate of AFB1 reached 82.4% at 40 °C. The degradation rate of AFB1 decreased rapidly because of the high temperatures ranging from 70 to 90 °C. However, the degradation rate of AFB1 exceeded 65% between 30 and 60 °C (Figure 6B). pH value had a strong influence on the degradation of AFB1. At pH 6.5, the degradation rate of AFB1 was the highest, reaching 84.2%. Under acidic conditions, the degradation rate of AFB1 was lower; at pH 4.0, the degradation rate was only 7.1%. From pH 6.0 to pH 8.0, the degradation rate of AFB1 exceeded 60% (Figure 6D). As AFB1 was efficiently degraded under strongly alkaline conditions, the effects of a pH greater than 9.0 on the degradation of AFB1 by the recombinant laccase were not considered in the present study.

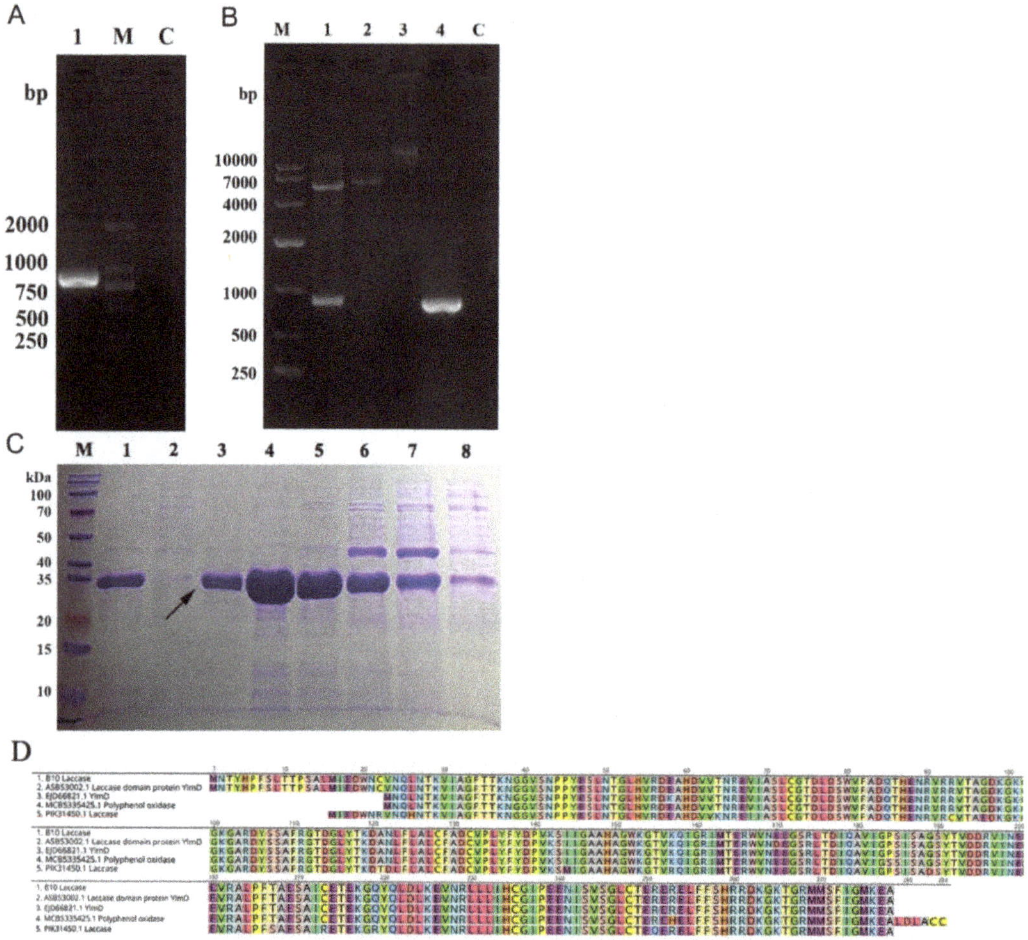

Figure 5. Gene cloning, expression, purification, and identification of B10 laccase: (**A**) Agarose gel electrophoresis to isolate the B10 laccase gene (M, marker; 1, B10 laccase PCR product; C, blank control). (**B**) Agarose gel electrophoresis to validate the results of pET-28a/B10 laccase recombinant vector (M, marker; 1, pET-28a/B10 laccase recombinant vector double digestion; 2, pET-28a double digestion; 3, pET-28a; 4, B10 laccase gene; C, blank control). (**C**) SDS-PAGE analysis of purified recombinant B10 laccase (M, marker; 1, beads; 2, 300 mM imidazole; 3, 200 mM imidazole; 4, 150 mM imidazole; 5, 100 mM imidazole; 6, 50 mM imidazole; 7, 10 mM imidazole; 8, imidazole-free buffer). (**D**) Amino acid sequence alignment of B10 laccase with *Bacillus amyloliquefaciens* 629 laccase, *Bacillus velezensis* RC218 laccase, and *Bacillus* 916 laccase.

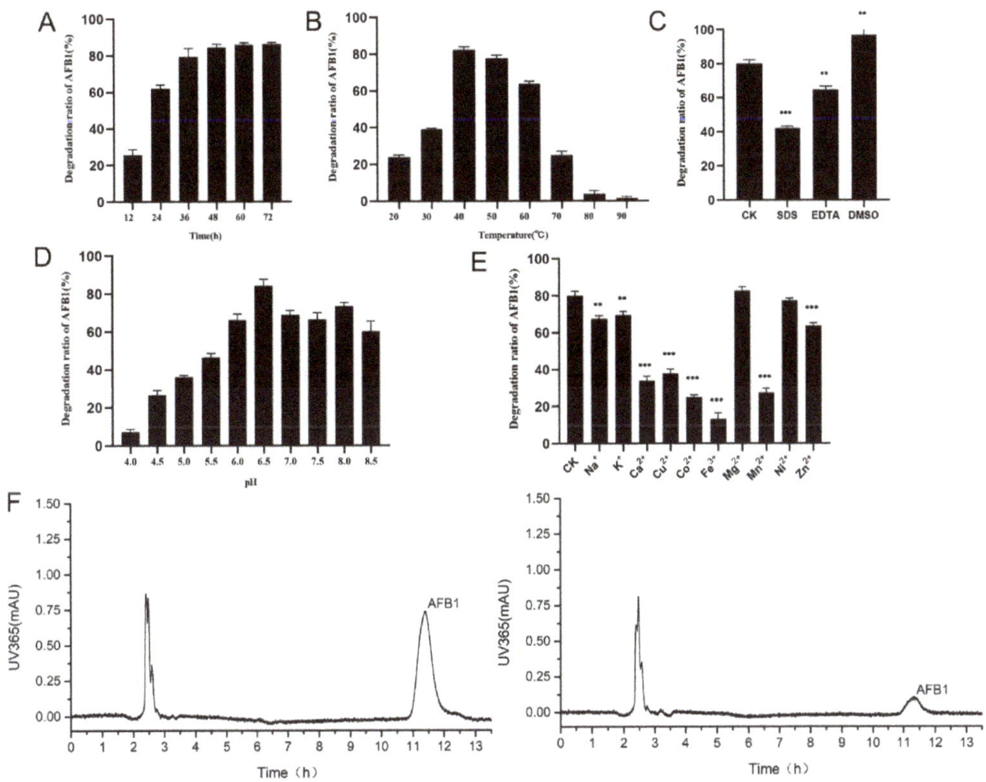

Figure 6. The efficiency of B10 laccase in degrading AFB1 and the effects of different conditions on the degradation: (**A**) Effect of co-culture time on the degradation of AFB1 by B10 laccase. (**B**) Effect of temperature on the degradation of AFB1 by B10 laccase. (**C**) Effect of common solutions on the degradation of AFB1 by B10 laccase (CK: In the control group, no substance was added). (**D**) Effect of pH on the degradation of AFB1 by B10 laccase. (**E**) Effect of metal ions on the degradation of AFB1 by B10 laccase (CK: In the control group, no metal ions was added). (**F**) Liquid chromatogram of AFB1 at 36 h co-culture (left, blank control; right, 36 h co-culture). **: $p < 0.05$; ***: $p < 0.001$.

The results with regard to the effects of different metal ions on the degradation of AFB1 by the recombinant laccase indicated that Ca^{2+}, Cu^{2+}, Co^{2+}, Fe^{3+}, Mn^{2+}, and Zn^{2+} significantly reduced the degradation activity of AFB1 by the recombinant laccase compared with the control group with an 80.3% degradation rate of AFB1 ($p < 0.001$), while Na^{+} and K^{+} slightly reduced the degradation rate of AFB1 ($p < 0.05$). The metal ion Ni^{2+} had no significant effect on the degradation activity of AFB1 (Figure 6E). It was also shown that the metal chelators EDTA ($p < 0.05$) and SDS ($p < 0.001$) significantly reduced the AFB1-degrading activity of the recombinant laccase (Figure 6C), and the AFB1-degrading efficiency was significantly increased with the addition of DMSO ($p < 0.05$).

2.7. Site-Specific Mutagenesis of B10 Laccases

The amino acid sequence of B10 laccase was matched to asb53002 by the NCBI protein database BLAST. The laccase domain protein YlmD score of 1 was 99.64%. After query, the annotation of the GenBank CDs region was directed to uniprotkb o31726. We found that B10 laccase contained three Zn^{2+} binding sites: namely, H-87, C-132, and H-149. Therefore, the roles of these three residues in laccase were assessed by site-specific mutagenesis. There were seven mutant proteins, single mutant H87, C132, and H149; double mutant H87/C132,

H87/H149, and C132/H149; and triple mutant H87/C132/H149. These proteins were obtained by the polymerase chain reaction (PCR) method to mutate these sites to alanine. The results showed that after expression and purification, single mutants H87, C132, and H149 could be analysed by measuring AFB1-degrading activity. The AFB1-degrading activity of the single and double mutants H87, C132, H149, H87/C132, H87/H149, and C132/H149 was significantly decreased compared with the wild type, while the AFB1-degrading activity was further weakened by the triple mutant H87/C132/H149. In contrast, the triple mutant H87/C132/H149 decreased the degradation rate of AFB1 to 6.7% (Figure 7).

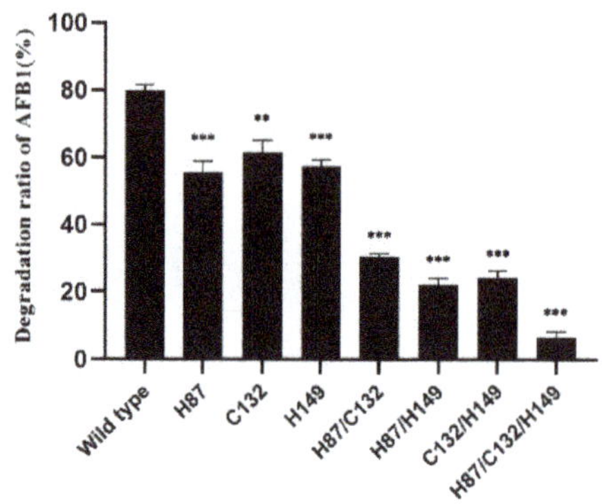

Figure 7. Effects on AFB1 degradation activity of B10 laccase mutants. **: $p < 0.05$; ***: $p < 0.001$.

3. Discussion

Aflatoxin is of widespread concern due to its high toxicity, with approximately 4.5 billion people worldwide chronically exposed to aflatoxin through contaminated food [28,29]. In addition, data show that the consumption of aflatoxin-contaminated food is expected to further increase in incidence due to the ongoing COVID-19 pandemic, which has made the management of food and feed more difficult and complex [30]. The detoxification of aflatoxins by microorganisms is a promising new technology with broad application prospects. The control of aflatoxin by microorganisms mainly includes the inhibition of aflatoxin production, the adsorption of aflatoxin, and the degradation of aflatoxin [11]. To date, strains that are highly efficient in degrading aflatoxins have been isolated from various environmental sample species, including fungi, bacteria, actinomycetes, and protozoa [31]. In this study, 74 strains were screened in the laboratory, and one strain (laboratory number B10), was found to be highly efficient in degrading AFB1. The strain was identified morphologically, characterized biochemically, and analysed in a phylogenetic tree as *Bacillus amyloliquefaciens*. The B10 strain was 99.8% homologous to the *Bacillus amyloliquefaciens* strain JX014631.1 on the same branch and 98.8% homologous to the *Bacillus amyloliquefaciens* strain GD4a. Xu et al. isolated a strain of *Bacillus shark ii* L7 from 43 strains of bacteria, which could degrade 92.1% of AFB1 at a final concentration of 100 μg/L for 72 h at 37 °C. Further experiments implied that the active substance degrading AFB1 of strain L7 was mainly present in the culture supernatant, which could degrade 77.9% of the AFB1 within 72 h [32]. This was similar to the results of this study, where the supernatant of strain B10 could degrade AFB1 by 72.9% at 24 h. In contrast, the degradation rate of AFB1 was lower in the high-temperature inactivation group, the intracellular fluid group, and the bacterial suspension group compared with the culture supernatant. The high-temperature inactivation group presented the degradation of AFB1 probably due to the physical adsorption capacity of the cell membrane of the B10 strain. The presence of active substances in culture supernatants for AFB1 degradation by *Bacillus* was also revealed in tests by Gayatri [33], Wang [34], and Shu [35]. In this study, we found that protein precipitated by an ammonium sulfate gradient had the highest AFB1-degrading efficiency, so the AFB1-degrading substance was likely to be an enzyme or several enzymes in the culture supernatant.

In this study, the protein precipitated by ammonium sulfate was further purified, and the active component was measured through separation by DEAE anion exchange chromatography and gel filtration chromatography. A novel laccase that can efficiently degrade AFB1 was identified by protein mass spectrometry. The laccase gene had an open reading frame of 837 bp, an apparent molecular weight of 35 kDa, and a predicted isoelectric point of 5.84, and encoded 278 amino acids. The laccase cloned from strain B10 showed 99.64% amino acid homology to *Bacillus amyloliquefaciens* 629 laccase [25], 99.28% to *Bacillus velezensis* RC218 laccase [26], and 99.22% to *Bacillus* 916 laccase [27]. Proteins with AFB1-degrading activity were previously isolated using similar methods from culture supernatants of *Aspergillus flavus* ANSM068 [36], and the edible fungus *Pleurotus ostreatus* [37].

Superior results in all previous applications were achieved when Alphafold2 was used to predict the structure of the B10 laccase protein, and the release of Deepmind's Alphafold2 software has ushered in a new revolution in high-quality three-dimensional (3D) protein structure prediction [38–40]. The predicted 3D model of B10 laccase was evaluated by structural refinement with a Rasch diagram, which allowed us to examine the model of the interaction of B10 and laccase with the substrate through molecular docking simulations without resolving the crystal structure. The active pocket of laccase appeared semi-open, with Zn^{2+} at the bottom of the pocket interacting with His-87, C-132, and H-149. This leads to speculation that Zn^{2+} may also be complexed with substrate or water molecules. In molecular docking, an implicit water model was used to approximate the solvent interaction, and a high degree of precision was employed to reveal the possible conformation of the ligand in the pocket. The docking results showed that both Lys-153 and Arg-268 could exert hydrogen bonding forces with aflatoxin, and therefore these two residues play an essential role in the binding of aflatoxin to B10 laccase. Notably, His-87 is known to play an essential role in catalysing the degradation of AFB1 by B10 laccase using a targeted mutation assay, and the docking results also indicated that His-87 can exert hydrophobic forces with aflatoxin, so His-87 is equally important in the binding of B10 laccase and substrate. The amino acid residues involved in hydrogen bonding were considered key residues for the interaction of laccase with specific ligands [41]. Lys-153 and Arg-268 will therefore be the focus of our future studies.

B10 laccase and aflatoxin have high docking scores. Therefore, we continue to verify the efficiency of B10 laccase in degrading aflatoxin. The efficiency of AFB1 degradation by B10 laccase was significantly different at different temperatures. The maximum degradation activity of AFB1 by this laccase reached 82.4% at 40 °C. Unlike the laccase identified by other researchers, the degradation activity of AFB1 decreased at temperatures ranging from 60 to 90 °C. The degradation rate of AFB1 by laccase isolated from *Stenotrophomonas* sp. CW117 was less than 60% at 40 °C but increased rapidly when treated at a high temperature for a short period; the degradation rate of AFB1 exceeded 84.6% at above 70 °C [31]. This may be due to evolutionary factors in the strain that led to the evolution of the laccase into a heat-resistant enzyme [42]. The highest AFB1-degradation activity of B10 laccase was achieved at pH 6.5, with a rate of degradation of 79.2%. The AFB1-degrading activity of this enzyme was deficient when pH was below 4.5, yielding a result more similar to that of Guo et al. [19]. The laccase isolated by Cai et al. exhibited low AFB1-degrading activity at pH 7–8 [31]. Since AFB1 was also efficiently degraded under strongly alkaline conditions, the effects of conditions above pH 9.0 on the degradation of AFB1 by recombinant laccase were not considered herein. Laccase activity could be increased at pH 9.8, but this condition was difficult to apply in food and feed detoxification [43]. In this study, we found that the addition of all metal ions exerted a negative effect on AFB1 degradation by B10 laccase, except for Mg^{2+}, which enhanced the AFB1-degrading activity of B10 laccase. Notably, the addition of Cu^{2+} significantly reduced the AFB1-degradation activity of B10 laccase. This finding was consistent with results that laccase activity can be enhanced by 0.4–1 mM Cu^{2+} but was inhibited by high concentrations of Cu^{2+} [44]. In the present study, the organic solvent DMSO significantly improved the degradation activity of AFB1, in line with the findings of Wu al. [45], which may be attributed to changes in enzymatic activity due to the altered microenvironment caused by the organic solvent [46].

The residues H87, C132, and H149 of B10 laccase are the three metal-binding sites of this enzyme. The metal-binding sites of laccase are associated with the catalytic degradation of AFB1. When supplemented with metal ion chaperones during the expression of laccase, there was an increase in the metal-ion content of laccase and an improvement in its specific activity [47]. Previous research showed that the ribose portion of such proteins was coordinated by the side chains of these three residues. One of the conserved triplets of B10 laccase was predicted to be the active site, with similar results to the YlmD protein [48,49]. In the presence or absence of excess metal ions, the conformation of the protein was found to be significantly altered, including the R-loop, MR helix, and metal site centre, which destabilized the structure of the protein [50]. Although this does not change the overall structure, it can significantly affect enzymatic activity [51,52]. The present study showed that these three residues are essential for the degradation of AFB1 by B10 laccase. When these residues were replaced by alanine, the AFB1-degrading activity of the mutants was significantly decreased compared with the wild type.

4. Conclusions

In this study, a strain of *Bacillus amyloliquefaciens* B10 with high AFB1-degradation capacity was obtained and the active component of AFB1 degradation was located in the culture supernatant of this strain. A new laccase with high AFB1 degradation activity was obtained by the crude extraction of culture supernatant proteins with ammonium sulfate, DEAE, and gel filtration chromatography. The molecular docking results showed higher scores. Meanwhile, it was speculated that Lys-153, Arg-268, and His-87 residues in this laccase played an essential role in the binding of aflatoxin to this laccase. The B10 laccase gene was cloned from the bacterial genome and was expressed in *E. coli*. The recombinant laccase showed the highest AFB1-degradation activity at 40 °C and pH 6.5. The mutation of three key metal sites of the laccase implied that the AFB1-degradation activity of the triple mutants was almost wholly lost. These results suggested that B10 laccase is a promising enzyme for aflatoxin degradation. Our laboratory will continue to explore B10 laccase to improve enzymatic activity and stability, with a view to prompting the application of this enzyme in practical production as soon as possible.

5. Materials and Methods

5.1. Chemicals and Reagents

Gram stain, TAE, nucleic acid dye, Komas Brilliant Blue R-250, and ammonium sulfate were purchased from Beijing Solarbio Science & Technology Co., Ltd. (Beijing, China); the plasmid extraction kit, gel recovery kit, and PCR product recovery kit were purchased from Thermo Fisher Scientific (Waltham, MA, USA); DH5α and BL21 (DE3) receptor cells were sourced from TransGen Biotech (Beijing, China); point mutation kits from TransGen Biotech (Beijing, China) were used; point-mutation kits were purchased from Vazyme Biotech Co., Ltd. (Beijing, China); HiTrap Capto DEAE (5 mL), Superdex™ 75 Increase 10/300 GL, and ÄKTA protein purification system were purchased from Cytiva (Washington, USA); and AFB1 was sourced from Sigma-Aldrich (Shanghai, China). All other reagents were at least of analytical purity.

5.2. Isolation of AFB1-Degrading Strains

Each of the 74 strains stored at −80 °C in the laboratory was inoculated into tubes containing 5 mL of LB liquid medium and incubated overnight at 37 °C on a shaker at 150 rpm. An inoculation loop was adopted to pick up a certain number of bacteria from the overnight culture of the strains to be tested on an LB agar gel plate for scribing. The strain was allowed to grow as a single colony for use, or the culture was purified if there were any stray colonies.

The purified strains to be tested were compared with the final concentration of 2.5 μg/mL AFB1 standard cultured in LB medium at 37 °C and 150 rpm for 24 h. The growth status of the strains in medium containing AFB1 was observed, and the strains that grew well were tested sequentially and at final concentrations of 5 and 10 μg/mL in AFB1 standard co-culture.

After initial screening, a laboratory strain labelled B10 grew well in AFB1 medium containing different concentrations of AFB1. All the components were incubated at 37 °C and 150 rpm for 12, 24, 48, and 96 h. The content of AFB1 was determined by high-performance liquid chromatography and ultraviolet (HPLC, UV) methods, and the degradation efficiency of strain B10 was calculated.

Chromatographic conditions: The chromatographic column was a C18 reverse adsorption column (4.6 mm × 250 mm, 5 μm), with an injection volume of 20 mL, a mobile phase of a 1:1 (v/v) water:methanol mixture, a flow rate of 1 mL/min, and a detection wavelength of 365 nm. The standard curve for the HPLC detection of AFB1 was $y = 0.7735x + 0.1521$, $R^2 = 0.9975$.

5.3. Localisation of AFB1-Degrading Active Substances by Strain B10

The degradation of active substances from the fermentation broth of strain B10 was localised in four fractions: 500 mL of fermentation broth was taken and centrifuged through a low-temperature high-speed centrifuge at 4 °C, 8000 rpm for 10 min. The separate of the supernatant of the fermentation broth was separated from the bacterial precipitation, and 3 mL was taken as the cell secretion group (extracellular fluid group). The bacterial sediment was resuspended with 20 mL of sterile PBS buffer; we centrifuged and poured some PBS and washed it thrice before resuspension in 20 mL of sterile PBS. Afterwards, 3 mL was taken as the live-cell group (bacterial suspension group). Of the bacterial suspension, 3 mL was placed in an autoclave at 121 °C for 30 min and allowed to cool before use (inactivation group). The remaining bacterial suspension was dosed with PMSF at a final concentration of 1%, crushed in a low-temperature ultrasonic cell crusher, then centrifuged at 4 °C, 13,000 rpm for 30 min. Each component was incubated with AFB1 standard at a final concentration of 2.5 μg/mL for 24 h in an incubator at 37 °C, and the degradation rate of each component against AFB1 was measured using HPLC.

5.4. Crude Extraction of Protein from Culture Medium

The B10 strain was inoculated in four vials of LB medium and incubated overnight at 37 °C. The supernatant was removed by high-speed low-temperature centrifugation, and ammonium sulfate was added to give final concentrations of 20, 40, 60, and 80%. After thorough mixing with a magnetic stirrer, the culture was kept at 4 °C overnight. The supernatant of each group was centrifuged at 13,000 rpm for 30 min at 4 °C. The supernatant was discarded, and the residue was dissolved in 10 mL of sterile PBS, transferred to a dialysis bag with a cut-off volume of 3.5 kDa, and dialysed in PBS buffer for 24 h at 4 °C.

One millilitre of crude protein solution of each component after dialysis was taken separately, filtered through a 0.22 μL needle filter, and incubated with 500 μL of AFB1 standard at a final concentration of 2.5 μg/mL for 24 h. The rate of degradation of each component was measured using HPLC.

5.5. Isolation and Identification of AFB1-Degrading Proteins

The crude extracted protein from the fermentation broth culture supernatant was concentrated and initially purified by passing through a DEAE ion-exchange column. The crude proteins were loaded onto a column pre-equilibrated with buffer (containing 20 mM/L Tris, 10 mM/L NaCl, pH 8.0) and eluted with a 0.01–2.0 M/L linear concentration gradient NaCl buffer at a flow rate of 1 mL/min. The eluate was collected separately according to the absorption peaks. Each component was incubated with 500 μL of AFB1 standard at a final concentration of 2.5 μg/mL for 24 h. The efficiency of each component in degrading AFB1 was obtained using HPLC.

The fraction with good degradation was purified by gel filtration chromatography. The solution to be measured was loaded onto a column pre-equilibrated with buffer (containing 20 mM/L Tris, 50 mM/L NaCl, pH 8.0) and eluted at a flow rate of 0.2 mL/min. The eluate was collected separately according to the absorption peaks. The degradation rate of each group was detected as described. The components with the best degradation effect were analysed by SDS-PAGE, and the proteins in the optimal components were identified by protein mass spectrometry.

5.6. Structural Modelling of the Recombinant Laccase and Molecular Docking

Through recombinant laccase amino acid sequencing using AlphaFold2 on the Beijing Supercomputing Cloud N22 partition with a full_dbs preset [25], a total of five initial models and five models relaxed by the Amber relaxation procedure were generated, after which the highest-ranked conformation was selected for subsequent docking analysis based on the average plDDT ranking. The recombinant laccase's optimal structure was plotted in Python 3.8 using the RamachanDraw package for quality and poor contact assessment between residues.

As retrieved from AFB1 (PubChem CID: 186907), AFB2 (PubChem CID: 2724360), AFG1 (PubChem CID:133065469), AFG2 (PubChem CID: 2724362), AFM1 (PubChem CID: 15558498), and AFM2 (PubChem CID: 23318), six molecules were downloaded and saved in SDF format, semi-empirically optimized using MOPAC2016 (http://openmopac.net/, accessed on 15 November 2020) under the PM7 PRNT = 2 parameter. Optimized structures were prepared using the MGLTools 1.57 suite of prepare_ligand.py which could be converted to pdbqt format to give the Gasteiger charge and retain the total hydrogen.

The prediction of the recombinant laccase Zn^{2+} binding site using the bioinformatics tool UniProt (EMBL-EBI, Cambridge, UK) was followed by use of the NCBI BLAST tool (National Center for Biotechnology Information, Bethesda, MD, USA). The PDB database was selected, the crystal structure of a purine nucleoside phosphorylase with 49.40% homology (PDB ID: 6T0Y) was retrieved, the two structures were fitted in pymol2.5, the template Zn^{2+} was retained, the amino acids in the Zn^{2+} binding site were adjusted, and the MGLTools 1.57 suite was used to create a recombinant laccase pdbqt file containing charge and H atoms.

Docking studies of the recombinant laccase were performed using watvina (https://github.com/biocheming/watvina, accessed on 30 November 2021, Ximing Xu, Qingdao, China), a deeply optimized offshoot of vina, using the getbox plugin in pymol to generate docking boxes centred on the metal-binding site of the recombinant laccase. The docking box was centred at center_x = 0.8, center_y = −3.1, center_z = −5.7; size_x = 19.5, size_y = 21.2, size_z = 18.5, after which the molecule and recombinant laccase were docked using watvina, and the solvent interaction was approximated by using an implicit water model. The docking results were plotted using pymol.

5.7. Gene Cloning, Protein Expression, and Purification of the Recombinant Laccase

Strain B10 was cultured overnight in LB medium, and DNA was extracted from the strain using a DNA extraction kit. Amino acid sequence results obtained from protein mass spectrometry identification in 5.5 were aligned by NCBI BLAST. Cloning primers were designed based on the CDs region base sequences of the most similar proteins. The restriction enzymes BamH I and Xho I were selected, and the expression vector pET-28a was designed with fragment amplification primers as follows:

Lac-F: 5′CGGGATCCATGAATACATATCACCCGTTTAGTCTT3′

Lac-R: 5′CCCTCGAGTTATGCCTCCTTCATTCCGATAAAG3′.

PCR reaction conditions consisted of denaturation at 98 °C for 10 s, annealing at 55 °C for 5 s, extension at 72 °C for 30 s, 34 cycles from denaturation to extension steps, and storage at 4 °C. PCR fragments were recovered using an agarose gel DNA extraction kit. The double-cleaved B10 laccase gene fragment was inserted into the expression vector pET-28a, transformed into Trans5α chemically competent cells, coated onto LB medium containing 100 mg/mL kanamycin, and incubated overnight. Three positive clones were selected for sequencing, and the plasmids from the positive clones corresponding exactly to the B10 laccase gene were extracted and transformed into BL21 (DE3) chemically competent cells. Positive clones coated on LB medium containing 100 mg/mL kanamycin were inoculated in LB liquid medium. Expressed cells were collected by centrifugation and resuspended in sterile PBS, fragmented by ultrasonic cell disruptor, and purified by Ni Sepharose 6 FF affinity chromatography packing with His-tagged laccase in the supernatant. The purity of the recombinant laccase was determined by sodium dodecyl sulfate–polyacrylamide gel electrophoresis (SDS-PAGE), and the protein concentration was detected by bicinchoninic acid (BCA) assay. BCA standard curve regression equation was $y = 0.7618x + 0.1371$.

5.8. Efficiency of AFB1 Degradation by a Recombinant Laccase and Its Effect on AFB1-Degrading Activity under Different Conditions

A recombinant laccase (20 μg/mL) was incubated with AFB1 standard at a final concentration of 2.5 μg/mL for 12, 24, 36, 48, 60, and 72 h at pH 7.2 and 37 °C. The samples were centrifuged in 1.5 mL centrifuge tubes at 13,000 rpm for 5 min at room temperature, and the supernatant was removed from the tubes. A total of 500 μL of the supernatant was transferred to a new 1.5 mL sterile centrifuge tube, and all mixed samples were withdrawn using a 1 mL syringe and filtered through a 0.22 μm organic phase needle filter. A total of 20 μL of the filtrate was injected into the HPLC detection system using a microsampling needle (the same treatment was applied to the control). The samples were assayed for AFB1 according to the AFB1 standard curve.

To evaluate the effects of different temperatures on the degradation of AFB1 by the recombinant laccase from strain B10, the recombinant laccase (20 μg/mL) and AFB1 standard at a final concentration of 2.5 μg/mL were dissolved in PBS buffer at pH 7.2 and incubated at 20, 30, 40, 50, 60, 70, and 80 °C for a total of 36 h.

To characterize the effects of different pH values on the degradation of AFB1 by recombinant laccase of strain B10, the recombinant laccase (20 μg/mL) and the AFB1 standard at a final concentration of 2.5 μg/mL were incubated at pH 4, 5, 6, 7, 8, 9, and 10 for 36 h at 37 °C.

To determine the effects of different metal ions on the degradation of AFB1 by the recombinant laccase from strain B10, the recombinant laccase (20 μg/mL) and 2.5 μg/mL of AFB1 standard were dissolved in PBS buffer at pH 7.2 and incubated for 36 h at 37 °C with 10 mM Na^+, K^+, Co^{2+}, Fe^{3+}, Mg^{2+}, Cu^{2+}, Ca^{2+}, Mn^{2+}, Zn^{2+}, and Ni^{2+} for 36 h.

To estimate the effects of other conditions on the degradation of AFB1 by the recombinant laccase of strain B10, the recombinant laccase (20 μg/mL) and the AFB1 standard at a final concentration of 2.5 μg/mL were dissolved in PBS buffer at pH 7.2 and incubated for 36 h at 37 °C in 10 mM ethylenediamine tetraacetic acid (EDTA), sodium dodecyl sulfate (SDS), and dimethyl sulfoxide (DMSO), respectively.

The PBS buffer was incubated with AFB1 standard at a final concentration of 2.5 μg/mL for the same length of time as the control group. HPLC was used to assay all such samples, and the content of AFB1 was determined according to the AFB1 standard curve.

5.9. Targeted Mutagenesis of the Recombinant Laccase

The amino acid sequence of the recombinant laccase was blasted through the UniProt database and the PDB database to determine proteins with a high matching score for this laccase, and its metalloid sites were recorded as H87, C132, and H149. Targeted mutations were achieved by the Fast Mutagenesis kit according to the instructions supplied with the kit. Mutation primers were as follows:

H87-F: 5′CCAGACAgctGAAAACCGCGTCCGGCGCGTGA3′

H87-R: 5′GGTTTTCagcTGTCTGGTCGGCGAACACCCAG3′

C132-F: 5′TGTTTTGCGGACgctGTGCCCTTGTATTTTTATGACCCG3′

C132-R: 5′ACagcGTCCGCAAAACAAAGGGCCAAAAAAAG3′

H149-F: 5′TTATCGGCGCTGCCgctGCCGGATGGAAGGGGACG3′

H149-R: 5′agcGGCAGCGCCGATAATGGATTTCACCGGGT3′.

The mutants were transformed into DH5α, and assays determined the correct sequence of the recombinant laccase mutant. The plasmid carrying the mutant was also transformed into *E. coli* BL21 receptor cells, and the purified mutant was expressed as described above, while its AFB1-degrading activity was determined.

Author Contributions: Methodology, M.L. and D.X.; software, D.X. and G.L.; validation, D.X., J.W. and T.L.; formal analysis, D.X. and J.W.; writing—original draft preparation, D.X.; writing—review and editing, M.L.; visualization, D.X. and G.L.; supervision, M.L. All authors have read and agreed to the published version of the manuscript.

Funding: This research was funded by the National Natural Science Foundation of China (Grant Nos. 31772809, 31972746, and 31872538) and through a Key Grant Project of Liaoning Provincial Department of Education (Grant No. LJKZ0632).

Institutional Review Board Statement: Not applicable.

Informed Consent Statement: Not applicable.

Data Availability Statement: The data presented in this study are available in this article.

Conflicts of Interest: The authors declare no conflict of interest.

References

1. Bennett, J.W.; Klich, M. Mycotoxins. *Clin. Microbiol. Rev.* **2003**, *16*, 497–516. [CrossRef] [PubMed]
2. Kensler, T.W.; Roebuck, B.D.; Wogan, G.N.; Groopman, J.D. Aflatoxin: A 50-Year Odyssey of Mechanistic and Translational Toxicology. *Toxicol. Sci.* **2011**, *120*, S28–S48. [CrossRef] [PubMed]
3. Peraica, M.; Radić, B.; Lucić, A.; Pavlović, M. Toxic effects of mycotoxins in humans. *Bull. World Health Organ.* **1999**, *77*, 754–766. [PubMed]
4. Nazhand, A.; Durazzo, A.; Lucarini, M.; Souto, E.B.; Santini, A. Characteristics, Occurrence, Detection and Detoxification of Aflatoxins in Foods and Feeds. *Foods* **2020**, *9*, 644. [CrossRef]
5. Zain, M.E. Impact of mycotoxins on humans and animals. *J. Saudi Chem. Soc.* **2011**, *15*, 129–144. [CrossRef]
6. Adebo, O.; Njobeh, P.; Gbashi, S.; Nwinyi, O.; Mavumengwana, V. Review on microbial degradation of aflatoxins. *Crit. Rev. Food Sci. Nutr.* **2017**, *57*, 3208–3217. [CrossRef] [PubMed]
7. Jard, G.; Liboz, T.; Mathieu, F.; Guyonvarc'H, A.; Lebrihi, A. Review of mycotoxin reduction in food and feed: From preven-tion in the field to detoxification by adsorption or transformation. *Food Addit. Contam. Part A Chem. Anal. Control. Expo. Risk Assess.* **2011**, *28*, 1590–1609. [CrossRef] [PubMed]
8. Mishra, H.N.; Das, C. A review on biological control and metabolism of aflatoxin. *Crit. Rev. Food Sci. Nutr.* **2003**, *43*, 245–264. [CrossRef] [PubMed]
9. Naresh, M.; Olsen, M. *Mycotoxins in food: Detection and control*; Woodhead Publishing: Oxford, UK, 2004; p. 289.
10. Kološová, A.; Stroka, J. Substances for reduction of the contamination of feed by mycotoxins: A review. *World Mycotoxin J.* **2011**, *4*, 225–256. [CrossRef]
11. Guan, Y.; Chen, J.; Nepovimova, E.; Long, M.; Wu, W.; Kuca, K. Aflatoxin Detoxification Using Microorganisms and Enzymes. *Toxins* **2021**, *13*, 46. [CrossRef] [PubMed]
12. Mate, D.; Alcalde, M. Laccase: A multi-purpose biocatalyst at the forefront of biotechnology. *Microb. Biotechnol.* **2016**, *10*, 1457–1467. [CrossRef]
13. Xu, F.; Shin, W.; Brown, S.H.; Wahleithner, J.A.; Sundaram, U.M.; Solomon, E.I. A study of a series of recombinant fungal laccases and bilirubin oxidase that exhibit significant differences in redox potential, substrate specificity, and stability. *Biochim. Biophys. Acta Protein Struct. Mol. Enzym.* **1996**, *1292*, 303–311. [CrossRef]
14. Morozova, O.V.; Shumakovich, G.P.; Gorbacheva, M.A.; Shleev, S.V.; Yaropolov, A.I. "Blue" laccases. *Biochemistry* **2007**, *72*, 1136–1150. [CrossRef] [PubMed]
15. Janusz, G.; Pawlik, A.; Swiderska-Burek, U.; Polak, J.; Sulej, J.; Jarosz-Wilkolazka, A.; Paszczynski, A. Laccase Properties, Physiological Functions, and Evolution. *Int. J. Mol. Sci.* **2020**, *21*, 996. [CrossRef] [PubMed]
16. Zhou, Z.; Li, R.; Ng, T.B.; Lai, Y.; Yang, J.; Ye, X. A New Laccase of Lac 2 from the White Rot Fungus Cerrena unicolor 6884 and Lac 2-Mediated Degradation of Aflatoxin B1. *Toxins* **2020**, *12*, 476. [CrossRef] [PubMed]

17. Hullo, M.-F.; Moszer, I.; Danchin, A.; Martin-Verstraete, I. CotA of Bacillus subtilis Is a Copper-Dependent Laccase. *J. Bacteriol.* **2001**, *183*, 5426–5430. [CrossRef] [PubMed]
18. Dwivedi, U.N.; Singh, P.; Pandey, V.P.; Kumar, A. Structure–function relationship among bacterial, fungal and plant laccases. *J. Mol. Catal. Enzym.* **2011**, *68*, 117–128. [CrossRef]
19. Guo, Y.; Qin, X.; Tang, Y.; Ma, Q.; Zhang, J.; Zhao, L. CotA laccase, a novel aflatoxin oxidase from Bacillus licheniformis, transforms aflatoxin B1 to aflatoxin Q1 and epi-aflatoxin Q1. *Food Chem.* **2020**, *325*, 126877. [CrossRef] [PubMed]
20. Loi, M.; Renaud, J.B.; Rosini, E.; Pollegioni, L.; Vignali, E.; Haidukowski, M.; Sumarah, M.W.; Logrieco, A.F.; Mule, G. Enzymatic transformation of aflatoxin B1 by Rh_DypB peroxidase and characterization of the reaction products. *Chemosphere* **2020**, *250*, 126296. [CrossRef] [PubMed]
21. Eaton, D.L.; Beima, K.M.; Bammler, T.K.; Riley, R.T.; Voss, K.A. 9.24—Hepatotoxic Mycotoxins. In *Comprehensive Toxicology*, 2nd ed.; Mcqueen, C.A., Ed.; Elsevier: Oxford, UK, 2010; pp. 527–569.
22. Hsieh, D.P.H.; Salhab, A.S.; Wong, J.J.; Yang, S.L. Toxicity of aflatoxin Q1 as evaluated with the chicken embryo and bacteri-al auxotrophs. *Toxicol. Appl. Pharm.* **1974**, *30*, 237–242. [CrossRef]
23. Martins, L.O.; Durao, P.; Brissos, V.; Lindley, P.F. Laccases of prokaryotic origin: Enzymes at the interface of protein science and protein technology. *Celll. Mol. Life Sci.* **2015**, *72*, 911–922. [CrossRef] [PubMed]
24. Jumper, J.; Evans, R.; Pritzel, A.; Green, T.; Figurnov, M.; Ronneberger, O.; Tunyasuvunakool, K.; Bates, R.; Žídek, A.; Potapenko, A.; et al. Highly accurate protein structure prediction with AlphaFold. *Nature* **2021**, *596*, 583–589. [CrossRef] [PubMed]
25. SantAnna, B.M.M.; Marbach, P.P.A.; Rojas-Herrera, M.; de Souza, J.T.; Roque, M.R.A.; Queiroz, A.T.L. High-Quality Draft Genome Sequence of Bacillus amyloliquefaciens Strain 629, an Endophyte from Theobroma cacao. *Genome Announc.* **2015**, *3*, e01325-15. [CrossRef] [PubMed]
26. Zheng, Z.; Zheng, J.; Zhang, Z.; Peng, D.; Sun, M. Nematicidal spore-forming Bacilli share similar virulence factors and mechanisms. *Sci. Rep.* **2016**, *6*, 31341. [CrossRef]
27. Wang, X.; Luo, C.; Chen, Z. Genome Sequence of the Plant Growth-Promoting Rhizobacterium Bacillus sp. Strain 916. *J. Bacteriol.* **2012**, *194*, 5467–5468. [CrossRef] [PubMed]
28. Hamid, A.S.; Tesfamariam, I.G.; Zhang, Y.; Zhang, Z.G. Aflatoxin B1-induced hepatocellular carcinoma in developing coun-tries: Geographical distribution, mechanism of action and prevention. *Oncol. Lett.* **2013**, *5*, 1087–1092. [CrossRef] [PubMed]
29. Williams, J.H.; Phillips, T.D.; Jolly, P.E.; Stiles, J.; Jolly, C.M.; Aggarwal, D. Human aflatoxicosis in developing countries: A review of toxicology, exposure, potential health consequences, and interventions. *Am. J. Clin. Nutr.* **2004**, *80*, 1106–1122. [CrossRef] [PubMed]
30. Singh, U.; Gupta, S.; Gupta, M. A review study on biological ill effects and health hazards of aflatoxins. *Asian J. Adv. Med. Sci.* **2021**, *3*, 1–8.
31. Cai, M.; Qian, Y.; Chen, N.; Ling, T.; Wang, J.; Jiang, H.; Wang, X.; Qi, K.; Zhou, Y. Detoxification of aflatoxin B1 by Stenotrophomonas sp. CW117 and characterization the thermophilic degradation process. *Environ. Pollut.* **2020**, *261*, 114178. [CrossRef] [PubMed]
32. Xu, L.; Eisa, A.M.; Sangare, L.; Zhao, Y.; Selvaraj, J.N.; Xing, F.; Wang, Y.; Yang, H.; Liu, Y. Novel Aflatoxin-Degrading En-zyme from Bacillus shackletonii L7. *Toxins* **2017**, *9*, 36. [CrossRef] [PubMed]
33. Suresh, G.; Cabezudo, I.; Pulicharla, R.; Cuprys, A.; Rouissi, T.; Brar, S.K. Biodegradation of aflatoxin B1 with cell-free ex-tracts of Trametes versicolor and Bacillus subtilis. *Res. Vet. Sci.* **2020**, *133*, 85–91. [CrossRef] [PubMed]
34. Wang, Y.; Zhang, H.; Yan, H.; Yin, C.; Liu, Y.; Xu, Q.; Liu, X.; Zhang, Z. Effective Biodegradation of Aflatoxin B1 Using the Bacillus licheniformis (BL010) Strain. *Toxins* **2018**, *10*, 497. [CrossRef] [PubMed]
35. Shu, X.; Wang, Y.; Zhou, Q.; Li, M.; Hu, H.; Ma, Y.; Chen, X.; Ni, J.; Zhao, W.; Huang, S.; et al. Biological Degradation of Aflatoxin B1 by Cell-Free Extracts of Bacillus velezensis DY3108 with Broad PH Stability and Excellent Thermostability. *Toxins* **2018**, *10*, 330. [CrossRef] [PubMed]
36. Zhao, L.; Guan, S.; Gao, X.; Ma, Q.; Lei, Y.; Bai, X.; Ji, C. Preparation, purification and characteristics of an aflatoxin degradation enzyme from Myxococcus fulvus ANSM068. *J. Appl. Microbiol.* **2010**, *110*, 147–155. [CrossRef]
37. Motomura, M.; Toyomasu, T.; Mizuno, K.; Shinozawa, T. Purification and characterization of an aflatoxin degradation enzyme from Pleurotus ostreatus. *Microbiol. Res.* **2003**, *158*, 237–242. [CrossRef] [PubMed]
38. Del Alamo, D.; Govaerts, C.; Mchaourab, H.S. AlphaFold2 predicts the inward-facing conformation of the multidrug transporter LmrP. *Proteins Struct. Funct. Bioinform.* **2021**, *89*, 1226–1228. [CrossRef] [PubMed]
39. Cramer, P. AlphaFold2 and the future of structural biology. *Nat. Struct. Mol. Biol.* **2021**, *28*, 704–705. [CrossRef] [PubMed]
40. Robertson, A.J.; Courtney, J.M.; Shen, Y.; Ying, J.; Bax, A. Concordance of X-ray and AlphaFold2 Models of SARS-CoV-2 Main Protease with Residual Dipolar Couplings Measured in Solution. *J. Am. Chem. Soc.* **2021**, *143*, 19306–19310. [CrossRef] [PubMed]
41. Liu, Y.; Mao, H.; Hu, C.; Tron, T.; Lin, J.; Wang, J.; Sun, B. Molecular docking studies and in vitro degradation of four afla-toxins (AFB1, AFB2, AFG1, and AFG2) by a recombinant laccase from Saccharomyces cerevisiae. *J. Food Sci.* **2020**, *85*, 1353–1360. [CrossRef]
42. Jardon-Xicotencatl, S.; Díaz-Torres, R.; Marroquin-Cardona, A.; Villarreal-Barajas, T.; Méndez-Albores, A. Detoxification of Aflatoxin-Contaminated Maize by Neutral Electrolyzed Oxidizing Water. *Toxins* **2015**, *7*, 4294–4314. [CrossRef]

43. Novoa, C.; Dhoke, G.V.; Mate, D.; Martinez, R.; Haarmann, T.; Schreiter, M.; Eidner, J.; Schwerdtfeger, R.; Lorenz, P.; Davari, M.D.; et al. KnowVolution of a Fungal Laccase toward Alkaline pH. *ChemBioChem* **2019**, *20*, 1458–1466. [CrossRef] [PubMed]
44. Shi, X.; Liu, Q.; Ma, J.; Liao, H.; Xiong, X.; Zhang, K.; Wang, T.; Liu, X.; Xu, T.; Yuan, S.; et al. An acid-stable bacterial laccase identified from the endophyte Pantoea ananatis Sd-1 genome exhibiting lignin degradation and dye decolor-ization abilities. *Biotechnol. Lett.* **2015**, *37*, 2279–2288. [CrossRef] [PubMed]
45. Wu, M.H.; Lin, M.C.; Lee, C.C.; Yu, S.M.; Wang, A.H.; Ho, T.D. Enhancement of laccase activity by pre-incubation with or-ganic solvents. *Sci. Rep.* **2019**, *9*, 9754. [CrossRef] [PubMed]
46. Klibanov, A.M. Improving enzymes by using them in organic solvents. *Nature* **2001**, *409*, 241–246. [CrossRef]
47. Kim, Y.; Maltseva, N.; Dementieva, I.; Collart, F.; Holzle, D.; Joachimiak, A. Crystal Structure of Hypothetical Protein YfiH From Shigella flexneri at 2 Å Resolution. *Proteins Struct. Funct. Bioinform.* **2006**, *63*, 1097–1101. [CrossRef] [PubMed]
48. Gunne, M.; Al-Sultani, D.; Urlacher, V.B. Enhancement of copper content and specific activity of CotA laccase from Bacillus licheniformis by coexpression with CopZ copper chaperone in E. coli. *J. Biotechnol.* **2013**, *168*, 252–255. [CrossRef] [PubMed]
49. Cader, M.Z.; de Almeida, R.R.; West, J.A.; Sewell, G.W.; Md-Ibrahim, M.N.; Reikine, S.; Sirago, G.; Unger, L.W.; Iglesias-Romero, A.B.; Ramshorn, K.; et al. FAMIN Is a Multifunctional Purine Enzyme Enabling the Purine Nucle-otide Cycle. *Cell* **2020**, *180*, 278–295. [CrossRef] [PubMed]
50. Wang, H.; Liu, X.; Zhao, J.; Yue, Q.; Yan, Y.; Gao, Z.; Dong, Y.; Zhang, Z.; Fan, Y.; Tian, J.; et al. Crystal struc-tures of multicopper oxidase CueO G304K mutant: Structural basis of the increased laccase activity. *Sci. Rep.* **2018**, *8*, 14252. [CrossRef] [PubMed]
51. Djoko, K.Y.; Chong, L.X.; Wedd, A.G.; Xiao, Z. Reaction Mechanisms of the Multicopper Oxidase CueO from Escherichia coli Support Its Functional Role as a Cuprous Oxidase. *J. Am. Chem. Soc.* **2010**, *132*, 2005–2015. [CrossRef] [PubMed]
52. Kataoka, K.; Komori, H.; Ueki, Y.; Konno, Y.; Kamitaka, Y.; Kurose, S.; Tsujimura, S.; Higuchi, Y.; Kano, K.; Seo, D.; et al. Structure and Function of the Engineered Multicopper Oxidase CueO from Escherichia coli—Deletion of the Methionine-Rich Helical Region Covering the Substrate-Binding Site. *J. Mol. Biol.* **2007**, *373*, 141–152. [CrossRef] [PubMed]

toxins

MDPI

Article

The Antagonistic Effect of Glutamine on Zearalenone-Induced Apoptosis via PI3K/Akt Signaling Pathway in IPEC-J2 Cells

Tianhu Wang †, Jingjing Wang †, Tong Zhang, Aixin Gu, Jianping Li * and Anshan Shan *

Institute of Animal Nutrition, Northeast Agricultural University, Harbin 150030, China; wthqh123@163.com (T.W.); jf_jing@163.com (J.W.); neauzt@outlook.com (T.Z.); aixingu@hotmail.com (A.G.)
* Correspondence: ljpneau@neau.edu.cn (J.L.); asshan@neau.edu.cn (A.S.); Tel.: +86-0451-5519-1439 (J.L.)
† These authors contributed equally to this work.

Abstract: Zearalenone (ZEN) is a non-steroidal estrogen mycotoxin produced by *Fusarium* fungi, which inevitably exists in human and animal food or feed. Previous studies indicated that apoptosis seems to be a key determinant of ZEN-induced toxicity. This experiment aimed to investigate the protective effects of Glutamine (Gln) on ZEN-induced cytotoxicity in IPEC-J2 cells. The experimental results showed that Gln was able to alleviate the decline of cell viability and reduce the production of reactive oxygen species and calcium (Ca^{2+}) induced by ZEN. Meanwhile, the mRNA expression of antioxidant enzymes such as glutathione reductase, glutathione peroxidase, and catalase was up-regulated after Gln addition. Subsequently, Gln supplementation resulted in the nuclear fission and Bad-fluorescence distribution of apoptotic cells were weakened, and the mRNA expression and protein expression of pro-apoptotic genes and apoptotic rates were significantly reduced. Moreover, ZEN reduced the phosphorylation Akt, decreased the expression of Bcl-2, and increased the expression of Bax. Gln alleviated the above changes induced by ZEN and the antagonistic effects of Gln were disturbed by PI3K inhibitor (LY294002). To conclude, this study revealed that Gln exhibited significant protective effects on ZEN-induced apoptosis, and this effect may be attributed to the PI3K/Akt signaling pathway.

Keywords: zearalenone; glutamine; PI3K/Akt pathway; apoptosis; IPEC-J2 cells

Key Contribution: The addition of Gln (2 mM) alleviated the negative effects resulting from ZEN (160 μM) in IPEC-J2 cells. Gln (2 mM) exerts an antagonistic effect on apoptosis by activating PI3K/Akt signaling pathway.

Citation: Wang, T.; Wang, J.; Zhang, T.; Gu, A.; Li, J.; Shan, A. The Antagonistic Effect of Glutamine on Zearalenone-Induced Apoptosis via PI3K/Akt Signaling Pathway in IPEC-J2 Cells. *Toxins* **2021**, *13*, 891. https://doi.org/10.3390/toxins13120891

Received: 15 November 2021
Accepted: 10 December 2021
Published: 12 December 2021

Publisher's Note: MDPI stays neutral with regard to jurisdictional claims in published maps and institutional affiliations.

1. Introduction

Zearalenone (ZEN), a mycotoxin produced by *Fusarium*, is one of the vital sources of food contamination. Its ubiquity in food and feed poses a threat to humans and animals health [1]. Studies have shown that after ingesting ZEN-contained foods, the toxic compound was absorbed through the gastrointestinal tract (GIT), metabolized, and distributed to different parts of the body [2]. Reports indicated that ZEN can induce hepatotoxicity, immunotoxicity, hematotoxicity, and genotoxicity, and lead to cell death by inducing oxidative stress, mitochondrial damage, and apoptosis [3–5]. Several toxicological models of ZEN's effects in the body and cells have been carried out in the past years. For instance, previous studies from this lab have shown that ZEN increased the levels of reactive oxygen species (ROS) and repressed the activity and expression of anti-oxidative enzymes in porcine kidney cells (PK15) or porcine intestinal epithelial cells (SIEC02), resulting in cell apoptosis [6,7]. However, little is known and it is worthy to further investigate ways to detoxify for ZEN-poisoned cells and organs.

Glutamine (Gln) is an α-amino acid and the most abundant free amino acid in the body [8]. As a precursor for nucleotide biosynthesis, Gln is one of the crucial substances for intestinal epithelial cell proliferation and integrity repair [9,10]. It was reported that the dietary addition of Gln reduced weaning stress caused intestinal dysfunction by cell proliferation and increased expression of tight junction proteins in weaned pups [11,12]. Similarly, in vitro studies also indicated that Gln could promote the proliferation of intestinal porcine epithelial cell lines [13,14]. In addition, Gln was reported toxic protection effects on the intestinal damage and the intestinal epithelial cell apoptosis caused by clostridium *difficile* toxin-A in the rabbit model [15]. At present, the mechanism of Gln that protects cells from ZEN-induced apoptosis is rarely reported and needs further exploration.

The gastrointestinal tract is a multifunctional and complex organ [16]. It is not only an organ for digestion and absorption of nutrients, but the first barrier to protect animal health from ingested chemicals, food contaminants, and natural toxins. Furthermore, intestinal homeostasis depends on the diverse functions of intestinal epithelial cells [17,18]. Hence, porcine jejunal epithelial cells (IPEC-J2) were selected for the study.

It was hypothesized that Gln might protect the cells against ZEN-induced apoptotic, and it may work via PI3K/Akt signaling pathway. Therefore, the IPEC-J2 cell line was studied as a model to investigate the detoxification of Gln addition on ZEN-induced cells in this study.

2. Results

2.1. Effects of ZEN and Gln on Cell Viability

To determine the suitable concentration of Gln in subsequent experiments, the viability of the IPEC-J2 cells (Figure 1) was measured by the CCK-8 method at first. As shown in Figure 1, compared with the control group, exposure to 160 μM ZEN for 48 h, the cell viability was reduced significantly ($p < 0.001$). The addition of 2 mM Gln significantly increased the cell viability compared with the ZEN group ($p < 0.001$).

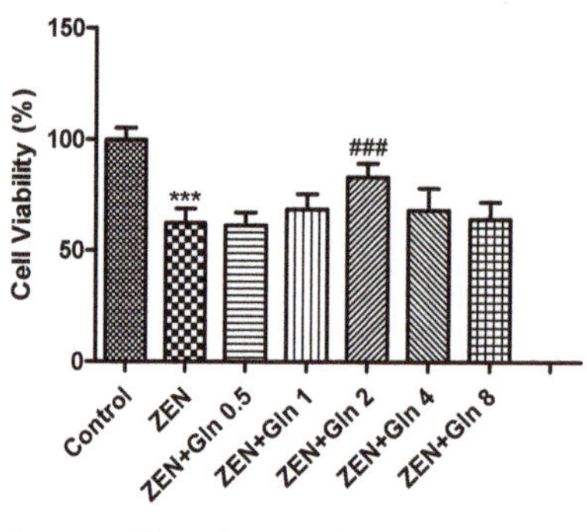

Figure 1. Effects of ZEN and Gln on the viability of the IPEC-J2 cells. Values are expressed as means ± SD of three independent experiments. *** $p < 0.001$ ZEN vs. control. ### $p < 0.001$ ZEN vs. ZEN + Gln 2.

2.2. Effects of ZEN and Gln on the Activities of Enzymes

When IPEC-J2 cells were exposed to ZEN and different concentrations of Gln for 48 h. As shown in Figure 2, the three enzyme activities (glutathione reductase (GR), glutathione peroxidase (GPx), and catalase (CAT)) decreased significantly upon exposure to ZEN compared with the control group ($p < 0.05$). Compared with the ZEN treatment, no differences were observed in the three enzyme activities at 0.5 mM Gln; however, the level of 1 and 2 mM Gln were showed significant increases ($p < 0.05$) in these three kinds of enzymes. Meanwhile, the concentration of 4 and 8 just observed improvements in one kind of enzyme (Gpx and CAT), respectively. Based on these data, a protective concentration of Gln (2 mM) was selected and incubated with ZEN for 48 h in subsequent experiments.

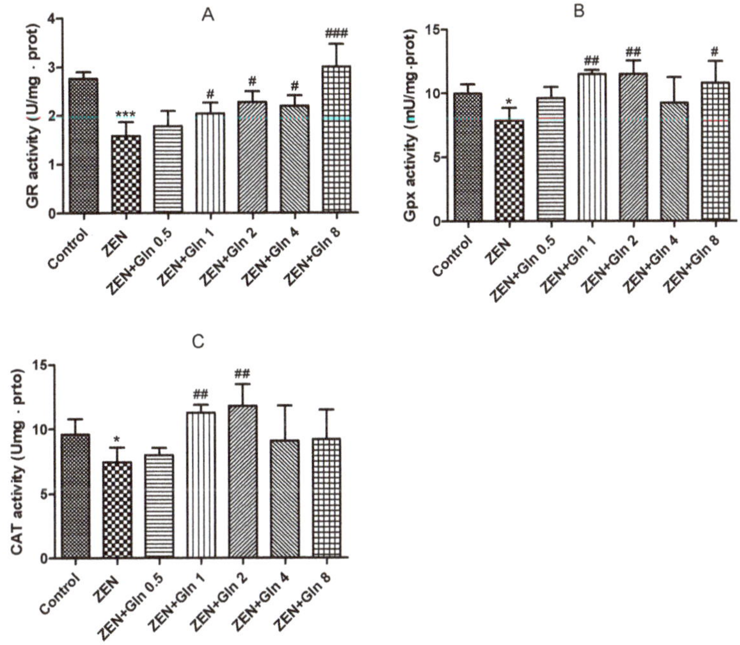

Figure 2. Effects of ZEN and Gln on the activities of the enzymes (GR, Gpx and CAT) in the IPEC-J2 cells. Values are expressed as means ± SD of three independent experiments. * $p < 0.05$ and *** $p < 0.001$ ZEN vs. control. # $p < 0.05$, ## $p < 0.01$, and ### $p < 0.001$ ZEN vs. ZEN + Gln. (**A**) Effects of ZEN and Gln on the activity of GR; (**B**) Effects of ZEN and Gln on the activity of Gpx; (**C**) Effects of ZEN and Gln on the activity of CAT.

2.3. Intracellular ROS Generation

To determine changes in oxidative damage, IPEC-J2 cells were exposed to different drugs for 48 h. The ROS production results were obtained by the fluorescein assay (Figure 3). The figure showed that the level of ROS was significantly higher in the ZEN group than that in the control group ($p < 0.001$). Compared with the ZEN group, the intracellular ROS production was decreased significantly after Gln addition ($p < 0.001$).

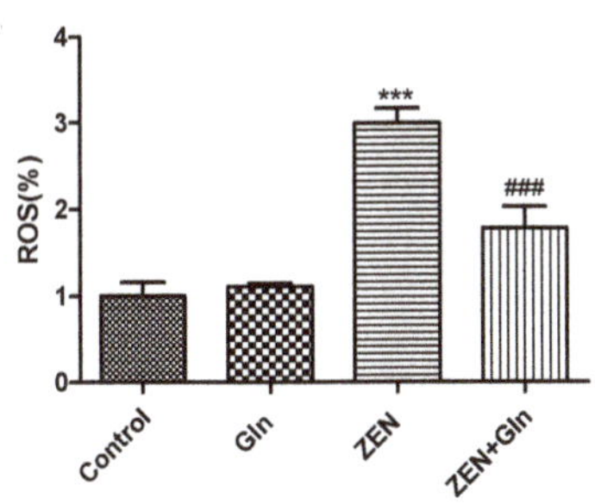

Figure 3. Effects of ZEN and Gln on intracellular ROS production. Values are expressed as means ± SD of three independent experiments. *** $p < 0.001$ ZEN vs. control. ### $p < 0.001$ ZEN vs. ZEN + Gln.

2.4. Intracellular Ca^{2+}

IPEC-J2 cells were incubated with Fluo-4 AM, bound to Ca^{2+} to produce strong fluorescence. As shown in Figure 4, compared with the control group, ZEN-induced levels of intracellular Ca^{2+} increased significantly ($p < 0.001$). Gln supplementation significantly reduced intracellular levels of Ca^{2+} in the ZEN-induced cells ($p < 0.001$).

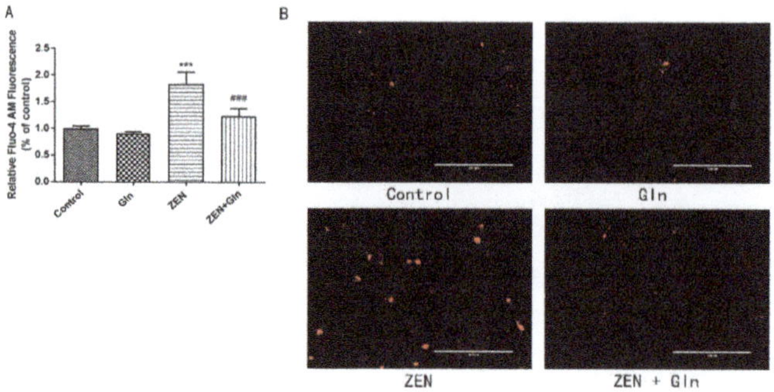

Figure 4. Effects of ZEN and Gln on intracellular Ca^{2+} production. (**A**) Values are expressed as means ± SD of three independent experiments. (**B**) Fluorescence microscopy observation of intracellular Ca^{2+} fluorescence intensity, Scale bar: 200 μm. *** $p < 0.001$ ZEN vs. control. ### $p < 0.001$ ZEN vs. ZEN + Gln.

2.5. Immunofluorescence Staining of Cells

The morphologic changes of apoptotic nuclei were observed by fluorescence microscopy with Hoechst-33258 staining (Figure 5). In control group cells, the nuclei displayed uniformly blue-stained with a smooth appearance. However, uneven nuclear staining, nuclear condensation, and fragmentation of nuclei were shown clearly in the ZEN group. In Comparison with the ZEN group, although Gln addition reduced nuclear shrinkage and rupture, pretreatment with LY294002 that did not reduce nuclear apoptosis.

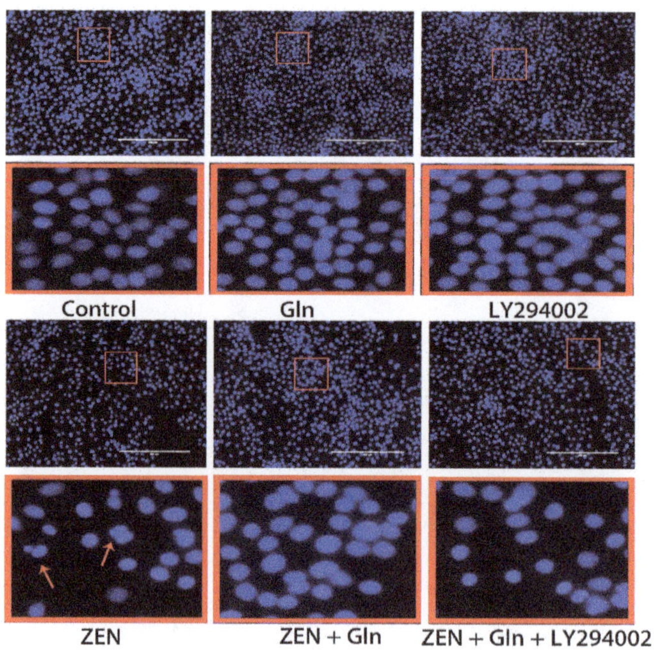

Figure 5. Effects of ZEN, Gln, and LY294002 on the apoptotic nuclei (Hoechst 33258 staining). The red color frame in the figure indicates the obvious change area. The arrow represents a change in nuclear morphology. Scale bar: 200 μm.

2.6. Apoptosis Rate in IPEC-J2 Cells

The Annexin V/FITC/PI apoptosis kit was used to analyze different drugs effects on apoptosis of IPEC-J2 cells. As shown in Figure 6A, ZEN induced a significant increase in the number of early apoptotic cells (Q2), as well as in the number of late apoptotic cells (Q4) in IPEC-J2 cells. The total apoptotic cell proportion was increased by 56.8% (Figure 6B) compared with the control group. Compared with the ZEN group, Gln addition significantly reduced early apoptosis and late apoptosis, and the total apoptotic cell proportion (12.5%) was decreased by 44.3%. In addition, pretreatment with LY294002 significantly increased late apoptosis, and the total cell apoptotic rate (31.6%) was increased by 19.1% compared with the ZEN + Gln group. This result was consistent with the results of nuclear apoptosis staining.

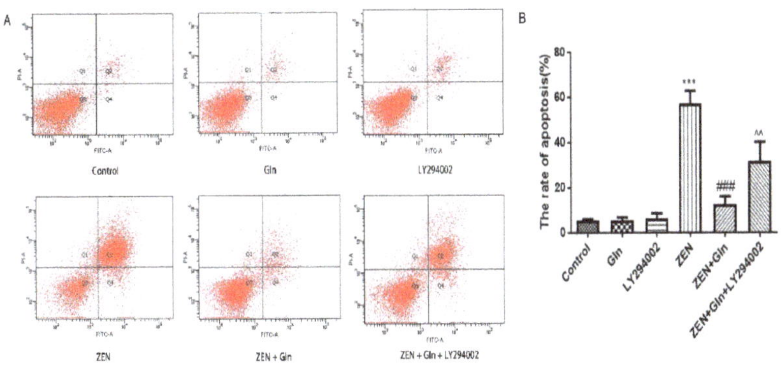

Figure 6. (**A**) The apoptotic cells were determined by annexin V-FITC/PI staining using flow cytometry. The Q1, Q2, Q3, and Q4, respectively, represented dead cells, the late cells apoptosis, normal cells, and the early cells apoptosis. Apoptosis was the sum of early apoptosis and late apoptosis. (**B**) The percentage of IPEC-J2 cells apoptosis was shown in statistical analysis. Each value represents the mean ± SD of the three independent experiments. *** $p < 0.001$ ZEN vs. control, ### $p < 0.001$ ZEN vs. ZEN + Gln. ^^ $p < 0.01$ ZEN + Gln vs. ZEN + Gln + LY294002.

2.7. The mRNA Expression of Apoptosis-Related Genes

To further investigate the effects of these drugs on cell apoptotic, the mRNA expression of apoptosis-related genes was measured. As shown in Figure 7, compared with the control group, ZEN induced a significant increase in the mRNA expression levels of pro-apoptotic genes: Caspase-3, Caspase-9, Cytochrome c (Cyto-c), and Bad ($p < 0.05$). Conversely, the mRNA expression of anti-apoptosis genes (Bcl-xl and Bcl-2) was significantly reduced ($p < 0.001$). After Gln addition, the mRNA expression of five pro-apoptotic genes (Caspase-3, Caspase-9, Cyto-c, Bax and Bad) were significantly down-regulated ($p < 0.001$), anti-apoptotic genes (Bcl-xl and Bcl-2) were significantly up-regulated ($p < 0.05$). Compared with the ZEN + Gln group, pretreatment with LY294002, the mRNA expression of three pro-apoptotic genes (Caspase-3, Caspase-9, and Bax) were increased significantly ($p < 0.05$) and had no significant effect on anti-apoptotic genes.

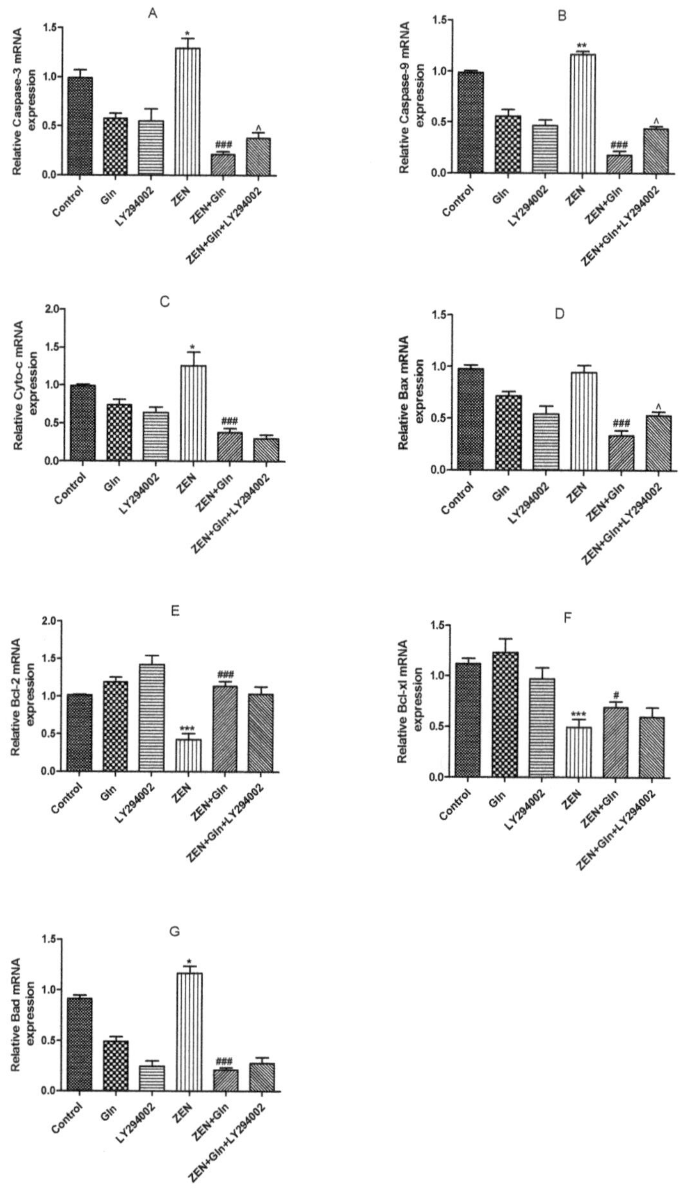

Figure 7. Effects of ZEN, Gln and LY294002 on the apoptosis-related genes in the IPEC-J2 cells. Values are expressed as means ± SD of three independent experiments. * $p < 0.05$, ** $p < 0.01$, and *** $p < 0.001$ ZEN vs. control. # $p < 0.05$ and ### $p < 0.001$ ZEN vs. ZEN + Gln. ^ $p < 0.05$ ZEN + Gln vs. ZEN + Gln + LY294002. (**A**) Effects of ZEN, Gln and LY294002 on the mRNA expression of Caspase-3; (**B**) Effects of ZEN, Gln and LY294002 on the mRNA expression of Caspase-9; (**C**) Effects of ZEN, Gln and LY294002 on the mRNA expression of Cyto-c; (**D**) Effects of ZEN, Gln and LY294002 on the mRNA expression of Bax; (**E**) Effects of ZEN, Gln and LY294002 on the mRNA expression of Bcl-2; (**F**) Effects of ZEN, Gln and LY294002 on the mRNA expression of Bcl-xl; (**G**) Effects of ZEN, Gln and LY294002 on the mRNA expression of Bad.

2.8. Immunofluorescence

The Bad protein is involved in initiating apoptosis. Next, we investigated the expression of Bad by immunofluorescence. As displayed in Figure 8, in the control group, the fluorescence of Bad was very weak. While in the ZEN group, Bad fluorescence expression was strongly positive, distributed around the nucleus in a spotted manner and decreased number of IPEC-J2 cells. In the Gln + ZEN group, the immunostaining of Bad was relatively weaker than the ZEN group, and there was no distribution pattern of aggregated spots. Moreover, in the LY294002 pretreatment group, the immunostaining of Bad was relatively stronger than that in the ZEN + Gln group.

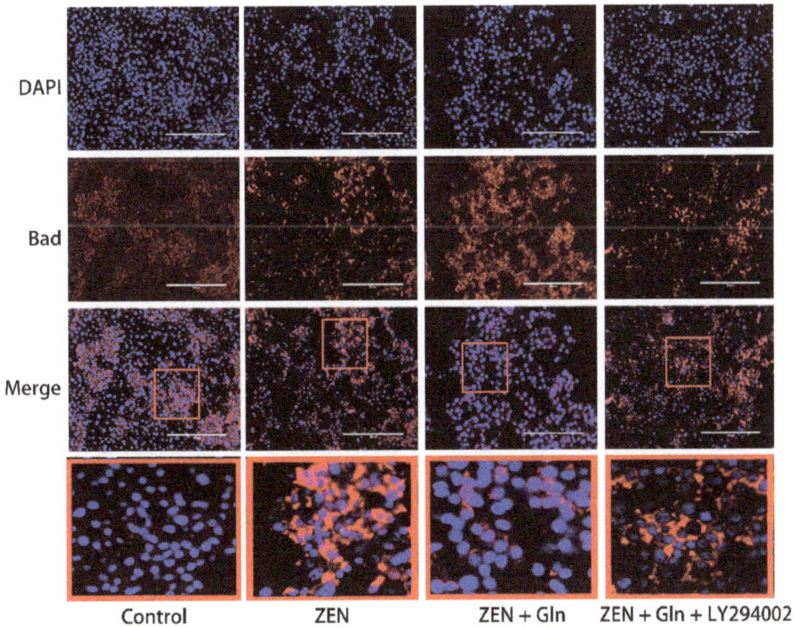

Figure 8. Expression of Bad in IPEC-J2 cells. Cells were stained with antibodies for Bad and detected by immunofluorescence after treatment. The images were collected by the nuclei showed blue fluorescence after counterstaining with DAPI. The red color frame in the figure indicates the obvious change area. Scale bar: 200 μm.

2.9. Western Blotting

To further verify that the PI3K/Akt signaling pathway is the mechanism by which Gln protects against ZEN-induced apoptosis, the related proteins of the PI3K/Akt signaling pathway and apoptosis-related proteins were measured by western blotting. As shown in Figure 9, there was no difference in the protein expression of Akt in the four treatment groups (Figure 9D). Compared with the control group, after ZEN-exposed, the protein expression of pro-apoptotic gene Bax and anti-apoptotic gene Bcl-2 were significantly increased ($p < 0.01$) and significantly decreased ($p < 0.001$), respectively. The addition of Gln significantly decreased the protein expression of Bax compared with the ZEN group ($p < 0.05$), conversely, the protein expression of Bcl-2 was significantly elevated ($p < 0.05$). Compared with the ZEN + Gln group, the pretreatment of LY294002 significantly decreased the protein expression of Bcl-2 ($p < 0.01$), and there was no significant change in Bax protein expression. Also, as shown in Figure 9E, after ZEN-exposed, a remarkable decrease in the protein expression of the p-Akt compared with the control group, the addition of Gln increased the protein expression of p-Akt but did not reach a significant level. In addition, pretreatment with LY294002 significantly reduced p-Akt protein expression.

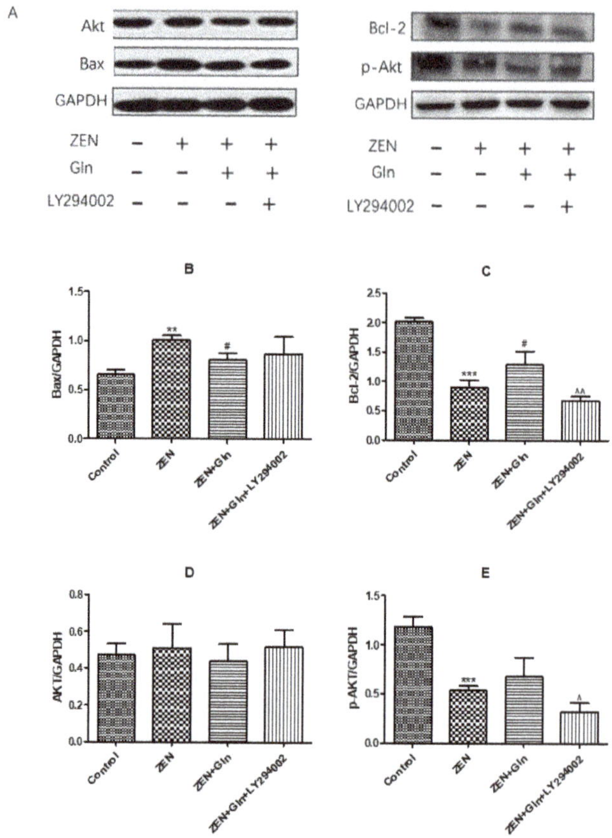

Figure 9. Effects of ZEN, Gln and LY294002 on the PI3K/Akt pathway- and apoptosis-related genes in the IPEC-J2 cells. Values are expressed as means ± SD of three independent experiments. ** $p < 0.01$ and *** $p < 0.001$ ZEN vs. control. # $p < 0.05$ ZEN vs. ZEN + Gln. ^ $p < 0.05$ and ^^ $p < 0.01$ ZEN + Gln vs. ZEN + Gln + LY294002. (**A**) Effects of ZEN, Gln and LY294002 on the protein expression of PI3K/Akt pathway; (**B**) Effects of ZEN, Gln and LY294002 on the protein expression of Bax; (**C**) Effects of ZEN, Gln and LY294002 on the protein expression of Bcl-2; (**D**) Effects of ZEN, Gln and LY294002 on the protein expression of Akt; (**E**) Effects of ZEN, Gln and LY294002 on the protein expression of p-Akt.

3. Discussion

ZEN is widely found in cereals and animal feed worldwide, which has a negative impact on human and animal health [4,19,20]. Intestinal epithelial cells are the first target of ZEN after ingestion of feed and foods contaminated with ZEN [21]. In vitro and in vivo studies found that oxidative damage was one of the crucial pathways by which ZEN induced cytotoxicity, resulting in cell apoptosis [22,23]. Oxidative damage is mainly caused by the mass production of ROS and free radicals [24]. Oxidative stress is caused by the excessive generation of ROS or the disruption of the oxidoreductase balance in the cell. It not only activated cell signaling but also induced apoptosis [25]. As described above, the results of this study found that ZEN exposure produced excess ROS (Figure 3), numerous uneven nuclear staining, nuclear fissures, mass cells apoptosis (Figure 6). To date, these antioxidant enzymes, comprising CAT, Gpx and GR, etc., are a source of protection against oxidative stress [26,27]. Overall, cells exposed to ZEN induced oxidative damage and reduced the intracellular antioxidant enzyme activities (Figure 2).

Gln, a major substrate utilized by intestinal cells, is not only a source of the main energy of the cell mitochondria, but it can eliminate some of the strong oxidants and protect cells from oxidative damage [28,29]. In gut physiology, the addition of Gln can promote enterocyte proliferation and protect against apoptosis under stress conditions [30]. Therefore, Gln was used to investigate the protective mechanism against ZEN-induced apoptosis in this study. It was observed that the addition of Gln increased cell survival (Figure 1), reduced nuclear shrinkage (Figure 5), and decreased apoptosis rate (Figure 6). The results showed Gln alleviated the apoptosis induced by ZEN. At the same time, the activities of antioxidant enzymes in the cells also increased, and the effect was the best when Gln concentration was 2 mM (Figure 2). Hence, 2 mM Gln was selected as the concentration for subsequent verification. Compared with 2mM Gln, the protective effect of Gln (4 and 8 mM) is weaker. The reason may be consistent with Curi's conclusion that although Gln supplementation can bring obvious benefits in many cases, the adverse effects of long-term use of high concentrations of glutamic acid might not be completely ruled out [31]. Further, the results of pretreatment with LY294002 that did not reduce nuclear apoptosis are consistent with earlier studies because it has cytostatic, but no cytotoxicity effects on cells [32–34].

ZEN exposure can induce IPEC-J2 cells apoptosis by mitochondrial damage [6,7,35]. The literature demonstrated that mitochondria-dependent apoptotic pathways involved a variety of events, such as the production of ROS, the release of Cyto-c in mitochondria, Bcl-2 family members, and activation of caspases-9 and caspases-3 [36]. We found that ZEN induced apoptosis via the mitochondrial pathway of IPEC-J2 cells. The expression of Cyto-c, caspases-9, caspases-3, and pro-apoptotic genes (Bax and Bad) were increased, while anti-apoptotic genes (Bcl-2 and Bcl-xl) were reduced (Figures 7 and 9B,C). In addition, mitochondria are a storage room for intracellular calcium [37]. Recently, with the in-depth discussion of the apoptotic process, it was suggested that intracellular Ca^{2+} and ROS surges were vital mediators of cell death [38,39]. The current study showed that ZEN exposure increased intracellular ROS and Ca^{2+} levels (Figures 3 and 4). Oppositely, the addition of Gln reduced the content of ROS and Ca^{2+} in cells (Figures 3 and 4). These results were consistent with the present study. Overall, Gln improved cell survival rate and protected cells from ZEN-induced mitochondrial apoptosis.

The PI3K/Akt pathway is an important regulator of cellular homeostasis in vivo [40,41]. In addition, it is a vital anti-apoptosis/proliferation signaling pathway that plays a key role in cellular functioning [42,43]. Recently, it was found that Gln increased the antioxidant capacity by activating PI3K/Akt signaling pathway in Parkinson's disease [44]. Phosphatidylinositol-3 kinases (PI3Ks) are a family of lipid kinases that regulate various metabolic activities in the cell [45]. Activated Akt is a downstream effector of PI3K, which can inhibit apoptosis by regulating multiple targets such as MPTP, ATPase, NF-κB, and the Bcl-2 family proteins [46,47]. We speculated that Gln could exert the effect of anti-apoptosis via PI3K/Akt signaling pathway and improve cell survival. The Bad protein is a downstream substrate of Akt, Akt-phosphorylate (p-Akt) can activate it [48]. In the presence of survival factors, the Bad protein is phosphorylated at two serine sites (Ser-112 and Ser-136) and sequestered in the form of inactive molecules in the cytosol, receiving the death signal, Bad dephosphorylates and interacts with Bcl-xl–Bcl-2 to form dimers that accumulate in mitochondria [49,50]. Therefore, to examine the position of Bad in the case of cell survival and death, immunofluorescence staining was used to observe the fluorescence change of the Bad gene. The results clearly showed that the strong accumulation of Bad gene fluorescence occurred in ZEN-induced cell apoptosis (Figure 8), the mRNA expression of Bad was increased simultaneously. However, there was no accumulation of Bad gene fluorescence occurred, but the mRNA expression of Bad was decreased (Figure 7G), and decreased cell apoptotic rate (Figure 6) when treated with Gln. Importantly, after pretreatment with inhibitor (LY294002) of the PI3K/Akt signaling pathway, the Bad gene strong accumulation of fluorescence occurred (Figure 8). This result suggested that Gln may play an anti-apoptotic effect via the PI3K/Akt signaling pathway. The same result was found in primary liver cancer cells increases of Bad-expressing caused by Akt-knocked [43].

The literature suggested that the activation of the PI3K/Akt signaling pathway could suppress apoptosis [46]. Western blotting results in this study showed that Gln increased the expression of p-Akt protein and anti-apoptotic protein Bcl-2 (Figure 9C,E), these results were consistent with the activation of the PI3K/Akt pathway leading to increased expression of Bcl-2 [51]. Our results also showed that Gln treatment stimulated phosphorylation of Akt, and reduced the apoptosis rates (Figure 6), consistent with a previous study showing that growth factor receptor Akt activation prevented apoptosis [52]. Thus, the addition of Gln activated the PI3K/Akt signaling pathway and protected cells from ZEN-induced apoptosis. Meanwhile, pretreatment with LY294002 reduced p-Akt protein expression, suggesting that the anti-apoptotic pathway of PI3K/Akt was inhibited. This result was supported by an early study that LY294002 inhibited the PI3K/Akt signaling pathway and significantly enhanced bufalin-induced apoptosis [53]. As described above, these results suggested that Gln protected cells from ZEN-induced apoptosis, and activation of the PI3K/Akt signaling pathway was one of the factors.

4. Conclusions

In conclusion, ZEN exposure induced the excessive generation of ROS, increased intracellular Ca^{2+} concentration, induced oxidative damage, and activated the intrinsic apoptotic cascade reaction in IPEC-J2 cells. However, Gln addition increased the activities of intracellular antioxidant enzymes, increased the expression of anti-apoptotic genes and p-Akt, reduced the expression of pro-apoptotic genes and caspase cascade enzymes. Overall, these findings suggested that Gln antagonized ZEN-induced apoptosis, possibly via the PI3K/Akt signaling pathway in IPEC-J2 cells.

5. Materials and Methods

5.1. Chemicals and Reagents

The Zearalenone (ZEN) and Glutamine (Gln) were obtained from Sigma-Aldrich (St. Louis, MO, USA) and were dissolved in ethanol and deionized water to a stock solution of 100 mM and 200 mM, respectively. DMEM-F:12 cell culture medium was purchased from Thermo Fisher (Hyclone, Beijing, China). Fetal bovine serum (FBS) was supplied by Gibco-Life Technology (Grand Island, NY, USA). Trypsin/EDTA, Penicillin/streptomycin, the activity assay kits of Catalase (CAT), Glutathione reductase (GR), Glutathione peroxidase (Gpx), and Alkaline phosphatase (AP), ECL detection kit, the Annexin V-FITC/PI apoptosis detection kit, Hoechst-33258 Staining Kit, Reactive Oxygen Species Assay Kit, Fluo-4 AM and BCA Assay Kit were supplied by Beyotime Biotechnology (Nantong, China). Cell Counting Kit-8 (CCK-8) was supplied by Dojindo (Kumamoto, Japan). Phosphate buffered saline (PBS) was purchased from Biotopped (Beijing, China).

5.2. Cell and Cell Culture

IPEC-J2 cells were donated by China Agricultural University. The cells were cultured in a complete medium composed of DMEM-F:12 medium (Hyclone, Beijing, China), 10% FBS (GIBCO, Grand Island, NY, USA), and 1% penicillin and streptomycin (Beyotime Biotechnology, Nantong, China). Cells were cultured in an incubator at 37 °C, with a continual supply of 5% CO_2.

5.3. Cell Viability Assay

Based on our previous findings (not yet published), we selected ZEN (160 μM) to infect IPEC-J2 cells. IPEC-J2 cells (0.8–1.0 × 10^5 cells/mL) were seeded in 96-well culture plates; culture medium was changed every 24 h. When cells became monolayer, cells were washed twice with PBS, then treated with ZEN (160 μM) and different concentrations of Gln (0.5, 1, 2, 4, and 8 mM) for 48 h. IPEC-J2 cells were washed three times with PBS after the cell culture medium was removed, then CCK-8 (10 μL) was added and incubated at 37 °C for

3 h. Cell viability was measured by absorbance on a microplate reader at 450 nm emission wavelength (Tecan Austria GmbH Untersbergatr, Austria). The microplate readers used in this study were from one manufacturer.

5.4. Determination of IPEC-J2 Cellular the Activities of Enzymes

IPEC-J2 cells (2.0–2.5 × 10^6 cells/mL) were grown in 6-well culture plates and treated with drugs as described in the previous section. The CAT, GR, and Gpx activities were determined according to the manufacturer's instructions.

5.5. Detection of ROS Generation

Changes in intracellular ROS was detected with dichlorofluorescein diacetate (DCFH-DA). IPEC-J2 cells (4.0–5.0 × 10^5 cells/mL) were grown in 24-well culture plates and treated with drugs as described in the previous section. After treatment, cells were washed thrice with PBS and incubated with 10 μM DCFH-DA at 37 °C for 20 min. Finally, cells were washed thrice with PBS and left a small amount of PBS. Intracellular production of ROS was measured by a microplate reader (Ex = 488 nm and Em = 525 nm). ROS production was expressed as a percentage of the control.

5.6. Measurement of Intracellular Calcium (Ca^{2+}) Levels

Changes in intracellular Ca^{2+} were detected by using the intracellular Ca^{2+} indicator Fluo-4 AM. Fluo-4 AM, an acetyl methyl ester derivative of Fluo-4, can penetrate cell membranes. Upon entering the cell, Fluo-4 AM can be cleaved by intracellular esterase to form Fluo-4, which retain in the cell. After treatment, cells were incubated with Fluo-4 AM (1 μM) at 37 °C for 30 min. Intracellular Ca^{2+} was measured by microplate reader (Ex = 488 nm and Em = 520 nm). Ca^{2+} images were obtained using a fluorescence microscope (Life Technologies Crop Bothell, Bothell, WA, USA).

5.7. Hoechst-33258 Staining

After treatment, cells were washed 2–3 times with PBS and added 200 μL Hoechst-33258 at room temperature for 3–5 min in the dark. Lastly, aspirated Hoechst-33258 staining solution and washed with PBS 2–3 times, 3–5 min each time. The stained cells were visualized and photographed under a fluorescence microscope (Life Technologies Crop Bothell, Bothell, WA, USA).

5.8. Apoptosis Detection

According to the manufacturer's protocol, the apoptosis rate of IPEC-J2 cells was measured by the Annexin V-FITC/PI apoptosis detection kit. After treatment, the cells were rinsed 2–3 times using PBS, trypsinized, and collected. Next, cells were resuspended in 195 μL of binding buffer. Then, 5 μL of Annexin V- FITC and 10 μL of propidium iodide (PI) were added to the tubes and gently vortexed. Lastly, the cells were analyzed by flow cytometry (Becton, Dickinson and Company, Franklin Lakes, NJ, USA).

5.9. Realtime PCR (RT-PCR) Assay

TRIzol reagent (Invitrogen, Shanghai, China) was used to isolate Total RNA from the cells. The concentration of RNA (A260/A280 ratio) was measured by using a Nano Photometer P-Class (IMPLEN, München, Germany). Reverse transcription of 5 μL RNA was performed using the PrimeScript™ RT reagent kit and the concentration of total RNA was 300 ng μL^{-1}. SYBR Green I RT-PCR kit (Takara, Dalian, China) was performed in a reaction volume of 10 μL using RT-PCR. Relative quantification of gene expression was calculated using the $2^{-\Delta\Delta Ct}$ method and normalized to GAPDH in each sample. The gene-specific primers are shown in Table 1.

Table 1. Primers used for RT-PCR.

Genes	Accession Number	Orientation	Sequence (5′-3′)	Fragments Size (bp)	Tm (°C)
GAPDH	NM_ 001206359.1	Forward Reversed	GATGGTGAAGGTCGGAGTGAAC TGGGTGGAATCATACTGGAACA	153	60.9
Caspase-3	NM_ 214131.1	Forward Reversed	GACACTCGCTCAACTTCTTGG TTGGACTGTGGGATTGAGAC	121	54.5
Caspase-9	XM_ 013998997.1	Forward Reversed	GGACATTGGTTCTGGAGGATT TGTTGATGATGAGGCAGTGG	116	52.3
Cyto-c	NM_ 001129970.1	Forward Reversed	CTCTTACACAGATGCCAACAA TTCCCTTTCTCCCTTCTTCT	139	56.1
Bax	XM_ 003127290.3	Forward Reversed	TTTGCTTCAGGGTTTCATCC GACACTCGCTCAACTTCTTGG	113	54.4
Bcl-2	AB 271960.1	Forward Reversed	GCGACTTTGCCGAGATGT CACAATCCTCCCCCAGTTC	116	55.9
Bcl-xl	XM_ 021077298.1	Forward Reversed	GCAGGTAGTGAACGAACTCTTCCG CCATCCAAGTTGCGATCCGACTC	140	60.08
Bad	XM_ 021082883.1	Forward Reversed	CTTACCCAGAGGGGACCGAG AGGAACCCTGGAACTCGTCA	153	58.39

5.10. Western Blotting Analyses

The density of each protein was detected by BCA Assay kit. Total protein was loaded onto 6–15% SDS–PAGE gel electrophoresis, separated by electrophoresis, and then was transferred to PVDF membranes. The membranes were blocked with 5% BSA in TBST at room temperature for 2 h, and probed with the indicated primary antibodies: Akt (1:2000, Sangon Biotech, Shanghai, China), p-Akt (1:1000, Cell Signaling Technology, Danvers, MA, USA), Bax, and Bcl-2 (1:1000, Beyotime Institute of Biotechnology, Nantong, China) at 4 °C overnight. Then, the members were washed in TBST three times, incubated with goat anti-rabbit/mouse secondary antibodies (1:1000; Beyotime Institute of Biotechnology, Nantong, China) at room temperature for 2 h and visualized using ECL Plus detection system (P1010, Applygen, Beijing, China). The density of the bands was analyzed using Image J software (National Institutes of Health, Bethesda, Rockville, MD, USA) and normalized to GAPDH.

5.11. Immunofluorescence Staining of Cells

IPEC-J2 cells were seeded in a polylysine-coated confocal dish (2.0–2.5 × 10^6 cells/mL). After treatment, the cells were washed with PBS three times, fixed with 4% polyoxymethylene for 30 min, and 0.2% Triton X-100 for 10 min. Subsequently, 2% BSA-PBS was added dropwise and blocked for 60 min at 37 °C. Cells were stained with primary rabbit anti-Bad (1:1000, Abcam, Cambridge, UK) antibody for one night at 4 °C, followed by incubation with Alexa Fluor 555-conjugated anti-rabbit secondary antibody (1:200; Beyotime Institute of Biotechnology, Nantong, China) for 120 min at 37 °C. Cell nuclei were stained with DAPI (Beyotime Institute of Biotechnology, Nantong, China) for 30 s at room temperature in the dark. 100 µL of PBS was added dropwise and photographed under a fluorescence microscope (Life Technologies Crop Bothell, Bothell, WA, USA).

5.12. Statistical Analyses

Data of three independent experiments were expressed as means ± Standard Deviation (SD) and performed using GraphPad Prime 6.0 software (GraphPad Software, Inc, CA, USA). Statistical figures were analyzed using SPSS 19.0 software (SPSS Inc, Chicago, IL, USA). All the experimental data were analyzed for variance uniformity, then analyzed by a one-way ANOVA and groups were compared using LSD's test. A p-value less than 0.05 was considered to indicate statistical significance.

Author Contributions: Conceptualization, J.L.; methodology, T.W. and J.W.; software, A.G. and T.Z.; validation, T.W., J.W., and T.Z.; formal analysis, T.W. and J.W.; investigation, J.L.; data curation, T.W. and T.Z.; writing—original draft preparation, T.W. and J.W.; writing—review and editing, J.W.;

supervision, J.L. and A.S.; project administration, J.L.; funding acquisition, J.L. and A.S. All authors have read and agreed to the published version of the manuscript.

Funding: This work was supported by the Natural Science Foundation of Heilongjiang Province of China (LC2018007) and the National Key R&D Program (2016YFD0501207).

Data Availability Statement: The data presented in this study are openly available in this article.

Conflicts of Interest: The authors have declared that no competing interest exist.

References

1. Lei, Y.; Zhao, L.; Ma, Q.; Zhang, J.Y.; Zhou, T.; Gao, C.; Ji, C. Degradation of zearalenone in swine feed and feed ingredients by *Bacillus subtilis* ANSB01G. *World Mycotoxin J.* **2014**, *7*, 143–151. [CrossRef]
2. Rai, A.; Das, M.; Tripathi, A. Occurrence and toxicity of a fusarium mycotoxin, zearalenone. *Crit. Rev. Food Sci. Nutr.* **2020**, *60*, 2710–2729. [CrossRef]
3. Zheng, W.; Wang, B.; Li, X.; Wang, T.; Zou, H.; Gu, J.; Yuan, Y.; Liu, X.; Bai, J.; Bian, J.; et al. Zearalenone Promotes Cell Proliferation or Causes Cell Death? *Toxins* **2018**, *10*, 184. [CrossRef]
4. Wang, N.; Wu, W.; Pan, J.; Long, M. Detoxification Strategies for Zearalenone Using Microorganisms: A Review. *Microorganisms* **2019**, *7*, 208. [CrossRef]
5. Wang, Y.C.; Deng, J.L.; Xu, S.W.; Peng, X.; Zuo, Z.C.; Cui, H.M.; Wang, Y.; Ren, Z.H. Effects of Zearalenone on IL-2, IL-6, and IFN-γmRNA Levels in the Splenic Lymphocytes of Chickens. *Sci. World J.* **2012**, *2012*, 567327. [CrossRef] [PubMed]
6. Zhang, W.; Zhang, S.; Zhang, M.; Yang, L.; Cheng, B.; Li, J.; Shan, A. Individual and combined effects of Fusarium toxins on apoptosis in PK15 cells and the protective role of N-acetylcysteine. *Food Chem. Toxicol.* **2018**, *111*, 27–43. [CrossRef] [PubMed]
7. Wang, J.; Li, M.; Zhang, W.; Gu, A.; Dong, J.; Li, J.; Shan, A. Protective Effect of N-Acetylcysteine against Oxidative Stress Induced by Zearalenone via Mitochondrial Apoptosis Pathway in SIEC02 Cells. *Toxins* **2018**, *10*, 407. [CrossRef] [PubMed]
8. De Oliveira, D.C.; Lima, F.D.S.; Sartori, T.; Santos, A.C.A.; Rogero, M.M.; Fock, R.A. Glutamine metabolism and its effects on immune response: Molecular mechanism and gene expression. *Nutrire* **2016**, *41*, 14. [CrossRef]
9. Santos, A.A.Q.A.; Braga-Neto, M.B.; Oliveira, M.R.; Freire, R.S.; Barros, E.B.; Santiago, T.M.; Rebelo, L.M.; Mermelstein, C.; Warren, C.A.; Guerrant, R.L.; et al. Glutamine and Alanyl-Glutamine Increase RhoA Expression and Reduce Clostridium difficile Toxin-A-Induced Intestinal Epithelial Cell Damage. *BioMed Res. Int.* **2012**, *2013*, 567327. [CrossRef]
10. Xing, S.; Zhang, B.; Lin, M.; Zhou, P.; Li, J.; Zhang, L.; Gao, F.; Zhou, G. Effects of alanyl-glutamine supplementation on the small intestinal mucosa barrier in weaned piglets. *Asian-Australas. J. Anim. Sci.* **2016**, *30*, 236–245. [CrossRef]
11. Wang, H.; Zhang, C.; Wu, G.; Sun, Y.; Wang, B.; He, B.; Dai, Z.; Wu, Z. Glutamine Enhances Tight Junction Protein Expression and Modulates Corticotropin-Releasing Factor Signaling in the Jejunum of Weanling Piglets. *J. Nutr.* **2014**, *145*, 25–31. [CrossRef]
12. Chen, S.; Xia, Y.; Zhu, G.; Yan, J.; Tan, C.; Deng, B.; Deng, J.; Yin, Y.; Ren, W. Glutamine supplementation improves intestinal cell proliferation and stem cell differentiation in weanling mice. *Food Nutr. Res.* **2018**, *62*, 1439. [CrossRef]
13. Jiang, Q.; Chen, J.; Liu, S.; Liu, G.; Yao, K.; Yin, Y. l-Glutamine Attenuates Apoptosis Induced by Endoplasmic Reticulum Stress by Activating the IRE1α-XBP1 Axis in IPEC-J2: A Novel Mechanism of l-Glutamine in Promoting Intestinal Health. *Int. J. Mol. Sci.* **2017**, *18*, 2617. [CrossRef]
14. Wang, B.; Wu, Z.; Ji, Y.; Sun, K.; Dai, Z.; Wu, G. L-Glutamine Enhances Tight Junction Integrity by Activating CaMK Kinase 2–AMP-Activated Protein Kinase Signaling in Intestinal Porcine Epithelial Cells. *J. Nutr.* **2016**, *146*, 501–508. [CrossRef] [PubMed]
15. Carneiro, B.A.; Fujii, J.; Brito, G.A.C.; Alcantara, C.; Oriá, R.; Lima, A.; Obrig, T.; Guerrant, R.L. Caspase and Bid Involvement in Clostridium difficile Toxin A-Induced Apoptosis and Modulation of Toxin A Effects by Glutamine and Alanyl-Glutamine In Vivo and In Vitro. *Infect. Immun.* **2006**, *74*, 81–87. [CrossRef] [PubMed]
16. Pluske, J.R.; Miller, D.W.; Sterndale, S.O.; Turpin, D. Associations between gastrointestinal-tract function and the stress response after weaning in pigs. *Anim. Prod. Sci.* **2019**, *59*, 2015. [CrossRef]
17. Goossens, J.; Pasmans, F.; Verbrugghe, E.; Vandenbroucke, V.; De Baere, S.; Meyer, E.; Haesebrouck, F.; De Backer, P.; Croubels, S. Porcine intestinal epithelial barrier disruption by the Fusarium mycotoxins deoxynivalenol and T-2 toxin promotes transepithelial passage of doxycycline and paromomycin. *BMC Veter.-Res.* **2012**, *8*, 245. [CrossRef] [PubMed]
18. Peterson, L.W.; Artis, D. Intestinal epithelial cells: Regulators of barrier function and immune homeostasis. *Nat. Rev. Immunol.* **2014**, *14*, 141–153. [CrossRef]
19. Li, X.; Zhao, L.; Fan, Y.; Jia, Y.; Sun, L.; Ma, S.; Ji, C.; Ma, Q.; Zhang, J. Occurrence of mycotoxins in feed ingredients and complete feeds obtained from the Beijing region of China. *J. Anim. Sci. Biotechnol.* **2014**, *5*, 37. [CrossRef]
20. Zhao, L.; Lei, Y.; Bao, Y.; Jia, R.; Ma, Q.; Zhang, J.; Chen, J.; Ji, C. Ameliorative effects of Bacillus subtilisANSB01G on zearalenone toxicosis in pre-pubertal female gilts. *Food Addit. Contam. Part A* **2014**, *32*, 617–625. [CrossRef]
21. Braicu, C.; Selicean, S.; Cojocneanu-Petric, R.; Lajos, R.; Balacescu, O.; Taranu, I.; Marin, D.E.; Motiu, M.; Jurj, A.; Achimas-Cadariu, P.; et al. Evaluation of cellular and molecular impact of zearalenone and *E. coli* co-exposure on IPEC-1 cells using microarray technology. *BMC Genom.* **2016**, *17*, 576. [CrossRef]

22. Ben Salem, I.; Prola, A.; Boussabbeh, M.; Guilbert, A.; Bacha, H.; Abid-Essefi, S.; Lemaire, C. Crocin and Quercetin protect HCT116 and HEK293 cells from Zearalenone-induced apoptosis by reducing endoplasmic reticulum stress. *Cell Stress Chaperones* **2015**, *20*, 927–938. [CrossRef]
23. Long, M.; Yang, S.; Zhang, Y.; Li, P.; Han, J.; Dong, S.; Chen, X.; He, J. Proanthocyanidin protects against acute zearalenone-induced testicular oxidative damage in male mice. *Environ. Sci. Pollut. Res.* **2016**, *24*, 938–946. [CrossRef]
24. Yang, D.; Jiang, X.; Sun, J.; Li, X.; Li, X.; Jiao, R.; Peng, Z.; Li, Y.; Bai, W. Toxic effects of zearalenone on gametogenesis and embryonic development: A molecular point of review. *Food Chem. Toxicol.* **2018**, *119*, 24–30. [CrossRef] [PubMed]
25. Venkataramana, M.; Nayaka, S.C.; Anand, T.; Rajesh, R.; Aiyaz, M.; Divakara, S.; Murali, H.; Prakash, H.; Rao, P.L. Zearalenone induced toxicity in SHSY-5Y cells: The role of oxidative stress evidenced by N-acetyl cysteine. *Food Chem. Toxicol.* **2014**, *65*, 335–342. [CrossRef] [PubMed]
26. Yin, L.; Mano, J.; Tanaka, K.; Wang, S.; Zhang, M.; Deng, X.; Zhang, S. High level of reduced glutathione contributes to detoxification of lipid peroxide-derived reactive carbonyl species in transgenic Arabidopsis overexpressing glutathione reductase under aluminum stress. *Physiol. Plant.* **2017**, *161*, 211–223. [CrossRef] [PubMed]
27. Kopp, T.I.; Vogel, U.; Dragsted, L.O.; Tjonneland, A.; Ravn-Haren, G. Association between single nucleotide polymorphisms in the antioxidant genes CAT, GR and SOD1, erythrocyte enzyme activities, dietary and life style factors and breast cancer risk in a Danish, prospective cohort study. *Oncotarget* **2017**, *8*, 62984–62997. [CrossRef] [PubMed]
28. Kvamme, E.; Roberg, B.; Torgner, I.A. Glutamine transport in brain mitochondria. *Neurochem. Int.* **2000**, *37*, 131–138. [CrossRef]
29. Marques, C.; Mauriz, J.L.; Simonetto, D.; Marroni, C.A.; Tuñon, M.J.; González-Gallego, J.; Marroni, N.P. Glutamine prevents gastric oxidative stress in an animal model of portal hypertension gastropathy. *Ann. Hepatol.* **2011**, *10*, 531–539. [CrossRef]
30. Kim, M.-H.; Kim, H. The Roles of Glutamine in the Intestine and Its Implication in Intestinal Diseases. *Int. J. Mol. Sci.* **2017**, *18*, 1051. [CrossRef]
31. Curi, R.; Newsholme, P.; Procopio, J.; Lagranha, C.; Gorjão, R.; Pithon-Curi, T.C. Glutamine, gene expression, and cell function. *Front. Biosci.* **2007**, *12*, 344–357. [CrossRef]
32. Gu, X.; Guo, W.; Zhao, Y.; Liu, G.; Wu, J.; Chang, C. Deoxynivalenol-Induced Cytotoxicity and Apoptosis in IPEC-J2 Cells through the Activation of Autophagy by Inhibiting PI3K-AKT-mTOR Signaling Pathway. *ACS Omega* **2019**, *4*, 18478–18486. [CrossRef]
33. Marone, R.; Erhart, D.; Mertz, A.C.; Bohnacker, T.; Schnell, C.; Cmiljanovic, V.; Stauffer, F.; Garcia-Echeverria, C.; Giese, B.; Maira, S.-M.; et al. Targeting Melanoma with Dual Phosphoinositide 3-Kinase/Mammalian Target of Rapamycin Inhibitors. *Mol. Cancer Res.* **2009**, *7*, 601–613. [CrossRef]
34. Gharbi, S.I.; Zvelebil, M.J.; Shuttleworth, S.J.; Hancox, T.; Saghir, N.; Timms, J.F.; Waterfield, M.D. Exploring the specificity of the PI3K family inhibitor LY294002. *Biochem. J.* **2007**, *404*, 15–21. [CrossRef] [PubMed]
35. Fan, W.; Shen, T.; Ding, Q.; Lv, Y.; Li, L.; Huang, K.; Yan, L.; Song, S. Zearalenone induces ROS-mediated mitochondrial damage in porcine IPEC-J2 cells. *J. Biochem. Mol. Toxicol.* **2017**, *31*, e21944. [CrossRef] [PubMed]
36. Gu, Z.T.; Li, L.; Wu, F.; Zhao, P.; Yang, H.; Liu, Y.S.; Geng, Y.; Zhao, M.; Su, L. Heat stress induced apoptosis is triggered by transcription-independent p53, Ca2+ dyshomeostasis and the subsequent Bax mitochondrial translocation. *Sci. Rep.* **2015**, *5*, 11497. [CrossRef]
37. Hyun, D.-H.; Hunt, N.D.; Emerson, S.S.; Hernandez, J.O.; Mattson, M.P.; de Cabo, R. Up-regulation of plasma membrane-associated redox activities in neuronal cells lacking functional mitochondria: Plasma Membrane-Associated Redox Activities in Neuronal Cells. *J. Neurochem.* **2006**, *100*, 1364–1374. [CrossRef]
38. Hempel, N.; Trebak, M. Crosstalk between calcium and reactive oxygen species signaling in cancer. *Cell Calcium* **2017**, *63*, 70–96. [CrossRef] [PubMed]
39. Lee, H.-J.; Oh, S.-Y.; Jo, I. Zearalenone Induces Endothelial Cell Apoptosis through Activation of a Cytosolic Ca^{2+}/ERK1/2/p53/Caspase 3 Signaling Pathway. *Toxins* **2021**, *13*, 187. [CrossRef] [PubMed]
40. Weichhart, T.; Saemann, M.D. The PI3K/Akt/mTOR pathway in innate immune cells: Emerging therapeutic applications. *Ann. Rheum. Dis.* **2008**, *67*, iii70–iii74. [CrossRef]
41. Kumar, D.; Shankar, S.; Srivastava, R.K. Rottlerin induces autophagy and apoptosis in prostate cancer stem cells via PI3K/Akt/mTOR signaling pathway. *Cancer Lett.* **2013**, *343*, 179–189. [CrossRef]
42. Xu, X.; Li, H.; Hou, X.; Li, D.; He, S.; Wan, C.; Yin, P.; Liu, M.; Liu, F.; Xu, J. Punicalagin Induces Nrf2/HO-1 Expression via Upregulation of PI3K/AKT Pathway and Inhibits LPS-Induced Oxidative Stress in RAW264.7 Macrophages. *Mediat. Inflamm.* **2015**, *2015*, 1–11. [CrossRef] [PubMed]
43. Hu, B.; Sun, M.; Liu, J.; Hong, G.; Lin, Q. The preventative effect of Akt knockout on liver cancer through modulating NF-?B-regulated inflammation and Bad-related apoptosis signaling pathway. *Int. J. Oncol.* **2016**, *48*, 1467–1476. [CrossRef]
44. Zhao, Y.; Wang, Q.; Wang, Y.; Li, J.; Lu, G.; Liu, Z. Glutamine protects against oxidative stress injury through inhibiting the activation of PI3K/Akt signaling pathway in parkinsonian cell model. *Environ. Health Prev. Med.* **2019**, *24*, 4. [CrossRef]
45. Sreenivasulu, K.; Nandeesha, H.; Dorairajan, L.N.; Ganesh, R.N. Over expression of PI3K-AkT reduces apoptosis and increases prostate size in benign prostatic hyperplasia. *Aging Male* **2018**, *23*, 440–446. [CrossRef]
46. Xuan, F.; Jian, J.; Qin, F.; Huang, R. Bauhinia championii flavone inhibits apoptosis and autophagy via the PI3K/Akt pathway in myocardial ischemia/reperfusion injury in rats. *Drug Des. Dev. Ther.* **2015**, *9*, 5933–5945. [CrossRef]
47. Plas, D.R.; Talapatra, S.; Edinger, A.L.; Rathmell, J.C.; Thompson, C.B. Akt and Bcl-xL Promote Growth Factor-independent Survival through Distinct Effects on Mitochondrial Physiology. *J. Biol. Chem.* **2001**, *276*, 12041–12048. [CrossRef]

48. Stiles, B.L. PI-3-K and AKT: Onto the mitochondria. *Adv. Drug Deliv. Rev.* **2009**, *61*, 1276–1282. [CrossRef] [PubMed]
49. Zha, J.; Harada, H.; Yang, E.; Jockel, J.; Korsmeyer, S.J. Serine Phosphorylation of Death Agonist BAD in Response to Survival Factor Results in Binding to 14-3-3 Not BCL-XL. *Cell* **1996**, *87*, 619–628. [CrossRef]
50. Siddiqui, W.A.; Ahad, A.; Ahsan, H. The mystery of BCL2 family: Bcl-2 proteins and apoptosis: An update. *Arch. Toxicol.* **2015**, *89*, 289–317. [CrossRef] [PubMed]
51. Qin, J.; Xie, L.-P.; Zheng, X.-Y.; Wang, Y.-B.; Bai, Y.; Shen, H.-F.; Li, L.-C.; Dahiya, R. A component of green tea, (−)-epigallocatechin-3-gallate, promotes apoptosis in T24 human bladder cancer cells via modulation of the PI3K/Akt pathway and Bcl-2 family proteins. *Biochem. Biophys. Res. Commun.* **2007**, *354*, 852–857. [CrossRef] [PubMed]
52. Zhu, H.; Zhang, Y.; Shi, Z.; Lu, D.; Li, T.; Ding, Y.; Ruan, Y.; Xu, A.-D. The Neuroprotection of Liraglutide Against Ischaemia-induced Apoptosis through the Activation of the PI3K/AKT and MAPK Pathways. *Sci. Rep.* **2016**, *6*, 26859. [CrossRef] [PubMed]
53. Li, D.; Qu, X.; Hou, K.; Zhang, Y.; Dong, Q.; Teng, Y.; Zhang, J.; Liu, Y. PI3K/Akt is involved in bufalin-induced apoptosis in gastric cancer cells. *Anti-Cancer Drugs* **2009**, *20*, 59–64. [CrossRef] [PubMed]

toxins

MDPI

Article

Comparison of Ameliorative Effects between Probiotic and Biodegradable *Bacillus subtilis* on Zearalenone Toxicosis in Gilts

Wenqiang Shen [1,†], Yaojun Liu [1,†], Xinyue Zhang [1], Xiong Zhang [1], Xiaoping Rong [1], Lihong Zhao [1], Cheng Ji [1], Yuanpei Lei [1], Fengjuan Li [2], Jing Chen [2] and Qiugang Ma [1,*]

1 State Key Laboratory of Animal Nutrition, China Agricultural University, Beijing 100193, China; Shenwenqiang1996@163.com (W.S.); yaojunliu@outlook.com (Y.L.); zhangxinyue090@163.com (X.Z.); Zhangxiongcau@163.com (X.Z.); rxp409291@163.com (X.R.); zhaolihongcau@cau.edu.cn (L.Z.); jicheng@cau.edu.cn (C.J.); lypcau@126.com (Y.L.)

2 FuQing Fengze Agricultural Science and Technology Development Co., Ltd., Fuzhou 350011, China; lifengjuan-1986@163.com (F.L.); chenjing101488@163.com (J.C.)

* Correspondence: maqiugang@cau.edu.cn

† These authors contributed equally to this work.

Abstract: This study was conducted to compare the potential ameliorative effects between probiotic *Bacillus subtilis* and biodegradable *Bacillus subtilis* on zearalenone (ZEN) toxicosis in gilts. Thirty-six Landrace×Yorkshire gilts (average BW = 64 kg) were randomly divided into four groups: (1) Normal control diet group (NC) fed the basal diet containing few ZEN (17.5 μg/kg); (2) ZEN contaminated group (ZC) fed the contaminated diet containing an exceeded limit dose of ZEN (about 300 μg/kg); (3) Probiotic agent group (PB) fed the ZC diet with added 5×10^9 CFU/kg of probiotic *Bacillus subtilis* ANSB010; (4) Biodegradable agent group (DA) fed the ZC diet with added 5×10^9 CFU/kg of biodegradable *Bacillus subtilis* ANSB01G. Results showed that *Bacillus subtilis* ANSB010 and ANSB01G isolated from broiler intestinal chyme had similar inhibitory activities against common pathogenic bacteria. In addition, the feed conversion ratio and the vulva size in DA group were significantly lower than ZC group ($p < 0.05$). The levels of IgG, IgM, IL-2 and TNFα in the ZC group were significantly higher than PB and DA groups ($p < 0.05$). The levels of estradiol and prolactin in the ZC group was significantly higher than those of the NC and DA groups ($p < 0.05$). Additionally, the residual ZEN in the feces of the ZC and PB groups were higher than those of the NC and DA groups ($p < 0.05$). In summary, the ZEN-contaminated diet had a damaging impact on growth performance, plasma immune function and hormone secretion of gilts. Although probiotic and biodegradable *Bacillus subtilis* have similar antimicrobial capacities, only biodegradable *Bacillus subtilis* could eliminate these negative effects through its biodegradable property to ZEN.

Keywords: zearalenone; degradable *Bacillus subtilis*; probiotic *Bacillus subtilis*; gilts

Key Contribution: The alleviating effects of biodegradable *Bacillus subtilis* ANSB01G on the ZEN-poisoned gilts were compared to probiotic *Bacillus subtilis* ANSB010, which was isolated from the same source and with similar bacteriostatic activity. Results showed that the improvement of ANSB01G on ZEN-poisoned gilts comes from its biodegradation activity to ZEN; not from its antibacterial activity.

Citation: Shen, W.; Liu, Y.; Zhang, X.; Zhang, X.; Rong, X.; Zhao, L.; Ji, C.; Lei, Y.; Li, F.; Chen, J.; et al. Comparison of Ameliorative Effects between Probiotic and Biodegradable *Bacillus subtilis* on Zearalenone Toxicosis in Gilts. *Toxins* **2021**, *13*, 882. https://doi.org/10.3390/toxins13120882

Received: 16 November 2021
Accepted: 8 December 2021
Published: 10 December 2021

1. Introduction

Zearalenone (ZEN), known as an F-2 mycotoxin, is a powerful estrogenic metabolite produced by certain species of *Fusarium* and *Erysipelas* spp. [1]. Several results showed that many feedstuffs for animals have been seriously contaminated with ZEN around the world [2,3], which could cause hyperestrogenism and fertility disorder in sows [4]. Moreover, a previous study found that exposure of post-weaning gilts to ZEN could increase the oxidative stress and had a negative impact on genital organs [5]. In addition, recent study found that ZEN can interfere with immune mediators at the spleen level and induce an intense inflammatory response [6]. Therefore, it is critical to find the appropriate and effective detoxifying strategies to prevent contamination by ZEN in animal husbandry.

Numerous researches have reported that physical, chemical and biological methods can be used to counter mycotoxicosis [7–9]. However, most of these methods are impractical or potentially unsafe because of losses in the nutritional value, high equipment costs and formation of toxic residues or derivatives [10]. Biodegradation is eco-friendly and highly efficient in minimizing the harmfulness of mycotoxins in feeds [11,12]. Previous work from our laboratory reported that *Bacillus subtilis* ANSB01G, which has both a biodegradable effect against ZEN and probiotic activities against pathogenic bacteria, can alleviate toxicosis of ZEN in pre-pubertal female gilts [13,14]. However, the article did not clarify whether the reduced toxicity was due to its probiotic or biodegradable properties.

Therefore, the aim of this study was to investigate the effects of biodegradable and probiotic *Bacillus subtilis* on growth performance, serum biochemical indexes and hormone, serum antioxidant, immune indicators and mycotoxin residue in gilts exposed to ZEN for 25 d in vivo, as well as the inhibitory activity of common harmful bacteria in vitro.

2. Results

2.1. Biochemical and Physiological Characteristics of Bacillus subtilis ANSB010 and ANSB01G

The colony morphologies showed that the surfaces of *Bacillus subtilis* ANSB010 and ANSB01G colonies are rough, opaque and milky white (Figure 1A,B). Under the microscope, the cells were found to be short, thin rods, positive for Gram staining and capable of forming spores (Figure 1C,D). As shown in Table 1, the physiological and biochemical results revealed ANSB010 and ANSB01G had typical characteristics of *Bacillus* spp., such as growing well at 37 °C but not at 10 and 55 °C; and being able to utilize cellulose, glucose and maltose as the only carbon source. Moreover, a phylogenetic tree based on 16s rDNA sequences suggested that both ANSB010 and ANSB01G have a close evolutionary relationship to *Bacillus subtilis* (Figure 1E).

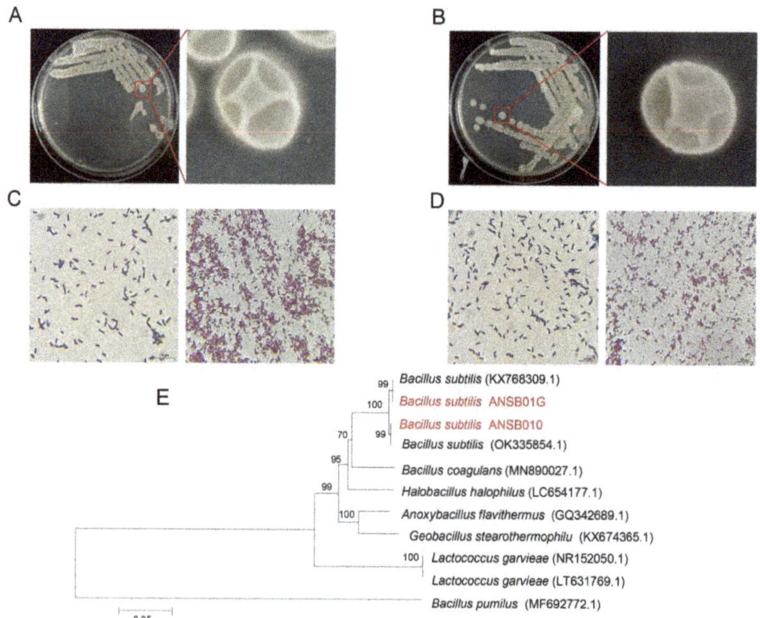

Figure 1. Colony characteristics of ANSB010 (**A**) and ANSB01G (**B**); cell (**left**) and spore (**right**) morphology of ANSB010 (**C**) and ANSB01G (**D**), scale bar, 100 µm; (**E**) the phylogenetic tree of *Bacillus subtilis* ANSB010 and ANSB01G, the GenBank accession numbers of sequences are shown in round brackets.

Table 1. Biochemical and physiological characteristics of *Bacillus subtilis* ANSB010 and ANSB01G.

Experimental Projects	ANSB010 [1]	ANSB01G [2]	Experimental Projects	ANSB010	ANSB01G
Gram	+	+	Glucose	+	+
Cell shape	Rod-shape	Rod-shape	Maltose	+	+
Cell diameter > 1 μm	+	+	Sucrose	+	+
Spore forming	+	+	D-xylose	+	+
Spore dilation	−	−	L-xylose	−	−
Round spores	−	−	D-arabinose	−	−
Glycerol	+	+	L-arabinose	+	+
Cellulose utilization	+	+	D-mannitol	+	+
Catalase	+	+	Gas production using glucose	−	−
Oxidase test	+	+	Citrate utilization	+	+
Anaerobic	−	−	Growth at 10 or 50 °C	−	−
Voges-Proskauer (VP) test	+	+	Growth at 37 °C	+	+
VP < pH 6	+	+	Growth at pH 5.7	+	+
VP > pH 7	−	−	Growth on 7% NaCl	+	+
Methyl red test	−	−	Hydrolysis of starch	+	+
Gluconate	−	−	Decomposition of casein	+	+
Xylitol	−	−	Nitrate reduction	+	+

[1] '+', '−' and 'w' mean positive, negative and weak response, respectively. [2] The results of ANSB01G after domestication were consistent with Lei et al. [13].

2.2. Bacteriostatic and ZEN-Degrading Effects of ANSB010 and ANSB01G

As shown in Figure 2 and Table 2, probiotic *Bacillus subtilis* ANSB010 and biodegradable *Bacillus subtilis* ANSB01G have a visible antibacterial effect on *Escherichia coli* (*E. coli*), *Salmonella choleraesuis* (*S. choleraesuis*) and *Staphylococcus aureus* (*S. aureus*) compared to the control group (Con) ($p < 0.05$, Table 2), while there was no significant difference in the antibacterial effect between ANSB010 and ANSB01G ($p > 0.05$, Table 2). Importantly, we noticed that ANSB01G could degrade 65.13%, 92.57% and 100.00% of ZEN in the fermentation broth at 6 h, 24 h and 48 h, respectively, but ANSB010 could not ($p < 0.05$, Table 3). Additionally, Supplementary Material Figures S1 and S2 show the representative chromatograms of degradation tests.

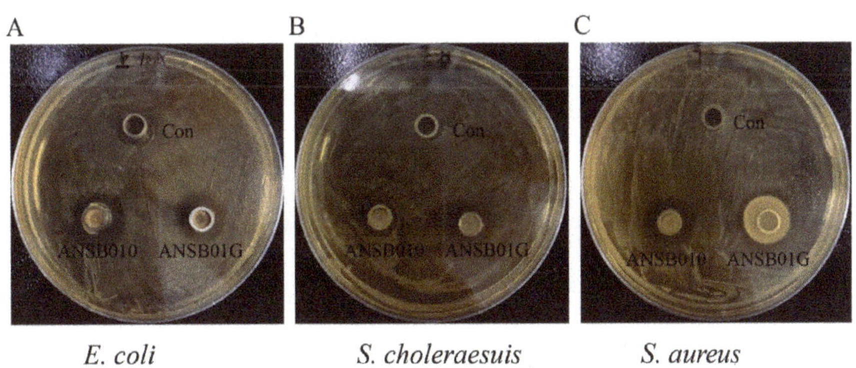

Figure 2. The antibacterial effects of probiotic *Bacillus subtilis* ANSB010 and biodegradable *Bacillus subtilis* ANSB01G on *E. coli* (**A**), *S. choleraesuis* (**B**) and *S. aureus* (C). *E. coli*, *Escherichia coli*; *S. choleraesuis*, *Salmonella choleraesuis*; *S. aureus*, Staphylococcus aureus. Con: MRS medium; ANSB010, probiotic *Bacillus subtilis* ANSB010; ANSB01G, biodegradable *Bacillus subtilis* ANSB01G.

Table 2. Antibacterial effects of *Bacillus subtilis* ANSB010 and ANSB01G.

Indicator Bacteria	Antibacterial Circle Diameter (cm)			SEM	*p*-Value
	Con	ANSB010	ANSB01G		
E. coli	1.10 ^{b}	1.76 ^{a}	1.68 ^{a}	0.07	0.00
S. choleraesuis	0.86 ^{b}	1.62 ^{a}	1.71 ^{a}	0.09	0.00
S. aureus	1.07 ^{b}	1.74 ^{a}	1.71 ^{a}	0.08	0.00

Con: MRS medium; ANSB010, probiotic *Bacillus subtilis* ANSB010; ANSB01G, biodegradable *Bacillus subtilis* ANSB01G. Different superscript letters represent significant difference.

Table 3. The degradation rate (%) of ANSB010 and ANSB01G on zearalenone in fermentation medium.

Time	ANSB010	ANSB01G	SEM	*p*-Value
6 h	0.05 ^{b}	65.13 ^{a}	12.23	0.00
24 h	0.38 ^{b}	92.57 ^{a}	4.81	0.00
48 h	−0.26 ^{b}	100.00 ^{a}	4.22	0.00

ANSB010, probiotic *Bacillus subtilis* ANSB010; ANSB01G, biodegradable *Bacillus subtilis* ANSB01G. Different superscript letters represent significant difference.

2.3. Growing Performance

As shown in Table 4, no significant difference were observed for the initial weight, terminal weight and average daily gain (ADG) ($p > 0.05$), while there was a decreasing trend of average daily feed intake (ADFI) in probiotic *Bacillus subtilis* ANSB010 agent (PB) and biodegradable *Bacillus subtilis* ANSB01G agent (DA) groups ($p = 0.10$) compared to ZEN contaminated (ZC) group. Of the four groups, the ZC group had the highest feed conversion ratio (F/G) value ($p < 0.05$), and was 1.07- and 1.13-fold higher than the PB and DA group, respectively.

Table 4. Comparison of growth performance among different treatment groups.

Items	NC	ZC	PB	DA	SEM	*p*-Value
Initial weight, kg	63.43	64.14	64.11	64.4	1.78	0.99
Terminal weight, kg	84.05	84.98	85.12	86.00	1.72	0.93
ADG, g	825	834	841	864	29.69	0.85
ADFI, g	2151	2244	2114	2047	81.44	0.10
F/G	2.61 ^{abc}	2.69 ^{ab}	2.52 ^{bc}	2.38 ^{c}	0.08	<0.01

ADG, average daily gain; ADFI, average daily feed intake; F/G, feed conversion ratio; NC, normal control diet group; ZC, Zearalenone (ZEN)-contaminated group; PB, probiotic *Bacillus subtilis* group; DA, biodegradable *Bacillus subtilis* group. Different superscript letters represent significant difference.

2.4. Vulva Size

The effects of the four diets on the vulva size are shown in Table 5. The vaginal length and area of the DA group was significantly lower compared to the ZC group ($p < 0.05$), and there was no significant difference between the ZC and PB groups ($p > 0.05$). Interestingly, there was a decreasing trend of vaginal width among four groups ($p = 0.10$). The vaginal volume of the DA group was dramatically lower compared to all other groups ($p < 0.05$), and that of the PB group was significantly lower than that of the ZC group ($p < 0.05$), but no significant difference existed between the ZC and normal control diet (NC) groups, or PB and NC groups ($p > 0.05$). There was no significant difference in the vaginal height among the four groups ($p > 0.05$).

2.5. Serum Biochemical Indicators, Antioxidant and Immunology Parameters

There was no significant difference in the serum biochemical indicators (e.g., total protein (TP), albumin (ALB), alkaline phosphatase (ALP), aspartate aminotransferase (AST), alanine aminotransferase (ALT), creatinine (CRE) and urea nitrogen (BUN)) among these groups (Figure 3A–G, $p > 0.05$), nor in the levels of glutathione peroxidase (GSH-Px) (Figure 4A, $p > 0.05$) and malondialdehyde (MDA) (Figure 4C, $p > 0.05$), which were indicators of antioxidant activities. Interestingly, the level of superoxide dismutase (SOD) in the PB group was significantly lower than that of the ZC group (Figure 4B, $p < 0.05$).

Table 5. The values of vulva indexes in different treatment groups.

Items	NC	ZC	PB	DA	SEM	p Value
Length, cm	3.12 [ab]	3.34 [a]	2.93 [ab]	2.85 [b]	0.66	0.03
Width, cm	2.51	2.53	2.42	2.21	0.48	0.10
Height, cm	2.36	2.30	1.90	1.95	0.75	0.43
Area, cm^2	6.15 [ab]	6.64 [a]	5.58 [ab]	4.99 [b]	0.21	0.03
Volume, cm^3	4.83 [ab]	5.11 [a]	3.54 [b]	3.23 [c]	0.24	0.00

NC, normal control diet group; ZC, ZEN-contaminated group; PB, probiotic *Bacillus subtilis* group; DA, biodegradable *Bacillus subtilis* group. Different superscript letters represent significant difference.

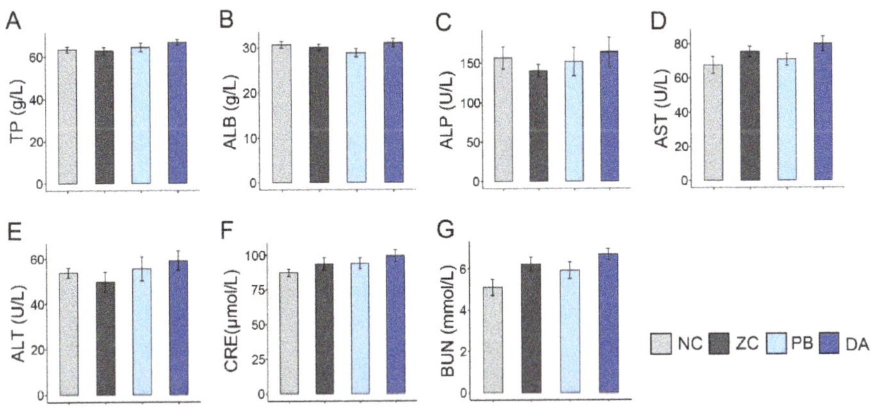

Figure 3. Comparison of serum levels of biochemical indicators among different treatment groups. (**A**) TP, total protein; (**B**) ALB, albumin; (**C**) ALP, alkaline phosphatase; (**D**) AST, aspartate aminotransferase; (**E**) ALT, alanine aminotransferase; (**F**) CRE, creatinine; (**G**) BUN, urea nitrogen. NC, normal control diet group; ZC, ZEN-contaminated group; PB, probiotic *Bacillus subtilis* group; DA, biodegradable *Bacillus subtilis* group.

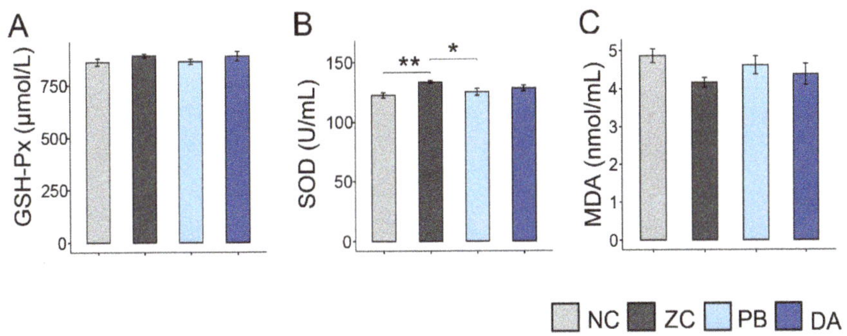

Figure 4. Comparison of serum levels of antioxidant parameters among different treatment groups. (**A**) GSH-Px, glutathione peroxidase; (**B**) SOD, superoxide dismutase; (**C**) MDA, malondialdehyde. NC, normal control diet group; ZC, ZEN-contaminated group; PB, probiotic *Bacillus subtilis* group; DA, biodegradable *Bacillus subtilis* group. *, $p < 0.05$; **, $p < 0.01$.

As shown in Figure 5, the levels of immunoglobulin G (IgG) and IgM in the PB and DA groups were dramatically lower than that of the ZC group (Figure 5B,C, $p < 0.05$). However, there was no significant difference in the levels of IgA (Figure 5A, $p < 0.05$). In addition, the levels of pro-inflammatory factors (e.g., interleukin 2 (IL-2) and tumor necrosis factor-α (TNFα)) in the serum of the ZC group was significantly higher than these of other groups (Figure 5E,G, $p < 0.05$). Additionally, the DA group had the lowest level of IL-2 (Figure 5E, $p < 0.05$). However, other pro-inflammatory factors (e.g., IL-1β and IL-6) were not significantly different (Figure 5D,F, $p > 0.05$).

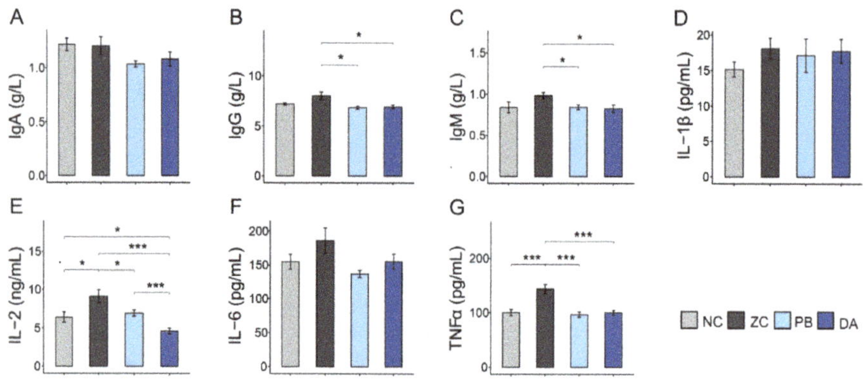

Figure 5. Comparison of serum levels of immune and inflammatory parameters in different treatment groups. (**A–C**) IgA, immunoglobulin A; IgG, immunoglobulin G; IgM, immunoglobulin M; (**D–G**) IL-1β, interleukin 1β; IL-2, interleukin 2; IL-6, interleukin 6; TNFα, tumor necrosis factor-α; NC, normal control diet group; ZC, ZEN-contaminated group; PB, probiotic *Bacillus subtilis* group; DA, biodegradable *Bacillus subtilis* group. *, $p < 0.05$; ***, $p < 0.001$.

2.6. Serum Hormone Parameters

The effects of diet supplemented with ZEN or *Bacillus subtilis* on the serum hormone of gilts were also shown in Figure 6. The level of estradiol (E2) in the NC and DA groups were significantly lower than that in the ZC groups (Figure 6C, $p < 0.05$), while no significant difference in the levels of follicle-stimulating hormone (FSH) and luteinizing hormone (LH) were observed in each treatment group (Figure 6A,B, $p > 0.05$). In addition, the level of prolactin (PRL) in the ZC group was the highest among the four groups (Figure 6D, $p < 0.05$). There was no difference in the PRL levels of the PB and DA groups compared to the NC group (Figure 6D, $p > 0.05$).

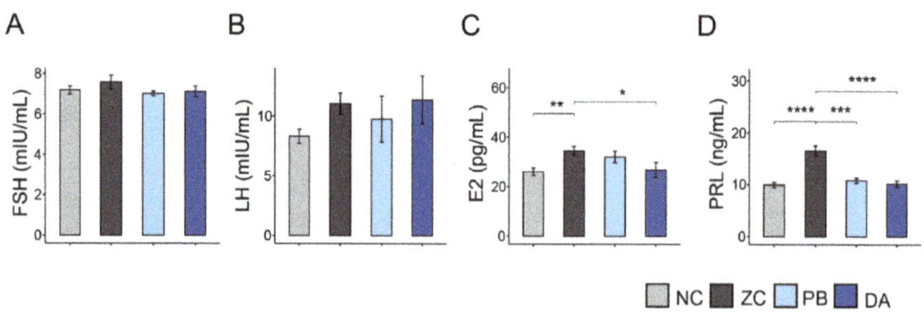

Figure 6. Comparison of serum levels of hormone parameters in different treatment groups. (**A–D**) FSH, follicle-stimulating hormone; LH, luteinizing hormone; E2, estradiol; PRL, prolactin. NC, normal control diet group; ZC, ZEN-contaminated group; PB, probiotic *Bacillus subtilis* group; DA, biodegradable *Bacillus subtilis* group. *, $p < 0.05$; **, $p < 0.01$; ***, $p < 0.001$; ****, $p < 0.0001$.

2.7. ZEN Residues

As shown in Table 6, mildewed maize enormously increased the content of ZEN in feed and feces. The NC group had very low levels of ZEN in feed (p = 0.06) and fecal samples ($p < 0.05$). In contrast, the ZC group had the highest levels of ZEN in feed and fecal samples ($p < 0.05$). Surprisingly, the content of ZEN in feces of the DA group were dramatically lower than the ZC and PB groups ($p < 0.05$). Additionally, the ZEN content in feces of the PB group was nearly same to that of the ZC group ($p > 0.05$). Intriguingly, although the ratio of ZEN contents between feces to diet was not significantly among the four groups ($p > 0.05$), the value in the DA group was indeed half that of the remaining three groups. In this study, both ZEN and its metabolites (α-zearalanol (α-ZAL), β-zearalanol (β-ZAL), α-zearalenol (α-ZOL), β-zearalenol (β-ZOL) and zearalanone (ZAN)) were not found in serum samples.

Table 6. Content of ZEN in feed and fecal samples.

Items	NC	ZC	PB	DA	SEM	*p*-Value
Content of ZEN in diet, μg/kg	17.50	304.80	297.30	307.70	55.42	0.06
Content of ZEN in feces, μg/kg	7.23 [b]	104.24 [a]	98.86 [a]	55.37 [b]	17.34	0.00
Ratio of ZEN contents between feces to diet, %	41.34	34.20	33.26	17.30	5.87	0.73

NC, normal control diet group; ZC, ZEN-contaminated group; PB, probiotic *Bacillus subtilis* group; DA, biodegradable *Bacillus subtilis* group. Different superscript letters represent significant difference.

3. Discussion

The recent years have witnessed growing interests in finding practical and effective methods to detoxify ZEN in contaminated cereals and feeds [15–17]. Previous studies had been focused on mycotoxin adsorbents used to control mycotoxins in animal feed [18]. However, these adsorbents would contribute to environmental pollution as they transfer mycotoxins to surrounding areas [19]. Biodegradation of mycotoxins was considered as an efficient and environmentally protective method for the treatment of contaminated diets in the livestock production [11]. Previous studies have shown that some strains of *Bacillus* spp. were able to prevent the toxicity of ZEN [13,14], while it is not well known whether these effects were due to their probiotic or degradative capacities. Previous reports from our laboratory suggested biodegradable *Bacillus subtilis* ANSB01G could degrade 84.58%, 83.04% and 66.34% of ZEN in naturally contaminated maize, swine complete feed and dried distillers' grains with solubles, respectively [13]. For these reasons, it is worth comparing the ameliorative effects between probiotic and biodegradable *Bacillus subtilis* in a ZEN-contaminated diet. The persuasion of the present study was ensured by the similarity of *Bacillus subtilis* ANSB010 and ANSB01G on the habitat source and the bacteriostatic activity against common pathogenic bacteria, including *E. coli, S. choleraesuis* and *S. aureus*.

It has been reported that the presence of ZEN reduced feed consumption, caused a subsequent growth depression, and increased susceptibility to diseases [20,21]. Although we did not observe significant difference in gilts weight in the present study, there is a decreasing trend in ADFI among these groups. In line with previous reports [22], our study showed that the gilts fed on diets containing ZEN significantly increased F/G. In our study, the contaminated diet contained about 300 μg/kg of ZEN, leading to an increase in ZEN levels in the feces. Our data revealed that biodegradable *Bacillus subtilis* ANSB01G alleviated the toxicity while significantly decreasing the F/G, but probiotic *Bacillus subtilis* ANSB010 did not.

Previous study has indicated that vulva swelling is the main clinical symptom of ZEN-induced toxicity in mammals [23,24], which has an adverse impact on the reproductive system and the breeding performance. Similarly, our data also indicated that ZEN significantly caused an increase in vulva size. Previous research found that the mycotoxin biodegradation agent composed of *Bacillus subtilis* ANSB01G and *Devosia* sp. ANSB714 can effectively reduce the estrogenic swelling of the vulva caused by ZEN in immature gilts [25].

In this study, we also discovered that only biodegradable *Bacillus subtilis* ANSB01G mediates vulva swelling caused by ZEN. Some available evidence also demonstrated the obvious adverse effects of ZEN on the secretion of these hormones and productivity of animals [26]. In addition, similar results were reflected by a fluctuation in hormone levels. Although there were no significant difference in the levels of FSH and LH, ZEN diets markedly increased the level of E2 in the gilts. It was well known that ZEN is a competitive substrate for endogenous estrogens, binding to estrogen receptors and thereby having a damaging effect on the function of gonads [27]. A recent long-term (48 d) study found that low doses of ZEN (20 μg/kg BW) induced changes in the concentrations of E2 levels in pre-pubertal gilts [28]. The addition of biodegradable *Bacillus subtilis* ANSB01G, in this study, restored serum E2 levels and modulated the function of gonads in gilts.

Van and his colleagues reported that ZEN ingestion partially induced oxidative stress in piglets, as it revealed an increased content of MDA and activity of SOD [29]. In addition, even low levels of ZEN (246 ug/kg) in the diet of gestation sows can lead to an increase in the level of serum MDA and cause cell apoptosis and moderate lesions of the liver, kidney, uterus, and ovary [30]. Importantly, our research showed that only biodegradable *Bacillus subtilis* ANSB01G reversed these increasing trends. However, no changes in the levels of antioxidant enzymes (e.g., GSH-Px and MDA) or SOD activity were observed in the serum of gilts. Moreover, we did not observe that diets treated with ZEN or both strains of *Bacillus subtilis* affected the serum biochemical indicators (e.g., ALB, AST, ALP and CRE). These data were inconsistent with the previous reports [31,32]. The difference in results might be attributed to the age difference of the animal model and different doses of ZEN contaminated. However, we noticed that ZEN increased IgG level in serum, but not in the levels of IgA and IgM. Gilts treated with biodegradable *Bacillus subtilis* ANSB01G and probiotic *Bacillus subtilis* ANSB010 decreased IgG levels in serum. IgG, IgM and IgA are the main components of immunoglobulins, of which the content of IgG is up to 70–75%. These results suggested that ZEN has antigenic activity which stimulated the immune system of gilts, and both biodegradable and probiotic *Bacillus subtilis* have the capacity to recover the stimulation caused by ZEN.

Moreover, it has been reported that ZEN could increase the synthesis and expression of pro-inflammatory factors through JNK signaling pathway activation [6]. Currently, few studies have focused on the effects of ZEN on the modulation of inflammation in gilts. In the present study, ZEN in the feed significantly increased the levels of TNFα and IL-2, and had no effects on IL-1β and IL-6 in serum. An increase in levels of TNFα, one of the most powerful pro-inflammation factors, might generate a risk of a more severe inflammatory response [33]. The inflammatory response in this trial was similar to previous results from our lab that showed an enormous increase in inflammatory cytokines (e.g., IL-2, IL-8 and IL-10) [25]. The elevation of cytokines caused by ZEN would impair the erythroid progenitor and red blood cells [34], which revealed a potential cancerotoxic effect of ZEN. Biodegradable *Bacillus subtilis* ANSB01G mitigated the acute inflammatory response, confirming the reliability of a biodegradation approach in myotoxin degradation. In comparison, the lack effect of probiotic *Bacillus subtilis* ANSB010 on the elevated inflammatory response demonstrated that probiotic *Bacillus subtilis* had no effects on alleviating the toxicity of ZEN. Therefore, we concluded that biodegradable *Bacillus subtilis* was more protective against ZEN toxicosis in gilts than the probiotic *Bacillus subtilis*. Taken together, this study provided further evidence that the specific strain of *Bacillus subtilis* ANSB01G can alleviate the toxicity of ZEN, mainly due to its biodegradable capacity.

4. Conclusions

This study demonstrated that a feeding diet contaminated with ZEN of 300 μg/kg had a damaging impact on the growth performance, plasma immune function and hormone secretion of gilts. Although probiotic *Bacillus subtilis* ANSB010 and biodegradable *Bacillus subtilis* ANSB01G have similar antimicrobial capacities and alleviate inflammatory responses, only biodegradable *Bacillus subtilis* ANSB010 could regulate estrogen levels, relieve swelling of the vulva, and reduce the F/G and fecal ZEN residues. Hence, the biodegradable *Bacillus subtilis* ANSB01G used in this study is considered to have great and promising potential for biodegradation of mycotoxin in feed industrial applications.

5. Materials and Methods

5.1. Source and Identification of Bacterial Strains

The two strains of *Bacillus subtilis* used in this study were isolated from healthy broiler intestinal chyme and identified and characterized using a standard method described by Holt et al. [35]. Gram-staining was performed using the Gram staining kit (G1060, Solarbio, Beijing, China). Briefly, the bacteria were activated in Luria-Bertani (LB) medium at 37 °C for 12 h. Then, a 2.5 μL sample was stained with crystal violet for 1 min, mordanted with iodine solution 1 min, decolorized for 30 s, and counterstained with safranine for 1 min. Spore-staining was performed using the Spore stain kit (G1133, Solarbio, Beijing, China). First, a 5 μL sample, cultured in LB medium at 37 °C for 48 h, was stained in malachite green solution for 10 min and counterstained with safranine for 3 min. Finally, the staining results were observed using a microscope. Then, DNA was extracted from the bacterial isolates using the Bacterial Genomic DNA kit (Beijing Zoman Biotechnology Co., Ltd., Beijing, China) according to the manufacturer's instructions. PCR amplification of 16S rDNA was performed with the primers 27F (5′-AGAGTTTGATCMTGGCTCAG-3′) and 1492R (5′-CGGTTACCTTGTTACGACTT-3′), and PCR products were purified and sequenced by Sangon Biotech (Beijing, China). Probiotic and biodegradable *Bacillus subtilis* were identified as strains of *Bacillus subtilis* named ANSB010 and ANSB01G, respectively. It has been shown that ANSB01G could degrade ZEN in naturally contaminated maize with high efficiency [13]. The *Bacillus subtilis* ANSB01G in our experiment was a domesticated strain based on the wild bacteria obtained by Lei et al. [13]. The method of microbial domestication was as follows: after activation, *Bacillus subtilis* ANSB01G was induced and cultured in a series of MRS mediums with gradually increasing concentrations of ZEN [36]. The efficiency of *Bacillus subtilis* ANSB01G in degrading ZEN was further improved after several domestications. The 16 s rDNA sequences of ANSB010 and domesticated ANSB01G are shown in the Supplementary Materials. A phylogenetic tree was drawn with the neighbor-joining method of 1000 bootstrap replications within Mega 5.0.

5.2. Antibacterial Activity and ZEN Degradation Tests

Selected indicated bacteria *E. coli* (No.10003), *S. choleraesuis* (No.21493) and *S. aureus* (No.10384) were purchased from the China Center of Industrial Culture Collection (CICC, Beijing, China). *E. coli, S. choleraesuis* and *S. aureus* were inoculated in MRS medium and incubated in MRS medium at 180 rpm for 24 h at 37 °C. After diluting twice in a gradient, 200 μL of the bacterial solution was added to MRS medium and spread evenly, then z sterilized Oxford cup was placed on the medium. Then, 200 μL of the supernatant of the probiotic and biodegradable *Bacillus subtilis* solution was aspirated into an Oxford cup and the non-inoculated MRS medium was used as a control. After incubating at 37 °C for 24 h, the antibacterial circle diameter (cm) was measured. For the ZEN degradation test, ZEN solution (100 μL, 2 μg/mL) was added into the LB medium of ANSB010 and ANSB01G (900 μL, 3.0×10^8 CFU/mL) and incubated with shaking for 6 h, 24 h, and 48 h at 37 °C in the dark, followed by measurement of ZEN levels using the HPLC method.

5.3. Animals and Experimental Treatments

The animal experiments were conducted according to the animal welfare requirements and approved by the Animal Protocol Review Committee of the China Agriculture University (Beijing, China).

Thirty-six healthy gilts (Landrace × Yorkshire, average body weight = 64 kg) were selected for the experiment. Then, these animals were randomly assigned to four treatments with nine replicates of one gilt per replicate for each group: (1) Normal control diet group (NC) fed the basal diet containing few ZEN (17.5 μg/kg diet) by controlling the quality of maize; (2) ZEN-contaminated group (ZC) fed the contaminated diet containing an exceeded limit dose of ZEN (about 300 μg/kg diet) by replacing normal maize in the basal diet with moldy maize; (3) Probiotic agent group (PB) fed the ZC diet with added 5×10^9 CFU/kg of probiotic *Bacillus subtilis* ANSB010; (4) Biodegradable agent group (DA) fed the ZC diet with added 5×10^9 CFU/kg of biodegradable *Bacillus subtilis* ANSB01G. The contaminated maize was purchased in 2018 from a small family farm in Henan Province, China. The *Bacillus subtilis* ANSB010 and ANSB01G were incubated in LB medium for 24 h at 37 °C, followed by drying at 65 °C, and then evenly mixed into the diets. Diets are formulated to meet or exceed nutrient requirements (Table 7) recommended by the National Research Council for replacement gilts (NRC, 2012). The experimental period lasts for 25 d. The contaminated diets were prepared by replacing corn in the control with the naturally contaminated maize. During the supplementation period, all piglets were individually housed in temperature-controlled stainless steel metabolism pens (25 ± 2 °C), allowing free access to drinking water. Animal care and experimental procedures were in accordance with the guidelines of the National Institutes of Health Guide and the China Ministry of Agriculture for the care and use of laboratory animals.

Table 7. Ingredients and compositions of the basal diet, as fed basis.

Ingredient	%	Nutrition Component	Content [1]
Maize	57.00	DE, Kcal/Kg	3100
Soybean meal	23.00	Crude protein, %	17.00
Wheat bran	16.00	Calcium, %	0.76
Calcium hydrophosphate	1.00	Total phosphorus, %	0.61
Limestone	1.05	Non-phytate phosphorus, %	0.36
Salt	0.30	Lysine, %	0.85
Threonine	0.05	Methionine, %	0.52
Lysine 70%	0.60	Threonine, %	0.59
Choline chloride	0.12		
Chlortetracycline	0.05		
Compound-premix [2]	0.83		
Total	100.00		

[1] The value is calculated. [2] Supplied the following per kilogram of diet: vitamin A, 5000 IU; vitamin D3, 900 IU; vitamin E, 40 IU; vitamin K, 2.5 mg; vitamin B1, 1.5 mg; vitamin B2, 6.4 mg; vitamin B6, 2.5 mg; vitamin B12, 0.025 mg; pantothenate, 20 mg; nicotinic acid, 30 mg; choline, 0.50 g; Fe 100 mg; Cu, 6 mg; Zn, 50 mg; Mn, 20 mg; Se, 0.30 mg; I, 0.24 mg.

5.4. Growth Performances

The initial body weight and terminal body weight were recorded. Moreover, ADG, ADFI, and F/G (ADFI/ADG) were calculated.

5.5. Vulva Size Determination

The vulva length (a), width (b) and height (h) were measured and recorded at 0, 12d and 24 d. The determination of vulva area and volume was performed and results were calculated according to the method previously described by Zhao and his colleagues [14]. The area and volume of vulva are approximately elliptical and conical, respectively. Therefore, the area of the vulva is in accordance with the formula: $S = (\pi \times a \times b)/4$; and the volume of the vulva is calculated with the equation: $V = 1/3 \times S \times h$.

5.6. Serum Parametes

Blood was collected from the marginal ear vein at the end of the experiment period. Then, the blood samples were centrifuged at 3000× *g* for 10 min to obtain serum for further analysis. The serum biochemical indicators including the TP, ALB, ALT, AST and ALP, CRE and BUN were measured with an automatic biochemical analyzer (Hitachi 7160, Hitachi High-Technologies Corporation, Tokyo, Japan). Immunoglobulin A (IgA), IgG and IgM were also determined with an automatic biochemical analyzer (Hitachi 7160). In addition, interleukin 1β (IL-1β), IL-2, IL-6 and TNFα were measured by an enzyme-linked immune sorbent assay (ELISA) kit (YuanMu Biotechnology, Shanghai, China). Serum SOD, GSH-Px and MDA were detected according to the instructions of the manufacturer using microplate test kits (Nanjing Jiancheng Bioengineering Institute, Nanjing, China). The concentrations of serum E2, FSH, LH and PRL were determined by radioimmunoassay (RIA). The serum samples were treated with radioactive-125I according to the instructions of the RIA kit (Beijing Kemei Biotechnology Co., Ltd., Beijing, China). Then, a GC-1200 radio-immunity gamma-counter (KeDa Innovation Co., Ltd., Hefei, China) was used to measure hormone concentrations. For each RIA, the intra- and inter-assay coefficients of variation were less than 15% and less than 10%, respectively. Hormone concentration was determined according to each sample's level of radioactivity.

5.7. Determination of ZEN in the Feed, Feces, Broth and Serum

Two days before the end of the test, fresh fecal samples were taken to determine the content of ZEN in feces according to the description of Chinese certification GB/T 23504-2009 and Lei et al. [13]. For feed and feces samples, 50 g ground samples were extracted by acetonitrile-water (70:30, *v/v*, 200 mL), followed by filtration with Whatman 4 filter paper. After dilution with PBS solution (PH = 7.40), the mixing solution were filtered through a micro-filter. A volume of 20 mL of suspension was passed through the immunoaffinity column (Femdetection FD-C21, Nanjing, China) at a flow rate of 1.0 mL/min under gravity. After washing the column with distilled water, the ZEN was subsequently eluted with 2 mL of methanol into a centrifuge tube for HPLC analysis. For liquid medium, samples (1 mL) were extracted with acetonitrile (9 mL) at 180 rpm for 2 h. The mixed samples were filtered using glass fibre filter paper, and then collected for subsequent HPLC analysis. ZEN and its metabolites in serum were analyzed using the method described by Duca et al. [37]. Briefly, serum (2 mL) was mixed with buffer ammonium acetate solution (8 mL). The mixed solution was incubated with glucuronidase/arylsulfatase (50 μL) at 37 °C for 15 h. After the samples were centrifuged at 5000 rpm for 10 min, the supernatant was passed through the immunoaffinity column. The column was then rinsed with 20 mL of ultrapure water. The analytes were then eluted with acetonitrile (2 mL). Subsequently, the solution was dried using a Speed Vac concentration system after which 200 μL mobile phase was added. For HPLC analysis, 20 μL sample solution was injected into the HPLC system. Separation was in a C18 column (4.6 mm × 150 mm, 5 μm; Thermo Fisher Scientific, Waltham, MA, USA) with mobile phase (water: acetonitrile, 50:50, *v/v*) at a flow rate of 1.0 mL/min. The analytes were detected by a fluorescence detector (Waters, Milford, MA, USA), excitation and emission wavelengths were 274 and 440 nm, respectively. The retention time was 7–8 min.

5.8. Statistical Analysis

Data were analyzed statistically using SAS 9.4 software (SAS Institute Inc., Cary, NC, USA) and were presented as mean ± SEM. The significance of difference between groups of gilts were analyzed by one-way ANOVA. A normality test (Shapiro–Wilk) was performed to determine normality before one-way ANOVA analysis. Differences were regarded as statistically significant at a probability of $p < 0.05$, and p-values < 0.10 were regard as a trend.

Supplementary Materials: The following are available online at https://www.mdpi.com/article/10.3390/toxins13120882/s1.

Author Contributions: W.S. and Y.L. (Yaojun Liu): conceptualization, conducted the animal experiments; methodology, writing—review and editing. X.Z. (Xinyue Zhang), X.Z. (Xiong Zhang), F.L. and J.C.: assisted with the experiments; X.R. and Y.L. (Yuanpei Lei): conducted the ZEN degradation test; L.Z. and C.J.: supervision; Q.M.: conceptualization, funding acquisition, writing—review and editing. All authors have read and agreed to the published version of the manuscript.

Funding: This study was funded by the National Natural Science Foundation of China (Grant No. 31772637, 31301981), and the Special Fund for Agro-scientific Research in the Public Interest (201403047).

Institutional Review Board Statement: The study was conducted according to the guidelines of the Beijing Muncipality on the Review of Welfare and Ethics of Laboratory Animals, and approved by the China Agricultural University Animal Care and Use Committee (Approval No. Aw72011202-1-6).

Informed Consent Statement: Not applicable.

Data Availability Statement: Data are available from the first author.

Acknowledgments: The authors wish to thank the students who participated in our study and the workers of the Fengze Farm.

Conflicts of Interest: The authors declare no conflict of interest.

References

1. Kuiper-Goodman, T.; Scott, P.; Watanabe, H. Risk assessment of the mycotoxin zearalenone. *Regul. Toxicol. Pharmacol.* **1987**, *7*, 253–306. [CrossRef]
2. Zinedine, A.; Soriano, J.M.; Moltó, J.C.; Mañes, J. Review on the toxicity, occurrence, metabolism, detoxification, regulations and intake of zearalenone: An oestrogenic mycotoxin. *Food Chem. Toxicol.* **2007**, *45*, 1–18. [CrossRef] [PubMed]
3. Ma, R.; Zhang, L.; Liu, M.; Su, Y.-T.; Xie, W.-M.; Zhang, N.-Y.; Dai, J.-F.; Wang, Y.; Rajput, S.A.; Qi, D.-S.; et al. Individual and Combined Occurrence of Mycotoxins in Feed Ingredients and Complete Feeds in China. *Toxins* **2018**, *10*, 113. [CrossRef]
4. Döll, S.; Dänicke, S. The Fusarium toxins deoxynivalenol (DON) and zearalenone (ZON) in animal feeding. *Prev. Vet. Med.* **2011**, *102*, 132–145. [CrossRef] [PubMed]
5. Jiang, S.Z.; Yang, Z.B.; Yang, W.R.; Gao, J.; Liu, F.X.; Broomhead, J.; Chi, F. Effects of purified zearalenone on growth performance, organ size, serum metabolites, and oxidative stress in postweaning gilts1. *J. Anim. Sci.* **2011**, *89*, 3008–3015. [CrossRef]
6. Pistol, G.C.; Braicu, C.; Motiu, M.; Gras, M.A.; Marin, D.E.; Stancu, M.; Calin, L.; Israel-Roming, F.; Berindan-Neagoe, I.; Taranu, I. Zearalenone Mycotoxin Affects Immune Mediators, MAPK Signalling Molecules, Nuclear Receptors and Genome-Wide Gene Expression in Pig Spleen. *PLoS ONE* **2015**, *10*, e0127503. [CrossRef]
7. Awad, W.A.; Ghareeb, K.; Böhm, J.; Zentek, J. Decontamination and detoxification strategies for the Fusarium mycotoxin deoxynivalenol in animal feed and the effectiveness of microbial biodegradation. *Food Addit. Contam.* **2010**, *27*, 510–520. [CrossRef]
8. Rustom, I.Y.S. Aflatoxin in food and feed: Occurrence, legislation and inactivation by physical methods. *Food Chem.* **1997**, *59*, 57–67. [CrossRef]
9. Schatzmayr, G.; Zehner, F.; Täubel, M.; Schatzmayr, D.; Klimitsch, A.; Loibner, A.P.; Binder, E.M. Microbiologicals for deactivating mycotoxins. *Mol. Nutr. Food Res.* **2006**, *50*, 543–551. [CrossRef] [PubMed]
10. Jardon-Xicotencatl, S.; Díaz-Torres, R.; Marroquin-Cardona, A.; Villarreal-Barajas, T.; Méndez-Albores, A. Detoxification of Aflatoxin-Contaminated Maize by Neutral Electrolyzed Oxidizing Water. *Toxins* **2015**, *7*, 4294–4314. [CrossRef]
11. Jouany, J.P. Methods for preventing, decontaminating and minimizing the toxicity of mycotoxins in feeds. *Anim. Feed. Sci. Technol.* **2007**, *137*, 342–362. [CrossRef]
12. Doyle, M.P.; Applebaum, R.S.; Brackett, R.E.; Marth, E.H. Physical, Chemical and Biological Degradation of Mycotoxins in Foods and Agricultural Commodities. *J. Food Prot.* **1982**, *45*, 964–971. [CrossRef]
13. Lei, Y.; Zhao, L.; Ma, Q.; Zhang, J.Y.; Zhou, T.; Gao, C.; Ji, C. Degradation of zearalenone in swine feed and feed ingredients by Bacillus subtilis ANSB01G. *World Mycotoxin J.* **2014**, *7*, 143–151. [CrossRef]
14. Zhao, L.; Lei, Y.; Bao, Y.; Jia, R.; Ma, Q.; Zhang, J.; Chen, J.; Ji, C. Ameliorative effects ofBacillus subtilisANSB01G on zearalenone toxicosis in pre-pubertal female gilts. *Food Addit. Contam. Part A* **2014**, *32*, 617–625. [CrossRef] [PubMed]
15. Galvano, F.; Piva, A.; Ritieni, A.; Galvano, G. Dietary Strategies to Counteract the Effects of Mycotoxins: A Review. *J. Food Prot.* **2001**, *64*, 120–131. [CrossRef] [PubMed]
16. Binder, E.M. Managing the risk of mycotoxins in modern feed production. *Anim. Feed. Sci. Technol.* **2007**, *133*, 149–166. [CrossRef]
17. Zaki, M.M.; El-Midany, S.; Shaheen, H.; Rizzi, L. Mycotoxins in animals: Occurrence, effects, prevention and management. *J. Toxicol. Environ. Health Sciences* **2012**, *4*, 13–28. [CrossRef]
18. Huwig, A.; Freimund, S.; Käppeli, O.; Dutler, H. Mycotoxin detoxication of animal feed by different adsorbents. *Toxicol. Lett.* **2001**, *122*, 179–188. [CrossRef]
19. Kološová, A.; Stroka, J. Substances for reduction of the contamination of feed by mycotoxins: A review. *World Mycotoxin J.* **2011**, *4*, 225–256. [CrossRef]
20. Swamy, H.V.L.N.; Smith, T.K.; Macdonald, E.J.; Karrow, N.A.; Woodward, B.; Boermans, H.J. Effects of feeding a blend of grains naturally contaminated with Fusarium mycotoxins on growth and immunological measurements of starter pigs, and the efficacy of a polymeric glucomannan mycotoxin adsorbent. *J. Anim. Sci.* **2003**, *81*, 2792–2803. [CrossRef]

21. Pierzgalski, A.; Bryła, M.; Kanabus, J.; Modrzewska, M.; Podolska, G. Updated Review of the Toxicity of Selected *Fusarium* Toxins and Their Modified Forms. *Toxins* **2021**, *13*, 768. [CrossRef]
22. Cheng, Y.-H.; Weng, C.-F.; Chen, B.-J.; Chang, M.-H. Toxicity of different Fusarium mycotoxins on growth performance, immune responses and efficacy of a mycotoxin degrading enzyme in pigs. *Anim. Res.* **2006**, *55*, 579–590. [CrossRef]
23. Zhang, J.; Zheng, Y.; Tao, H.; Liu, J.; Zhao, P.; Yang, F.; Lv, Z.; Wang, J. Effects of *Bacillus subtilis* ZJ-2019-1 on Zearalenone Toxicosis in Female Gilts. *Toxins* **2021**, *13*, 788. [CrossRef]
24. Gao, X.; Xiao, Z.; Li, C.; Zhang, J.; Zhu, L.; Sun, L.; Zhang, N.; Khalil, M.M.; Rajput, S.A.; Qi, D. Prenatal exposure to zearalenone disrupts reproductive potential and development via hormone-related genes in male rats. *Food Chem. Toxicol.* **2018**, *116*, 11–19. [CrossRef] [PubMed]
25. Shi, D.; Zhou, J.; Zhao, L.; Rong, X.; Fan, Y.; Hamid, H.; Li, W.; Ji, C.; Ma, Q. Alleviation of mycotoxin biodegradation agent on zearalenone and deoxynivalenol toxicosis in immature gilts. *J. Anim. Sci. Biotechnol.* **2018**, *9*, 42. [CrossRef] [PubMed]
26. Hassan, A.A.; Rashid, M.; Koratum, K.M. Effect of aflatoxin B1, zearalenone and ochratoxin A on some hormones related to fertility in male rats. *Life Sci. J.* **2010**, *7*, 64–72.
27. Silva, J.R.; Hurk, R.V.D.; de Matos, M.H.T.; dos Santos, R.R.; Pessoa, C.; de Moraes, M.O.; de Figueiredo, J.R. Influences of FSH and EGF on primordial follicles during in vitro culture of caprine ovarian cortical tissue. *Theriogenology* **2004**, *61*, 1691–1704. [CrossRef] [PubMed]
28. Zielonka, L.; Gajęcka, M.; Lisieska-Żołnierczyk, S.; Dąbrowski, M.; Gajęcki, M.T. The Effect of Different Doses of Zearalenone in Feed on the Bioavailability of Zearalenone and Alpha-Zearalenol, and the Concentrations of Estradiol and Testosterone in the Peripheral Blood of Pre-Pubertal Gilts. *Toxins* **2020**, *12*, 144. [CrossRef] [PubMed]
29. Van Le Thanh, B.; Lemay, M.; Bastien, A.; Lapointe, J.; Lessard, M.; Chorfi, Y.; Guay, F. The potential effects of antioxidant feed additives in mitigating the adverse effects of corn naturally contaminated with Fusarium mycotoxins on antioxidant systems in the intestinal mucosa, plasma, and liver in weaned pigs. *Mycotoxin Res.* **2016**, *32*, 99–116. [CrossRef]
30. Zhou, J.; Ao, X.; Lei, Y.; Ji, C.; Ma, Q. Bacillus subtilis ANSB01G culture alleviates oxidative stress and cell apoptosis induced by dietary zearalenone in first-parity gestation sows. *Anim. Nutr.* **2020**, *6*, 372–378. [CrossRef]
31. Long, M.; Yang, S.; Zhang, W.; Zhang, Y.; Li, P.; Guo, Y.; Wang, Y.; Chen, X.; He, J. The Influence of Selenium Yeast on Hematological, Biochemical and Reproductive Hormone Level Changes in Kunming Mice Following Acute Exposure to Zearalenone. *Biol. Trace Element Res.* **2016**, *174*, 362–368. [CrossRef] [PubMed]
32. Su, Y.; Sun, Y.; Ju, D.; Chang, S.; Shi, B.; Shan, A. The detoxification effect of vitamin C on zearalenone toxicity in piglets. *Ecotoxicol. Environ. Saf.* **2018**, *158*, 284–292. [CrossRef] [PubMed]
33. Ben Salah-Abbès, J.; Belgacem, H.; Ezzdini, K.; Abdel-Wahhab, M.A.; Abbès, S. Zearalenone nephrotoxicity: DNA fragmentation, apoptotic gene expression and oxidative stress protected by Lactobacillus plantarum MON03. *Toxicon* **2020**, *175*, 28–35. [CrossRef]
34. Tiemann, U.; Brüssow, K.-P.; Dannenberger, D.; Jonas, L.; Pöhland, R.; Jäger, K.; Dänicke, S.; Hagemann, E. The effect of feeding a diet naturally contaminated with deoxynivalenol (DON) and zearalenone (ZON) on the spleen and liver of sow and fetus from day 35 to 70 of gestation. *Toxicol. Lett.* **2008**, *179*, 113–117. [CrossRef]
35. Cowan, S.T. Bergey's Manual of Determinative Bacteriology. *Nature* **1948**, *162*, 833. [CrossRef]
36. Ma, Q.G.; Zhao, L.H.; Ji, C.; Zhang, J.Y.; Li, M.L.; Zheng, R. A Strain Domestication Method for Improving the Degradation Efficiency of Mycotoxin-Degrading Bacteria. *Patent CN10804 8385A*, 2018.
37. Duca, R.C.; Bravin, F.; Delaforge, M.; Vladescu, L.; Badea, I.A.; Criste, R.D. Development of a New HPLC Method Used for Determination of Zearalenone and Its Metabolites in Broiler Samples. Influence of Zearalenone on the Nutritional Properties of Broiler Meat. *J. Agric. Food Chem.* **2009**, *57*, 10497–10504. [CrossRef]

toxins

MDPI

Article

Four PQQ-Dependent Alcohol Dehydrogenases Responsible for the Oxidative Detoxification of Deoxynivalenol in a Novel Bacterium *Ketogulonicigenium vulgare* D3_3 Originated from the Feces of *Tenebrio molitor* Larvae

Yang Wang [1], Donglei Zhao [2], Wei Zhang [1], Songshan Wang [1], Yu Wu [1], Songxue Wang [1], Yongtan Yang [1] and Baoyuan Guo [1,*]

1 Academy of National Food and Strategic Reserves Administration, Beijing 100037, China; wy@ags.ac.cn (Y.W.)
2 School of Health Science and Engineering, University of Shanghai for Science and Technology, Shanghai 200093, China
* Correspondence: gby@ags.ac.cn

Abstract: Deoxynivalenol (DON) is frequently detected in cereals and cereal-based products and has a negative impact on human and animal health. In this study, an unprecedented DON-degrading bacterial isolate D3_3 was isolated from a sample of *Tenebrio molitor* larva feces. A 16S rRNA-based phylogenetic analysis and genome-based average nucleotide identity comparison clearly revealed that strain D3_3 belonged to the species *Ketogulonicigenium vulgare*. This isolate D3_3 could efficiently degrade 50 mg/L of DON under a broad range of conditions, such as pHs of 7.0–9.0 and temperatures of 18–30 °C, as well as during aerobic or anaerobic cultivation. 3-keto-DON was identified as the sole and finished DON metabolite using mass spectrometry. In vitro toxicity tests revealed that 3-keto-DON had lower cytotoxicity to human gastric epithelial cells and higher phytotoxicity to *Lemna minor* than its parent mycotoxin DON. Additionally, four genes encoding pyrroloquinoline quinone (PQQ)-dependent alcohol dehydrogenases in the genome of isolate D3_3 were identified as being responsible for the DON oxidation reaction. Overall, as a highly potent DON-degrading microbe, a member of the genus *Ketogulonicigenium* is reported for the first time in this study. The discovery of this DON-degrading isolate D3_3 and its four dehydrogenases will allow microbial strains and enzyme resources to become available for the future development of DON-detoxifying agents for food and animal feed.

Keywords: DON biodegradation; 3-keto-DON; toxicity; *Ketogulonicigenium vulgare*; PQQ-dependent alcohol dehydrogenases

Key Contribution: A novel DON-oxidizing strain, *Ketogulonicigenium vulgare* D3_3, was isolated from yellow mealworm feces, with oxidizing ability under either aerobic or anaerobic conditions. The oxidation product, 3-keto-DON, exhibited a higher phytotoxicity compared to its parent DON. Furthermore, all four PQQ-dependent alcohol dehydrogenases in the DON degrader were found to possess DON-oxidizing activities.

Citation: Wang, Y.; Zhao, D.; Zhang, W.; Wang, S.; Wu, Y.; Wang, S.; Yang, Y.; Guo, B. Four PQQ-Dependent Alcohol Dehydrogenases Responsible for the Oxidative Detoxification of Deoxynivalenol in a Novel Bacterium *Ketogulonicigenium vulgare* D3_3 Originated from the Feces of *Tenebrio molitor* Larvae. *Toxins* **2023**, *15*, 367. https://doi.org/10.3390/toxins15060367

Received: 8 May 2023
Revised: 25 May 2023
Accepted: 26 May 2023
Published: 30 May 2023

1. Introduction

Deoxynivalenol (DON) is a natural toxin that is produced as a secondary metabolite by certain species of phytopathogenic fungi belonging to the genus *Fusarium* [1]. Its high prevalence and widespread occurrence in small cereal grains and their derivatives have been extensively documented in the literature [2,3]. The presence of DON in the food supply chain constitutes a notable risk to food safety and public health. A range of adverse health effects, including gastrointestinal disturbances, dermatological manifestations, and immunosuppression, have been associated with varying doses and exposure durations to

DON [4–7]. Given DON's chemical and thermal stability, salvaging DON-contaminated cereal grains and avoiding the health risks associated with DON is technically challenging. Over recent decades, extensive efforts have been made to seek cost-effective and efficacious DON mitigation strategies. For instance, a range of physicochemical detoxification approaches, such as ultraviolet irradiation [8], argon plasma exposure [9], ozonation treatment [10,11], sorting and separation [12], washing [13,14], and thermal treatment [15] have been explored. However, each of these DON-mitigation methods possesses inherent limitations regarding environmental impacts, nutritional quality, and potential generation of toxic byproducts. The biodegradation method, involving enzymatic transformation of mycotoxins into less toxic or non-toxic products via selected microorganisms, is characterized by mild reaction conditions and a relatively high specificity. Therefore, it was viewed as a promising alternative strategy to reduce DON contamination in agricultural products.

The main toxicity determinants in the molecular structure of DON are the three hydroxyl moieties located at C3, C7, and C15, as well as the epoxy ring connecting C12 and C13 [7,16], with C3-OH and the epoxy moiety as the primary targets for biodegradation. The oxidation of C3-OH can yield 3-keto-DON, which can further undergo epimerization to form 3-epi-DON. Furthermore, de-epoxidation of the epoxy moiety can generate deepoxy-deoxynivalenol (DOM-1). These three metabolites have demonstrated a lower toxicity than DON in vitro [17–19], and many species of microorganisms have been found to possess these metabolic capabilities. For instance, several soil-borne bacterial strains, including *Nocardioides* sp. WSN05-2 [20], *Devosia mutans* 17-2-E-8 [21], *Paradevosia shaoguanensis* DDB001 [22], and *Nocardioides* sp. ZHH-013 [23], can produce 3-epi-DON. The two-step enzymatic reaction mechanism converting DON to 3-epi-DON was first elucidated in *D. mutans* 17-2-E-8. The process involves a PQQ-dependent alcohol dehydrogenase (DepA) oxidizing DON to 3-keto-DON, followed by an NADPH dependent aldo-keto reductase (DepB) stereospecifically reducing the intermediate 3-keto-DON to 3-epi-DON [21,24]. Although DON can also be converted to 3-keto-DON by *Agrobacterium* E3-39 [17], *D. insulae* A16 [25], and *Pelagibacterium halotolerans* ANSP101 [26], these strains cannot further convert 3-keto-DON to 3-epi-DON, likely due to the absence of a functional enzyme such as DepB.

Despite numerous successful isolations of DON-degrading microorganisms from various sources, there have been no reports of such isolations from the intestines or feces of insects. As a kind of edible insect, *Tenebrio molitor* larvae (yellow mealworm) are an alternative protein source for food and feed, due to their low production costs and good nutritional characteristics [27]. They can be readily raised on cereal bran or flour; however, these diets may be contaminated with mycotoxins, particularly DON. Long-term feeding of mycotoxin-contaminated diets could thus create favorable conditions for the acclimation of mycotoxin-degrading microorganisms in yellow mealworm. Several investigations have explored the impacts of DON on the biological parameters of insect larvae, as well as the profiles of DON metabolism and excretion [28–30]. For example, Van Broekhoven et al. [30] found that DON and its metabolites generally do not remain in the bodies of insect larvae fed diets contaminated with mycotoxins, but can be detected in larval feces, implying that yellow mealworms and/or their gut microbiota may have the potential ability to metabolize DON. On the other hand, microbial resources capable of degrading DON under both anaerobic and aerobic conditions are limited, posing a challenge for the effective mitigation of DON contamination in various application scenarios. In this regard, there is an urgent need to screen for novel DON-degrading strains from diverse sources, to enrich the repository of DON-degrading strains. This can help to address specific application demands in different scenarios, such as agricultural production and animal feed.

In the current study, a novel microbial DON degrader D3_3 was isolated from a sample of yellow mealworm feces using conventional enrichment and isolation procedures. Genome-based ANI analysis and 16S rRNA-based phylogenetic tree analysis consistently revealed its taxonomic position. The impact of various conditions on the strain's DON-degrading activity was also investigated. Subsequently, the chemical structure of the DON metabolite was determined via mass spectrometry, and its animal cell cytotoxicity and phytotoxicity were evaluated using human gastric epithelial cells (GES-1) and duckweed (*Lemna minor*), respectively. Finally, four PQQ-dependent alcohol dehydrogenases (ADHs) in strain D3_3 were identified as being responsible for the catalysis of the oxidation of DON to 3-keto-DON.

2. Results

2.1. A Potent DON-Oxidizing Strain Ketogulonicigenium Vulgare D3_3 Isolated from Yellow Mealworm Feces

As demonstrated in Figure 1A, four successive subcultures of yellow mealworm excreta in MMFS (mineral medium supplemented fecal supernatant) liquid resulted in a 20.9% decrease in 50 mg/L DON at 30 °C and 220 rpm in a shaking incubator after 5 d. However, no DON reduction was observed for the sample of PYM that underwent the same enrichment procedure. Thus, the mixed culture with DON-degrading activity was spread on MMFS agar plates after serial dilution. After growing at 30 °C for 7 d, 40 colonies were examined for their ability to remove DON. This resulted in the isolation and selection of the DON-degrading strain D3_3. As demonstrated in Figure 1B, a pure culture of this isolate D3_3 grown in MMFS (initial pH 7) was capable of removing 50 mg/L of DON after 12 h of incubation. Furthermore, a peak in the MMFS+DON+D3_3 sample had the same retention time (Rt = 10.7 min) as the 3-keto-DON standard, which was not detected when D3_3 was incubated in a MMFS medium without DON (MMFS+D3_3), implying that it was a DON metabolite and possibly 3-keto-DON.

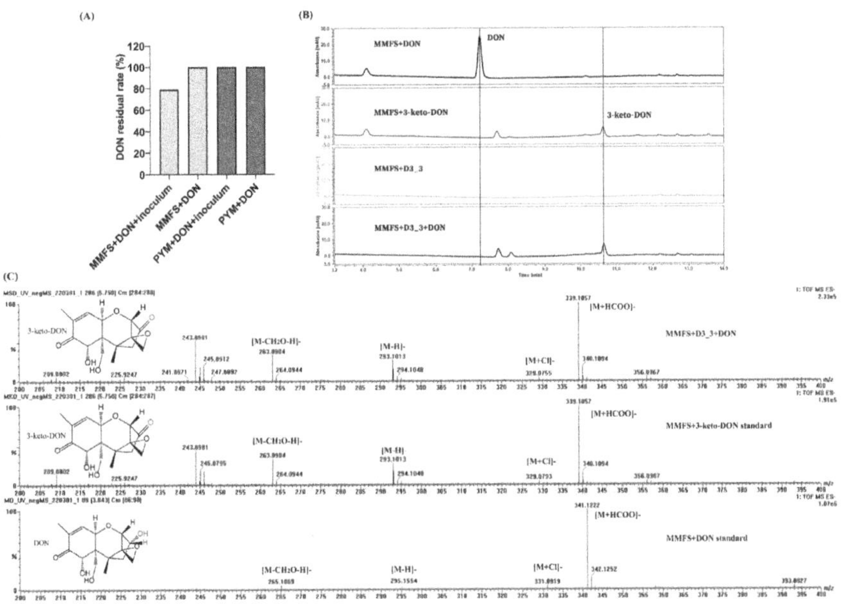

Figure 1. DON degradation profiles of the enrichment cultures and the isolate D3_3′ pure culture, as well as determination of the chemical structure of DON metabolite. (**A**) DON residual rate in MMFS and PYM media containing 50 mg/L of DON with or without insect fecal slurry after four instances of serial subcultivation. (**B**) LC profiles of DON degradation by the isolate D3_3 in MMFS medium containing 50 mg/L of DON. (**C**) MS profiles of DON, 3-keto-DON, and the putative metabolite 3-keto-DON generated by *K. vulgare* strain D3_3.

The chemical structure of possible 3-keto-DON (Rt = 5.75 min in UPLC-Q-TOF-MS analysis) was further ascertained via MS analysis. Figure 1C reveals that there were several peaks at *m*/*z* 339.1057, 329.0755, 293.1013, and 263.0904 in the negative mode, which were assigned to $[M+HCOO]^-$, $[M+Cl]^-$, $[M-H]^-$, and $[M-CH2O-H]^-$, respectively. These data were consistent with those of the 3-keto-DON standard. Furthermore, the MS data of DON (Rt = 3.84 min) showed that a number of peaks, such as m/z 341.1222 ($[M+HCOO]^-$), 331.0919 ($[M+Cl]^-$), 295.1154 ($[M-H]^-$), and 265.1069 ($[M-CH2O-H]^-$), were larger by 2 in comparison to the corresponding ions of the DON metabolites; such an increase was equal to the atomic weight of two hydrogen atoms. These findings conclusively determined that the DON metabolite produced by *K. vulgare* D3_3 was indeed the oxidation product of DON, namely 3-keto-DON.

The 16S rRNA gene sequence of strain D3_3 determined in this study comprised 1322 nt (GenBank accession no. OQ102971), exhibiting the highest sequence similarities with that of *Ketogulonicigenium vulgare* DSM 4025^T (99.85%) and *Ketogulonicigenium robustum* X6L^T (99.47%), respectively. Furthermore, phylogenetic analysis based on a 16S rRNA gene sequence also indicated that strain D3_3 was positioned within the genus *Ketogulonicigenium* and formed a phylogenetic clade with *K. vulgare* DSM 4025^T and *K. robustum* X6L^T, suggesting that it belonged to the genus *Ketogulonicigenium* (Figure 2A). However, these results could not definitively confirm which species strain D3_3 belonged to. To further clarify its taxonomic status, its genome sequence (obtained from the genome sequencing analysis) and several other genome sequences of *Ketogulonicigenium* available in the NCBI Genome database were subjected to pairwise average nucleotide identity (ANI) calculations using the FastANI algorithm [31]. As shown in Figure 2B, the heatmap indicated that the strain D3_3 shared an ANI of 98.33%, 98.36%, 98.40%, 98.43%, 98.44%, and 81.69% with *K. vulgare* Y25 (GCF_000164885), *K. vulgare* SKV (GCF_001693655), *K. vulgare* SPU B805 (GCF_001855295), *K. vulgare* Hbe602 (GCF_001399515), *K. vulgare* WSH-001 (GCF_000223375), and *K. robustum* SPU B003 (GCF_002117445), respectively. According to the classification criteria of >95% intra-species and <83% inter-species ANI values [31], the DON-degrading strain D3_3 was definitively identified as *K. vulgare*.

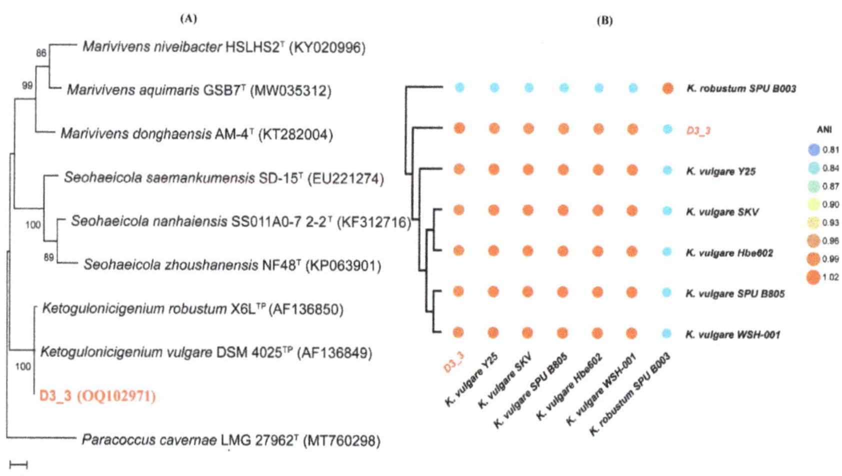

Figure 2. Taxonomic identification of DON-degrading strain D3_3. (**A**) The 16S rRNA-based phylogenetic tree of strain D3_3 reconstructed using the maximum likelihood method. The numbers near the nodes on the phylogenetic tree indicate that bootstrap values greater than 50%, while the content in parentheses are the GenBank accession numbers of the 16S rRNA sequences. Scale bar: 2 nucleotide substitutions per 100 positions. T: type strain, TP: use of patent strain as type strain. (**B**) ANI heatmap of seven *Ketogulonicigenium* strains. Heatmap generated based on ANI matrix obtained from *Ketogulonicigenium* genomes' average nucleotide identity (ANI) values, ranging from low (blue) to high (orange).

2.2. Effects of Different Growth Factors on the DON-Degrading Activity of Strain D3_3

As shown in Figure 3A, residual DON was not detected after 12 h in samples with an initial pH of 7, 8, and 9, meanwhile the DON residual rate was not significantly different for the pH 5 and 6 groups compared to the control. After 72 h, the DON residual rate of pH 6 decreased to 96.7 ± 0.6%, exhibiting a significant contrast with the control and the 12 h rate for the same group ($p < 0.05$). However, no such change was observed for the pH 5 group. Conclusively, strain D3_3 can degrade DON with a pH range of between 6 and 9, with greater activity between pH 7 and 9 than at pH 6, but no activity at pH 5.

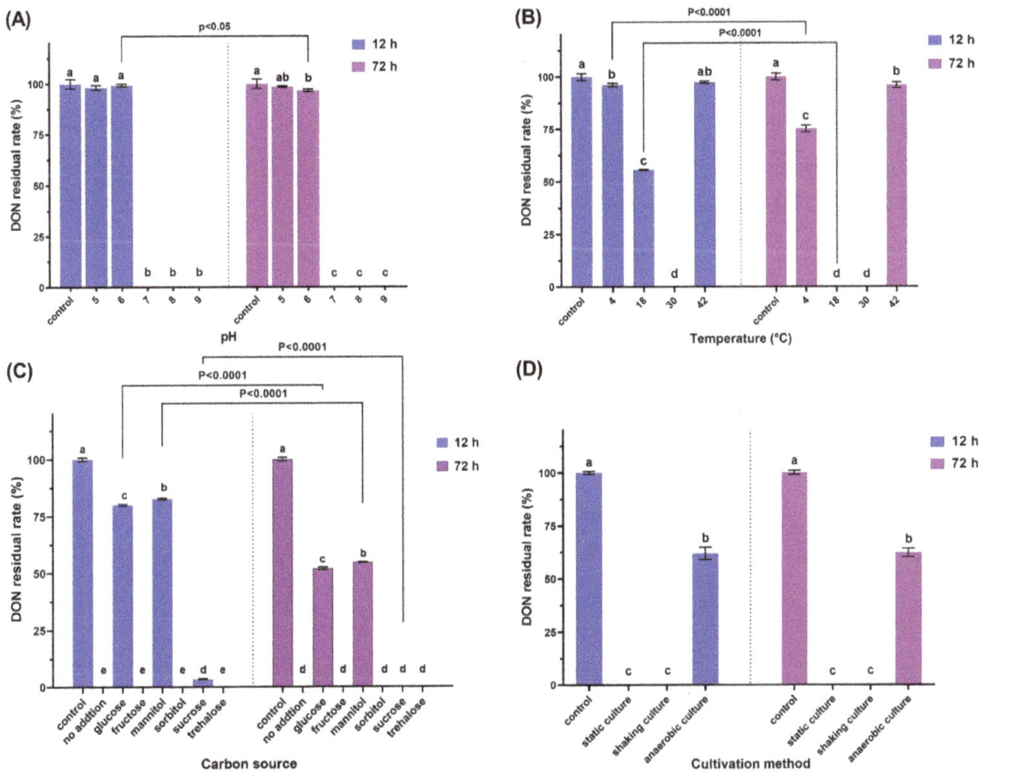

Figure 3. The impacts of different factors on the DON-degrading activity of strain D3_3 in the MMFS medium containing 50 mg/L of DON at incubation times of 12 and 72 h. (**A**) pH, (**B**) temperature, (**C**) carbon source, and (**D**) cultivation method. Significant differences ($p < 0.05$) in the degradation rates between the two groups, as determined via ordinary one-way ANOVA with a use of Tukey's multiple comparisons test, are indicated by the different letters (a–e) above the columns. "Control" stands for the MMFS medium containing 50 mg/L of DON, while "no addition" stands for the DON-added and inoculated MMFS medium.

As illustrated in Figure 3B, the 12 h DON residual rates for the samples at 4, 18, 30, 37, and 42 °C were 96.2 ± 0.8, 55.6 ± 0.3, 0, 97.6 ± 0.9, and 98.0 ± 0.6%, respectively. After 72 h, the DON residual rates at 4 and 18 °C decreased significantly to 75.2 ± 1.7 and 0% ($p < 0.0001$), respectively, while the rates at 37 and 42 °C remained statistically similar to their 12 h counterparts. Overall, strain D3_3 exhibited DON-degradation activity at temperatures ranging from 4 to 30 °C, with an optimal temperature of 30 °C. No DON-degradation activity was observed at 37 °C or higher, which is logically understandable considering that the strain does not grow at temperatures above 37 °C. Notably, even at 4 °C, D3_3 still displayed a 24.8 ± 1.7% DON-degradation rate after 72 h.

As shown in Figure 3C, the addition of glucose, mannitol, and sucrose significantly enhanced the DON residual rate at 12 h ($p < 0.05$), with glucose having the greatest effect, followed by mannitol and sucrose. Fructose, sorbitol, and trehalose had no significant effect on the DON residual rate. After 72 h, the DON residual rate for samples with glucose, mannitol, and sucrose declined significantly compared to the respective 12 h treatments ($p < 0.001$). Overall, the D3_3 strain's ability to degrade DON was inhibited by glucose, mannitol, and sucrose, but not by fructose, sorbitol, and trehalose.

As demonstrated in Figure 3D, the 12 h DON residual rates for samples under static, shaking, and anaerobic conditions were 0, 0, and 61.9 ± 2.9%, respectively. Despite extending the incubation duration to 72 h, the anaerobic DON residual rate remained virtually unchanged (62.1 ± 2.0%). These findings indicate that strain D3_3 possesses the capacity to degrade DON under both anaerobic and aerobic conditions, with the latter being more favorable for DON degradation. This expands this strain's potential application scenarios, such as being adapted to be used in agricultural practice in aerobic condition, in addition to being used as a feed additive administered to the anaerobic gastrointestinal tract of animals.

2.3. *Cytotoxicity and Phytotoxicity of 3-Keto-DON*

2.3.1. Effect of 3-Keto-DON on the Viability of GES-1 Cells

To gain a better understanding of the toxicity of 3-keto-DON, GES-1 was used as an in vitro model to investigate its toxicity. The results, as illustrated in Figure 4A,B, showed that both DON and 3-keto-DON reduced the viability of GES-1 cells in a dose-dependent manner, with GES-1 being more sensitive to DON than 3-keto-DON at the same treatment dose. The calculated IC_{50} values for DON and 3-keto-DON against GES-1 cells were 2.66 and 29.70 mg/L, respectively, indicating an 11.1-fold decrease in the toxicity of 3-keto-DON against GES-1 cells compared to the parent mycotoxin DON.

2.3.2. Effect of 3-Keto-DON on L. Minor

As shown in Figure 4C, 0.5 mg/L of DON standard significantly decreased the number of fronds by 41.4% when compared to the control (13.3 ± 2.3 vs. 22.7 ± 6.1 fronds/well), while the area of fronds was lowered by 49.4% (0.43 ± 0.07 vs. 0.85 ± 0.24 cm^2/well). Duckweed development was entirely repressed by DON at 1 mg/L or above, with no increase in both the area and number of fronds after 7 d of exposure.

As demonstrated in Figure 4D, 0.5 mg/L of the 3-keto-DON standard resulted in a considerable decrease of 72.2% in the number of fronds (6.3 ± 1.1 compared to 22.7 ± 6.1 fronds/well in the control) and a 77.7% decrease in area (0.19 ± 0.04 compared to 0.85 ± 0.24 cm^2/well in the control), both of which were statistically significant ($p < 0.05$). When compared to 0.5 mg/L of DON standard, the number and area of fronds were reduced by 52.6% (6.3 ± 1.1 vs. 13.3 ± 2.3 fronds/well) and 55.8% (0.19 ± 0.04 vs. 0.43 ± 0.07 cm^2/well), respectively, with statistical significance ($p < 0.05$). Furthermore, it was observed that only a small number of duckweed fronds were bleached when exposed to 3-keto-DON at 0.5 mg/L. In comparison, all of the fronds became bleached when exposed to 1 and 2 mg/L of the compound. However, the phenomenon of fronds becoming bleached did not occur at DON concentrations of 0.5 mg/L or higher.

Prior to the phytotoxicity experiment, UPLC analysis of MMSF+D3_3+DON revealed that DON had been completely converted to 3-keto-DON at a concentration of around 50 mg/L. Moreover, 100-fold, 50-fold, and 25-fold dilutions of the sample yielded 0.5, 1, and 2 mg/L of 3-keto-DON, respectively. Exposure of duckweed to the 100-fold dilution resulted in a 76.7% reduction in number of fronds (5.3 ± 2.3 vs. 22.7 ± 6.1 fronds/well in the control) and a 76.5% reduction in the area of fronds (0.2 ± 0.09 vs. 0.85 ± 0.24 cm^2/well in the control), while 50-fold and 25-fold dilutions completely inhibited growth. No significant differences in growth inhibition were observed between the 100-fold dilution of MMSF+D3_3+DON and 0.5 mg/L of the 3-keto-DON standard, or between the 50-fold and

25-fold dilutions and 1 and 2 mg/L of the 3-keto-DON standard, respectively ($p > 0.05$). According to these results, we can conclude that the phytotoxicity of 3-keto-DON seems to be greater than that of the same amount of DON.

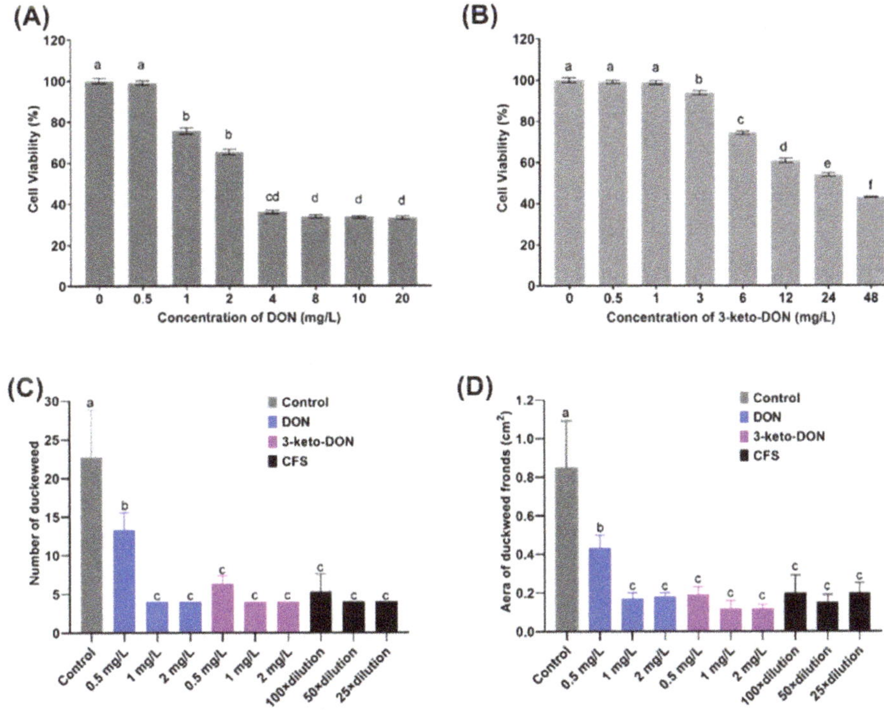

Figure 4. Toxicity evaluation of DON and 3-keto-DON. Effect of different concentrations of DON (**A**) and 3-keto-DON (**B**) on the viabilities of GES-1 cells. Phytotoxicity assessment of DON, 3-keto-DON, and cell-free supernatant of DON-degradation culture (CFS) on the number (**C**) and area (**D**) of duckweed fronds. The use of different lowercase letters (a–f) indicates that there are significant differences ($p < 0.05$) between the treatments being compared. Values that share the same letter are not significantly different.

2.4. Four PQQ-Dependent Alcohol Dehydrogenases Responsible for DON Transformation

To gain a greater understanding of *K. vulgare* strain D3_3 and screen its DON-oxidizing genes, its genome was sequenced, assembled, and then annotated. The analysis of D3_3's genome revealed that it comprised one circular chromosome and three plasmids, with a total length of 3,293,003 bp and a GC content of 61.36%. PGAP annotation predicted 3236 putative protein-coding genes, 5 rRNA operons, and 60 tRNAs (Table S1).

Currently, there are two different types of sequence-known enzymes that can oxidize the hydroxyl at C3 of DON into a keto group, namely an aldo-keto reductase AKR18A1 from *Sphingomonas* sp. S3-4 and two pyrroloquinoline quinone (PQQ)-dependent dehydrogenases DepA and QDAH from *D. mutans* 17-2-E-8 and *Devosia* sp. D6_9 [24,32,33]. Since DepA and QDAH share the same amino acid sequences, only DepA and AKR18A1 were used as query protein sequences to search for potential genes involved in DON oxidation against the resulting genome database. Based on BLASTp search outcomes, the genome contained fourteen potential genes for DON oxidation, comprising eight ADH-encoding genes and six AKR-encoding genes. The protein sequences encoded by these genes displayed amino acid sequence similarities to DepA and AKR18A1, ranging 24.8–57.8% and

28.2–34.7%, respectively (Tables S2 and S3). Furthermore, six of the eight dehydrogenase-encoding genes were located on a 2.8 Mb bacterial chromosome, whereas the other two were separately found on two 0.22 Mb megaplasmids of pP1 and pP2 (Table S2). Five of the six aldo-keto reductase-encoding genes were located on the bacterial chromosome, and the other one was found on megaplasmid pP1 (Table S3).

The signal peptide sequences were predicted and then deleted prior to heterogeneously expressing candidate enzymes. As predicted using SignalP 5.0, seven of the eight dehydrogenases possessed a signal sequence ranging from 21 to 24 amino acids in length, whereas all aldo/keto reductases did not possess it (Tables S2 and S3). Cloning, expression, and confirmation of enzymatic activity were performed on the fourteen candidate genes. As shown in Figure 5A, four of the eight ADHs, designated *Kv*ADH1, *Kv*ADH2, *Kv*ADH3, and *Kv*ADH4, yielded a single band at a molecular mass of 61.7, 61.2, 61.3, and 61.6 kDa, respectively. *Kv*ADH1, *Kv*ADH2, *Kv*ADH3, and *Kv*ADH4 showed 57.76%, 55.14%, 54.87%, and 55.86% amino acid similarity with DepA or QDDH, respectively; the amino acid sequence identities between the four ADHs obtained in this study ranged from 80.7 to 86.5% (Table S4). The four purified *Kv*ADHs exhibited catalytic activity toward DON in the presence of the cofactors PQQ and $CaCl_2$, as well as the artificial electron acceptor PMS (Figure 5B), but no activity was observed without PQQ being present. It is therefore suggested that PQQ is necessary for maintaining the enzymatic function of the four dehydrogenases. Additionally, none of the six AKRs had such enzymatic activity in the presence of NADH or NADPH. These results clearly demonstrated that the oxidation of DON to 3-keto-DON in the *K. vulgare* strain D3_3 was caused by the PQQ-dependent alcohol dehydrogenase rather than the aldo-keto reductase.

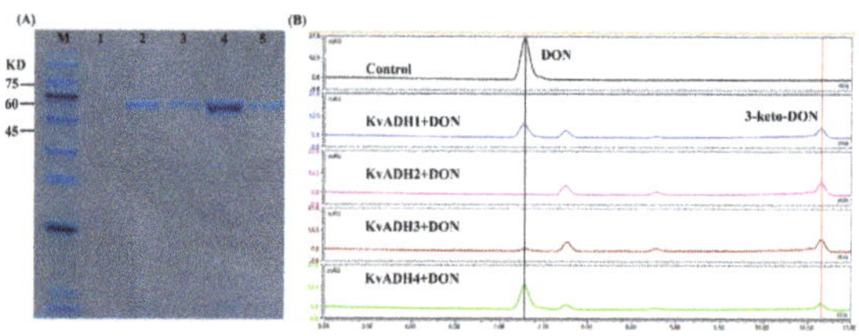

Figure 5. Molecular weight characterization and enzymatic activity confirmation of four recombinant PQQ-dependent alcohol dehydrogenases. (**A**) SDS-PAGE analysis of four Ni-affinity purified recombinant ADHs. M: molecular weight markers (10–180 KD); Lane 1: control; Lane 2: *Kv*ADH1; Lane 3: *Kv*ADH2; Lane 4: *Kv*ADH3; Lane 5: *Kv*ADH4. (**B**) LC profiles of 3-keto-DON produced via in vitro DON oxidation using four ADHs with PQQ, Ca^{2+}, and PMS present.

3. Discussion

Many species of DON-degrading microbes have so far been successfully isolated from a variety of sources, such as wheat leaf, wheat head, soil, lake water, seawater, various animal intestines, bovine rumen, and human milk [20,22,34–44]. Nevertheless, there have been no reports regarding the isolation of a DON-degrading strain from yellow mealworm. Herein, we successfully isolated a DON-degrading strain *K. vulgare* D3_3 from a single insect feces sample. Additionally, three DON-degrading *Devosia* strains were effectively screened from five other yellow mealworm feces samples collected from different geographical areas [45]. These results exemplify that yellow mealworm feces are a highly efficient source for the isolation of mycotoxin-degrading microorganisms.

Furthermore, this is the first report on a member of the genus *Ketogulonicigenium* as a potent DON-degrading microorganism. *K. vulgare*, the type species of the genus *Ketogulonicigenium*, was first isolated from soil and taxonomically characterized in 2001 [46]. It is currently used in the microbial production of a key intermediate for the industrial synthesis of vitamin C, namely 2-keto-L-gulonic acid [47]. Due to its unique nature and lack of many amino acid biosynthesis pathways, *K. vulgare* grew poorly, even on nutrient-rich media [48]. The isolate D3_3 also exhibited this poor growth phenotype, with a maximum optical density

(OD_{600}) of 0.13 (4×10^6 CFU/mL) in MMFS medium and only 0.25 in nutrient-rich TSB medium after 72 h incubation. However, it exhibited a highly efficient DON-degradation rate of 104.1 μg/h/per 10^7 cells, exceeding the degradation rates reported in earlier studies for *D. mutans* 17-2-E-8, *P. shaoguanens* is DDB001, and *Devosia* sp. D6-9, which had degradation rates of 0.3, 6.8, and 15.0 μg/h/per 10^7 cells, respectively [22,33,37]. In addition, the versatile DON-degrading capabilities of the facultative anaerobic strain D3_3 under aerobic or anaerobic conditions greatly expand its potential application scenarios.

3-keto-DON is a common intermediate or end metabolite in microbial DON degradation. Several strains of bacteria, including *Agrobacterium-Rhizobium* E3-39, *D. insulae* strain A16, and *P. halotolerans* ANSP101 can directly convert DON to 3-keto-DON [17,25,26]. Other strains, including *D. mutans* 17-2-E-8, *Devosia* sp. D6-9, *Lactobacillus rhamnosus* SHA113, *Nocardioides* sp. WSN05-2, *Nocardioides* sp. ZHH-013, and *Sphingomonas* sp. S3-4, could first convert DON to 3-keto-DON, which was then stereospecifically reduced to 3-epi-DON [23,32,33,37,42,43]. The absence of 3-epi-DON in the MMFS+DON+D3_3 sample indicated that *K. vulgare* D3_3 was incapable of reducing 3-keto-DON to 3-epi-DON, likely due to a lack of the responsible enzyme.

The bacterium *K. vulgare* D3_3 could transform DON to 3-keto-DON. However, before considering its potential use, it is essential to evaluate the toxicity of its transformation product. Currently, there is a scarcity of toxicity data for 3-keto-DON, with only two studies assessing its in vitro cytotoxicity on human colon cancer cells (Caco-2) and mouse spleen lymphocytes [17,49], and data on cytotoxicity against GES-1 cells are also lacking. In this study, the calculated IC_{50} values for DON and 3-keto-DON against GES-1 cells were 2.66 and 29.70 mg/L, respectively. However, in a previously described MTT (3-[4,5-dimethylthiazol-2-yl]-2,5 diphenyl tetrazolium bromide) bioassay assessing Caco-2 cell viability, the IC_{50} values for 3-keto-DON and DON were 1.24 mg/L and 0.409 mg/L, respectively [49]. It is considered that the discrepancy in IC_{50} values between the two studies mainly resulted from different cell lines (GES-1 vs. Caco-2) and assay methodologies (CKK-8 vs. MTT). In addition, taking into account the fact that the IC_{50} value (2.66 mg/L) of DON in this study is analogous to the IC_{50} value (2.99 mg/L) published in earlier studies [50] using the same cell lines and detection methods, the validity of the IC_{50} values of DON and 3-keto-DON in the current study should be verified. In summary, our study provides initial evidence that 3-keto-DON is less toxic to GES-1 compared to its parent mycotoxin DON, similarly to other cell lines.

PQQ-dependent alcohol dehydrogenases represent a class of oxidoreductases that utilize the cofactor PQQ to catalyze the oxidation of a variety of substrates, including alcohols and sugars [51–53]. Based on their tertiary structures, PQQ-dependent ADHs can be categorized into two families: six-bladed "propeller fold" and eight-bladed "propeller fold" [54]. The latter can be further classified into three types based on their location and structural characteristics: type I (periplasmic, dimeric), type II (periplasmic, monomeric, with a heme c group), and type III (membrane-bound, dimeric or trimeric, with multiple heme c groups) [55]. The crystal structures of many different types of PQQ-dependent ADHs have been determined, providing structural foundations for understanding the substrate–cofactor–enzyme binding and electron transfer mechanisms. Recently, the crystal structures of the first identified PQQ-dependent DON dehydrogenase, DepA (PDB:7WMD), and its complex with PQQ (PDB:7WMK), were determined, suggesting that DepA belongs to the Type I PQQ-dependent ADH [56]. These crystal structures, along with biochemical evidence, confirm the interactions between DepA, PQQ, and DON, and reveal a unique tyrosine residue crucial for substrate selection. In this study, four PQQ-dependent alcohol dehydrogenases (*Kv*ADHs) with DON oxidation activity in the genome of strain D3_3 were identified. These four enzymes exhibit high amino acid homology (54.87–57.76%) with DepA and 80.7–86.5% homology with each other. This suggests that the four *Kv*ADHs and DepA may share a similar catalytic mechanism, but with some differences in their amino acid sequences that could indicate differences in substrate specificity or other functional properties. Further research is required to elucidate the structure and function of the four

*Kv*ADHs, including studies which involve the crystallization and determination of their three-dimensional structures, as well as observations of their kinetic properties with DON and other alcohol substrates. In addition, some bacteria, such as *Pseudomonas*, methanotrophic, and methylotrophic bacteria, generally express different categories and even multiple PQQ-ADHs of the same type, demonstrating the importance of these enzymes for the metabolism of various alcohol substrates [52,57,58]. Within D3_3, there are eight putative PQQ-dependent dehydrogenases whose functional redundancy may enhance the adaptability of microorganisms in maintaining critical functions under fluctuating environmental conditions and diverse microbial communities.

Despite challenges such as low bacterial cell yields, high-cost cofactor PQQ requirements, and the weak robustness of wild enzymes, optimizing the cultivation conditions to improve bacterial cell yield and engineering the enzyme to enhance its robustness could help overcome these limitations and maximize the potential of the isolate and enzyme for industrial applications.

4. Conclusions

In this study, a novel DON-degrading bacterial strain, *K. vulgare* D3_3, was isolated from yellow mealworm feces and was found to completely degrade 50 mg/L of DON within 12 h under optimal conditions of pH 7.0–9.0, 30 °C, and anaerobic cultivation. The metabolite 3-keto-DON displayed lower cytotoxicity against GES-1 cells than its parent mycotoxin DON, but greater phytotoxicity toward duckweed. Additionally, four genes encoding PQQ-dependent dehydrogenases in the genome of isolate D3_3 were identified as being responsible for catalyzing the oxidation of DON. These findings suggest that the strain and enzyme have potential to be developed as detoxification agents to address DON contamination in food and animal feed, despite certain challenges, such as the low yields of bacterial cells, high-cost cofactor PQQ, and weak robustness of the wild-type enzyme.

5. Materials and Methods

5.1. Chemicals and Reagents

Standard DON and 3-keto-DON, both with a purity of 98%, were bought from Pribolab Pte. Ltd. (Qingdao, China) and TripleBond Corporation (Guelph, ON, Canada), respectively. The methanol and acetonitrile for ultra-performance liquid chromatography (UPLC) and UPLC coupled to quadrupole time-of-flight mass spectrometry (UPLC-Q-TOF-MS) analysis were of chromatographic grade and were purchased from Fisher Scientific International Inc. (Pittsburgh, PA, USA). All other chemicals and reagents were analytical-grade and obtained from various commercial sources.

5.2. Enrichment, Isolation, and Identification of DON-Degrading Microorganisms

5.2.1. Yellow Mealworm Feces Collection and Processing

Approximately 1 kg of yellow mealworms were bought from a pet supply market located in the Daxing District of Beijing. Upon arrival at the laboratory, they were placed in a pre-sterilized plastic container and fasted for 48 h, to empty their gastrointestinal tract of feed and produce an abundance of fecal pellets. These pellets were sifted from the yellow mealworms using a 40-mesh sieve and harvested with a sterile sampling bag. Then, 25 g of fresh feces was weighed and immersed in 500 mL of aseptic phosphate buffer (50 mM, pH 7) in a 2 L sterile Erlenmeyer flask. The mixture was homogenized via shaking for 30 min at 220 rpm and 30 °C, to make a 5% (w/v) fecal slurry. A small portion of this slurry served as the inoculum for subsequent enrichment culture experiments. The remaining slurry was centrifuged and subsequently filter sterilized. The resulting supernatant, namely the cell-free yellow mealworm fecal supernatant (FS), was stored at −20 °C for further use.

5.2.2. Enrichment and Isolation Procedures

For the enrichment and isolation of DON-degrading microorganisms, two different media, namely MMFS and PYM, were employed. The MMFS medium comprised

$Na_2HPO_4 \cdot 12H_2O$ (4.03 g/L), KH_2PO_4 (1 g/L), KNO_3 (0.59 g/L), $(NH_4)_2SO_4$ (0.5 g/L), $MgSO_4 \cdot 7H_2O$ (0.5 g/L), $CaCl_2$ (0.02 g/L), FS (10:90, *v*/*v*), and 1× trace metal mixture [59]; while the PYM medium contained peptone (10 g/L), yeast extract (2 g/L), and $MgSO_4 \cdot 7H_2O$ (1 g/L). For the enrichment procedure, 5 mL of each medium pre-spiked with 50ppm DON was inoculated with and without 500 μL fecal slurry as the treatment samples and control samples, respectively. These cultures were continuously shaken at 220 rpm under 30 °C for 5 d. After the initial incubation, two cultures were subcultured four times in the respective fresh media at an inoculum ratio of 1:100. Residual DON in the final enrichment cultures was assessed via UPLC, to ascertain DON-degrading capabilities. The positive culture was diluted and plated, and randomly picked single colonies were then cultivated individually in DON-containing liquid media for 3 d before being analyzed for residual DON via UPLC to obtain the DON-degrading microbial strain.

5.2.3. 16S rRNA-Based Phylogenetic Analysis of DON-Degrading Strain

A 16S rRNA-based phylogenetic analysis was conducted to determine the taxonomic position of the DON-degrading strain. Briefly, a partial 16S rRNA gene fragment was amplified via PCR using universal primers 27F (AGAGTTTGATYMTGGCTCAG) and 1492R (CGGYTACCTTGTTACGACTT). Then, the amplified product was sequenced and compared with sequences in the 16S ribosomal RNA database to obtain evolutionarily closely related 16S rRNA gene sequences using the BLASTn program. Finally, a phylogenetic tree was built using the maximum likelihood method with MEGA 11.0 software [60].

5.3. Effects of Various Factors on the DON-Degrading Activity of Strain D3_3

In a degradation system consisting of 2 mL of MMFS medium, 50 mg/L DON, and 8×10^5 CFU/mL of initial cell concentration (OD = 0.025), we investigated the effects on degradation activity as a result of four factors: incubation temperature (4, 18, 30, 37, and 42 °C); initial pH (5.0, 6.0, 7.0, 8.0, and 9.0); addition of 2% (*w*/*v*) sugar and sugar alcohol (glucose, fructose, mannitiol, sorbitol, sucrose, and trehalose); and cultivation conditions (shaking at 220 rpm, static, and anaerobic). During the investigation of the impact of a single factor, the other parameters were maintained constant at 30 °C, pH = 7.0, no sugar alcohol addition, and shaking at 220 rpm. A control group was established using MMFS with DON but without inoculum. Each treatment was conducted in triplicate and its residual DON was analyzed using the UPLC technique at 12 and 72 h. The DON residual rate (%) was calculated by measuring the peak area of DON (UV absorption at 220 nm) and using the following equation:

$$\text{DON residual rate} = \text{Asample}/\text{Acontorl} \times 100\% \quad (1)$$

where A_{sample} and $A_{contorl}$ are the values of the sample and control, respectively.

5.4. DON and Its Metabolite Analysis

5.4.1. DON Detection Using UPLC Technique

A Thermo Scientific Dionex UltiMate 3000 system comprising a quaternary RS pump, a column oven, an autosampler, a diode array detector, and a system controller software Chromeleon 7.2 was used for UPLC analysis. First, 2 μL of sample was injected into an Acquity BEH C18 column (1.7 μm, 100 mm × 2.1 mm) and analyzed under the conditions of 40 °C, 0.2 mL/min flow rate, and a 220 nm detection wavelength. The elution gradients of solvent A (water) and solvent B (acetonitrile) were used as follows: 0–6 min, gradient 5% to 25% B; 6–12 min, isocratic 25% B; 12–13 min gradient 25% to 5% B; 13–18 min, isocratic 5% B.

5.4.2. Analysis of DON Degradation Metabolite Using the UPLC-Q-TOF-MS Method

The DON metabolite was analyzed using a Waters Xevo G2-S quadrupole time-of-flight mass spectrometer with an electrospray ionization source (negative ion mode), coupled to a Waters Acquity UPLC system. Liquid chromatography was performed similarly to the DON detection method 5.4.1, with the exception of using a different gradient elution program with solvent A (water containing 5 mM ammonium formate) and solvent B (methanol) as follows: 0–1 min, isocratic 10% B; 1–19 min, gradient 10–90% B; 19–24 min, isocratic 90% B; 24–25 min, gradient 90–10% B; 25-30 min, isocratic 10% B. The ionization source conditions were set as follows: 2 kV capillary voltage; 450 °C desolvation temperature; 800 L/h desolvation gas (N_2); 120 °C source temperature; and 50 L/h cone gas (N_2).

5.5. *Toxicity Assay of DON and Its Metabolite*

5.5.1. In Vitro Cytotoxicity Assay Using GES-1

The in vitro toxicity of DON and its oxidation product (3-keto-DON) were investigated using GES-1 cells and a CKK-8 test. GES-1 cells were bought from Cobioer Biosciences Co., Ltd. (Nanjing, China), and maintained under the cultivation conditions reported by Yang et al. [50]. First, 10,000 cells/well were plated into 96-well plates and allowed to proliferate to 80% confluence before cells were collected and exposed to 3-keto-DON (0.5, 1, 3, 6, 12, 24, and 48 mg/L) and DON standard (0.5, 1, 2, 4, 8, 10, and 20 mg/L) for 24 h. Following that, the spent medium was exchanged with a fresh one containing 1 mg/mL of CCK-8, which was then incubated for 4 h. Finally, cell viability was determined using an ELISA reader to measure the optical density at 450 nm. The percentage inhibition compared to control-treated cells was calculated for each compound concentration. All analyses were performed in triplicate. Statistical differences were determined using one-way ANOVA, with $p < 0.05$ considered to be statistically significant. IC_{50} values were calculated using Dr Fit software version 1.042 [61] with default setting parameters.

5.5.2. Phytotoxicity Assay Using *Lemna minor*

To evaluate the phytotoxicity of 3-keto-DON, the growth of *Lemna minor* (common duckweed) was examined after exposure to test samples. Duckweed was collected from a pond in Daxing District, Beijing, China, disinfected, and acclimated to experimental conditions according to the method described by Megateli et al. [62]. Four duckweed fronds were aseptically inoculated into each well of 24-well plates preloaded with 2 mL of media supplemented with different test samples, including working concentrations of 100-fold, 50-fold, and 25-fold dilutions of MMFS+D3_3+DON (derived from cell-free culture supernatant of strain D3_3 grown for 3 d in MMFS medium supplemented with 50 mg/L of DON); 0.5, 1, and 2 mg/L of DON and 3-keto-DON standard solution; and no supplement. All treatments were performed in triplicate. After 7 d of incubation, the number of fronds was counted with a microscope. Each well was photographed, and the frond area was calculated using ImageJ software version 1.53p.

5.6. *Identification of DON-Oxidizing Enzyme in Strain D3_3*

5.6.1. Sequencing, Assembly and Annotation of D3_3' Genome, as well as Scouting Potential Genes Involved in DON Oxidation

Genome sequencing was carried out at Biomaker Technology Inc. (Beijing, China) using a combination of the Nanopore PromethION 48 system and the Illumina NovaSeq 6000 platform, and technical details on sequencing, assembly, and annotation can be found in our previous report [63].

To screen the candidate enzymes responsible for DON degradation, the two protein sequences of DepA (GenBank accession no. KFL25551.1) and AKR18A1 (GenBank accession no. ASY03293.1), which have been reported to have the catalytic function of oxidizing DON into 3-keto-DON, were individually used as queries against the genome sequence of D3_3 for homology search using BLASTp with an E-value of 10^{-6} using TBtools version 1.098769 [64]. Based on the results of the homology search, superfamily classifications for these candidate proteins were predicted using Superfamily 2.0 [65,66], and their signal peptides were predicted with SignalP-5.0 [67], prior to recombinant protein expression.

Multiple sequence alignment of all protein sequences was performed using Clustal Omega software version 1.2.2 [68].

5.6.2. Cloning, Expression, and Activity Assay for Potential DON-Oxidizing Enzymes

To verify whether the candidate enzymes suggested by the homology search results really had the function of degrading DON, we purified fourteen candidate enzymes heterologously overexpressed in *E. coli* and tested their DNA degradation function. For the DNA cloning experiment, the fourteen candidate genes encoding mature enzymes and linearized expression vector pET28a were amplified via PCR using Q5 High-Fidelity DNA Polymerase (NEB), and the detailed information of the PCR primers and cycling conditions are listed in Table S5. The fourteen DNA inserts were then individually cloned into linearized vector pET28a via the T5 exonuclease-dependent DNA assembly (TEDA) method [69], and the resulting ligation reaction mixtures were individually transformed into the expression host *E.coli* BL21(DE3)pLysS for expression as N-terminal 6-His-tag fusion proteins. Recombinant proteins were expressed in an autoinduction medium at 18 °C, according to the previously reported method [59], and purified using PureCube Ni-NTA Agarose (Cube Biotech), according to the manufacturer's instructions. Purified recombinant protein was assessed using SDS-PAGE and Coomassie Brilliant Blue R-250 staining. Regarding the enzymatic assay for alcohol dehydrogenase, 20 μL of each of the purified recombinant proteins was added to 180 μL of a reaction system containing 1mM Ca^{2+}, 100 μM PQQ·Na_2, 400 μM phenazine methosulfate (PMS), and 50 mg/L DON in 50 mM Tris-HCl buffer (pH 8). Aldo-keto reductase (AKR) activity was determined according to the method reported by He et al. [32]. After 1 h, enzyme-catalyzed reactions were stopped by adding 200 μL of methanol to the reaction systems followed by UPLC analysis.

Supplementary Materials: The following supporting information can be downloaded at: https://www.mdpi.com/article/10.3390/toxins15060367/s1, Table S1: Genomic characteristics of *Ketogulonicigenium vulgare* D3_3; Table S2: Eight candidate genes for oxidation of C3-OH group of DON; Table S3: Six candidate genes for oxidation of C3-OH group of DON; Table S4: Percent identity matrix of amino acid sequences created with the Clustal Omega program; Table S5: DNA primers and PCR conditions.

Author Contributions: Conceptualization, Y.W. (Yang Wang) and B.G.; methodology, Y.W. (Yang Wang), D.Z., W.Z., and S.W. (Songshan Wang); formal analysis, Y.W. (Yang Wang), W.Z., S.W. (Songshan Wang), and Y.W. (Yu Wu); investigation, Y.W. (Yang Wang); writing—original draft preparation, Y.W. (Yang Wang); writing—review and editing, Y.W. (Yang Wang), D.Z., and B.G.; supervision, S.W. (Songxue Wang), Y.Y., and B.G.; funding acquisition, Y.W. (Yang Wang) and B.G. All authors have read and agreed to the published version of the manuscript.

Funding: This research was funded by National Natural Science Foundation of China (Grant No. 31972605) and Key Research and Development Program of Ningxia, China (Grant No. 2022BBF03031).

Institutional Review Board Statement: Not applicable.

Informed Consent Statement: Not applicable.

Data Availability Statement: The full genomic sequence of *K. vulgare* D3_3 has been deposited in NCBI/GenBank under BioProject number (PRJNA875257) with the GenBank accession numbers (CP103997, CP103998, CP103999, and CP104000) and BioSample number (SAMN30609498). The 16S rRNA gene was deposited under the GenBank accession number OQ102971. Additionally, the data supporting the findings of this work are available within the paper and its Supplementary Information.

Conflicts of Interest: The authors declare no conflict of interest.

References

1. Kushiro, M. Effects of Milling and Cooking Processes on the Deoxynivalenol Content in Wheat. *Int. J. Mol. Sci.* **2008**, *9*, 2127–2145. [CrossRef]
2. Pestka, J.J. Deoxynivalenol: Toxicity, Mechanisms and Animal Health Risks. *Anim. Feed Sci. Technol.* **2007**, *137*, 283–298. [CrossRef]

3. Pestka, J.J.; Smolinski, A.T. Deoxynivalenol: Toxicology and Potential Effects on Humans. *J. Toxicol. Environ. Health Part B* **2005**, *8*, 39–69. [CrossRef] [PubMed]
4. Del Favero, G.; Janker, L.; Neuditschko, B.; Hohenbichler, J.; Kiss, E.; Woelflingseder, L.; Gerner, C.; Marko, D. Exploring the Dermotoxicity of the Mycotoxin Deoxynivalenol: Combined Morphologic and Proteomic Profiling of Human Epidermal Cells Reveals Alteration of Lipid Biosynthesis Machinery and Membrane Structural Integrity Relevant for Skin Barrier Function. *Arch. Toxicol.* **2021**, *95*, 2201–2221. [CrossRef] [PubMed]
5. Pestka, J.J. Deoxynivalenol: Mechanisms of Action, Human Exposure, and Toxicological Relevance. *Arch. Toxicol.* **2010**, *84*, 663–679. [CrossRef]
6. Rotter, B.A. Toxicology of Deoxynivalenol (vomitoxin). *J. Toxicol. Environ. Health* **1996**, *48*, 1–34. [CrossRef]
7. Sobrova, P.; Adam, V.; Vasatkova, A.; Beklova, M.; Zeman, L.; Kizek, R. Deoxynivalenol and Its Toxicity. *Interdiscip. Toxicol.* **2010**, *3*, 94–99. [CrossRef]
8. Murata, H.; Yamaguchi, D.; Nagai, A.; Shimada, N. Reduction of Deoxynivalenol Contaminating Corn Silage by Short-Term Ultraviolet Irradiation: A Pilot Study. *J. Vet. Med. Sci.* **2011**, *73*, 1059–1060. [CrossRef]
9. Park, B.J.; Takatori, K.; Sugita-Konishi, Y.; Kim, I.-H.; Lee, M.-H.; Han, D.-W.; Chung, K.-H.; Hyun, S.O.; Park, J.-C. Degradation of Mycotoxins Using Microwave-Induced Argon Plasma at Atmospheric Pressure. *Surf. Coat. Technol.* **2007**, *201*, 5733–5737. [CrossRef]
10. Savi, G.D.; Piacentini, K.C.; Scussel, V.M. Ozone Treatment Efficiency in *Aspergillus* and *Penicillium* Growth Inhi- bition and Mycotoxin Degradation of Stored Wheat Grains (*Triticum aestivum* L.): Ozone Treatment Efficiency on Fungi and Mycotoxins. *J. Food Process Pres.* **2015**, *39*, 940–948. [CrossRef]
11. Wang, L.; Luo, Y.; Luo, X.; Wang, R.; Li, Y.; Li, Y.; Shao, H.; Chen, Z. Effect of Deoxynivalenol Detoxification by Ozone Treatment in Wheat Grains. *Food Control* **2016**, *66*, 137–144. [CrossRef]
12. Zhang, C.; Zhuang, K.; Chen, L.; Ding, W.; Chen, X.; Shao, L.; Wang, G.; Wang, Y. Reducing Deoxynivalenol Con- tent in Wheat by a Combination of Gravity Separation and Milling and Characterization of the Flours Produced. *J. Cereal Sci.* **2022**, *104*, 103372. [CrossRef]
13. Matumba, L.; Van Poucke, C.; Njumbe Ediage, E.; Jacobs, B.; De Saeger, S. Effectiveness of Hand Sorting, Flotation/Washing, Dehulling and Combinations Thereof on the Decontamination of Mycotoxin-Contaminated White Maize. *Food Addit. Contam. Part A* **2015**, *32*, 960–969. [CrossRef]
14. Trenholm, H.L.; Charmley, L.L.; Prelusky, D.B.; Warner, R.M. Washing Procedures Using Water or Sodium Carbonate Solutions for the Decontamination of Three Cereals Contaminated with Deoxynivalenol and Zearalenone. *J. Agric. Food. Chem.* **1992**, *40*, 2147–2151. [CrossRef]
15. Yumbe-Guevara, B.E.; Imoto, T.; Yoshizawa, T. Effects of Heating Procedures on Deoxynivalenol, Nivalenol and Zearalenone Levels in Naturally Contaminated Barley and Wheat. *Food Addit. Contam.* **2003**, *20*, 1132–1140. [CrossRef] [PubMed]
16. Sundstøl Eriksen, G.; Pettersson, H.; Lundh, T. Comparative Cytotoxicity of Deoxynivalenol, Nivalenol, Their Acetylated Derivatives and de-Epoxy Metabolites. *Food Chem. Toxicol.* **2004**, *42*, 619–624. [CrossRef]
17. Shima, J.; Takase, S.; Takahashi, Y.; Iwai, Y.; Fujimoto, H.; Yamazaki, M.; Ochi, K. Novel Detoxification of the Trichothecene Mycotoxin Deoxynivalenol by a Soil Bacterium Isolated by Enrichment Culture. *Appl. Environ. Microb.* **1997**, *63*, 3825–3830. [CrossRef]
18. He, J.W.; Bondy, G.S.; Zhou, T.; Caldwell, D.; Boland, G.J.; Scott, P.M. Toxicology of 3-Epi-Deoxynivalenol, a De oxynivalenol-Transformation Product by *Devosia mutans* 17-2-E-8. *Food Chem. Toxicol.* **2015**, *84*, 250–259. [CrossRef] [PubMed]
19. Mayer, E.; Novak, B.; Springler, A.; Schwartz-Zimmermann, H.E.; Nagl, V.; Reisinger, N.; Hessenberger, S.; Schatzmayr, G. Effects of Deoxynivalenol (DON) and Its Microbial Biotransformation Product Deepoxy-Deoxynivalenol (DOM-1) on a Trout, Pig, Mouse, and Human Cell Line. *Mycotoxin Res.* **2017**, *33*, 297–308. [CrossRef]
20. Ikunaga, Y.; Sato, I.; Grond, S.; Numaziri, N.; Yoshida, S.; Yamaya, H.; Hiradate, S.; Hasegawa, M.; Toshima, H.; Koitabashi, M.; et al. Nocardioides Sp. Strain WSN05-2, Isolated from a Wheat Field, Degrades Deoxynivalenol, Producing the Novel Intermediate 3-Epi-Deoxynivalenol. *Appl. Microbiol. Biotechnol.* **2011**, *89*, 419–427. [CrossRef] [PubMed]
21. Carere, J.; Hassan, Y.I.; Lepp, D.; Zhou, T. The Identification of DepB: An Enzyme Responsible for the Final De- toxification Step in the Deoxynivalenol Epimerization Pathway in *Devosia mutans* 17-2-E-8. *Front. Microbiol.* **2018**, *9*, 1573. [CrossRef] [PubMed]
22. Wang, Y.; Zhang, H.H.; Zhao, C.; Han, Y.T.; Liu, Y.C.; Zhang, X.L. Isolation and Characterization of a Novel De- oxynivalenol-Transforming Strain *Paradevosia shaoguanensis* DDB001 from Wheat Field Soil. *Lett. Appl. Microbiol.* **2017**, *65*, 414–422. [CrossRef] [PubMed]
23. Zhang, H.; Zhang, H.; Qin, X.; Wang, X.; Wang, Y.; Bin, Y.; Xie, X.; Zheng, F.; Luo, H. Biodegradation of Deoxyni- valenol by *Nocardioides* sp. ZHH-013: 3-Keto-Deoxynivalenol and 3-Epi-Deoxynivalenol as Intermediate Products. *Front. Microbiol.* **2021**, *12*, 658421. [CrossRef] [PubMed]
24. Carere, J.; Hassan, Y.I.; Lepp, D.; Zhou, T. The Enzymatic Detoxification of the Mycotoxin Deoxynivalenol: Iden- tification of DepA from the DON Epimerization Pathway. *Microb. Biotechnol.* **2018**, *11*, 1106–1111. [CrossRef] [PubMed]
25. Wang, G.; Wang, Y.; Ji, F.; Xu, L.; Yu, M.; Shi, J.; Xu, J. Biodegradation of Deoxynivalenol and Its Derivatives by *Devosia insulae* A16. *Food Chem.* **2019**, *276*, 436–442. [CrossRef]
26. Zhang, J.; Qin, X.; Guo, Y.; Zhang, Q.; Ma, Q.; Ji, C.; Zhao, L. Enzymatic Degradation of Deoxynivalenol by a Novel Bacterium, *Pelagibacterium halotolerans* ANSP101. *Food Chem. Toxicol.* **2020**, *140*, 111276. [CrossRef]

27. Grau, T.; Vilcinskas, A.; Joop, G. Sustainable Farming of the Mealworm *Tenebrio molitor* for the Production of Food and Feed. *Z. Für Nat. C* **2017**, *72*, 337–349. [CrossRef]
28. Janković-Tomanić, M.; Petković, B.; Todorović, D.; Vranković, J.; Perić-Mataruga, V. Physiological and Behavioral Effects of the Mycotoxin Deoxynivalenol in *Tenebrio molitor* Larvae. *J. Stored Prod. Res.* **2019**, *83*, 236–242. [CrossRef]
29. Ochoa Sanabria, C.; Hogan, N.; Madder, K.; Gillott, C.; Blakley, B.; Reaney, M.; Beattie, A.; Buchanan, F. Yellow Mealworm Larvae (*Tenebrio molitor*) Fed Mycotoxin-Contaminated Wheat—A Possible Safe, Sustainable Protein Source for Animal Feed? *Toxins* **2019**, *11*, 282. [CrossRef]
30. Van Broekhoven, S.; Gutierrez, J.M.; De Rijk, T.C.; De Nijs, W.C.M.; Van Loon, J.J.A. Degradation and Excretion of the *Fusarium* Toxin Deoxynivalenol by an Edible Insect, the Yellow Mealworm (*Tenebrio molitor* L.). *World Mycotoxin J.* **2017**, *10*, 163–169. [CrossRef]
31. Jain, C.; Rodriguez-R, L.M.; Phillippy, A.M.; Konstantinidis, K.T.; Aluru, S. High Throughput ANI Analysis of 90K Prokaryotic Genomes Reveals Clear Species Boundaries. *Nat. Commun.* **2018**, *9*, 5114. [CrossRef]
32. He, W.-J.; Zhang, L.; Yi, S.-Y.; Tang, X.-L.; Yuan, Q.-S.; Guo, M.-W.; Wu, A.-B.; Qu, B.; Li, H.-P.; Liao, Y.-C. An Aldo-Keto Reductase Is Responsible for *Fusarium* Toxin-Degrading Activity in a Soil *Sphingomonas* Strain. *Sci. Rep.* **2017**, *7*, 9549. [CrossRef]
33. He, W.-J.; Shi, M.-M.; Yang, P.; Huang, T.; Zhao, Y.; Wu, A.-B.; Dong, W.-B.; Li, H.-P.; Zhang, J.-B.; Liao, Y.-C. A Quinone-Dependent Dehydrogenase and Two NADPH-Dependent Aldo/Keto Reductases Detoxify Deoxynivalenol in Wheat via Epimerization in a *Devosia* Strain. *Food Chem.* **2020**, *321*, 126703. [CrossRef]
34. Fuchs, E.; Binder, E.M.; Heidler, D.; Krska, R. Structural Characterization of Metabolites after the Microbial Deg- radation of Type A Trichothecenes by the Bacterial Strain BBSH 797. *Food Addit. Contam.* **2002**, *19*, 379–386. [CrossRef] [PubMed]
35. Gao, X.; Mu, P.; Wen, J.; Sun, Y.; Chen, Q.; Deng, Y. Detoxification of Trichothecene Mycotoxins by a Novel Bacterium, *Eggerthella* sp. DII-9. *Food Chem. Toxicol.* **2018**, *112*, 310–319. [CrossRef] [PubMed]
36. Gao, X.; Mu, P.; Zhu, X.; Chen, X.; Tang, S.; Wu, Y.; Miao, X.; Wang, X.; Wen, J.; Deng, Y. Dual Function of a Novel Bacterium, *Slackia* sp. D-G6: Detoxifying Deoxynivalenol and Producing the Natural Estrogen Analogue, Equol. *Toxins* **2020**, *12*, 85. [CrossRef] [PubMed]
37. He, J.W.; Hassan, Y.I.; Perilla, N.; Li, X.-Z.; Boland, G.J.; Zhou, T. Bacterial Epimerization as a Route for Deoxyni- valenol Detoxification: The Influence of Growth and Environmental Conditions. *Front. Microbiol.* **2016**, *7*, 572. [CrossRef] [PubMed]
38. Ito, M.; Sato, I.; Koitabashi, M.; Yoshida, S.; Imai, M.; Tsushima, S. A Novel Actinomycete Derived from Wheat Heads Degrades Deoxynivalenol in the Grain of Wheat and Barley Affected by *Fusarium* Head Blight. *Appl. Microbiol. Biotechnol.* **2012**, *96*, 1059–1070. [CrossRef]
39. Ito, M.; Sato, I.; Ishizaka, M.; Yoshida, S.; Koitabashi, M.; Yoshida, S.; Tsushima, S. Bacterial Cytochrome P450 System Catabolizing the *Fusarium* Toxin Deoxynivalenol. *Appl. Environ. Microb.* **2013**, *79*, 1619–1628. [CrossRef]
40. Jia, R.; Cao, L.; Liu, W.; Shen, Z. Detoxification of Deoxynivalenol by *Bacillus subtilis* ASAG 216 and Characteri- zation the Degradation Process. *Eur. Food Res. Technol.* **2021**, *247*, 67–76. [CrossRef]
41. Li, B.; Duan, J.; Ren, J.; Francis, F.; Li, G. Isolation and Characterization of Two New Deoxynivalenol-Degrading Strains, *Bacillus* sp. HN117 and *Bacillus* sp. N22. *Toxins* **2022**, *14*, 781. [CrossRef] [PubMed]
42. Qu, R.; Jiang, C.; Wu, W.; Pang, B.; Lei, S.; Lian, Z.; Shao, D.; Jin, M.; Shi, J. Conversion of DON to 3- *Epi* -DON in Vitro and Toxicity Reduction of DON *in Vivo* by *Lactobacillus rhamnosus. Food Funct.* **2019**, *10*, 2785–2796. [CrossRef] [PubMed]
43. Sato, I.; Ito, M.; Ishizaka, M.; Ikunaga, Y.; Sato, Y.; Yoshida, S.; Koitabashi, M.; Tsushima, S. Thirteen Novel Deox- ynivalenol-Degrading Bacteria Are Classified within Two Genera with Distinct Degradation Mechanisms. *FEMS Microbiol. Lett.* **2012**, *327*, 110–117. [CrossRef] [PubMed]
44. Wang, S.; Hou, Q.; Guo, Q.; Zhang, J.; Sun, Y.; Wei, H.; Shen, L. Isolation and Characterization of a Deoxynivalenol-Degrading Bacterium *Bacillus licheniformis* YB9 with the Capability of Modulating Intestinal Microbial Flora of Mice. *Toxins* **2020**, *12*, 184. [CrossRef]
45. Wang, Y.; Zhao, D.L.; Guo, B.Y. Biodegradation of Deoxynivalenol by Two-Member Bacterial Consortia Isolated from *Tenebrio molitor* Faeces. *Int. Biodeter. Biodegr.* **2023**. *in preparation*.
46. Urbance, J.W.; Bratina, B.J.; Stoddard, S.F.; Schmidt, T.M. Taxonomic Characterization of *Ketogulonigenium vulgare* Gen. Nov., Sp. Nov. and *Ketogulonigenium robustum* sp. Nov., Which Oxidize L-Sorbose to 2-Keto-L-Gulonic Acid. *Int. J. Syst. Evol. Micr.* **2001**, *51*, 1059–1070. [CrossRef]
47. Sugisawa, T.; Miyazaki, T.; Hoshino, T. Microbial Production of L-Ascorbic Acid from D-Sorbitol, L-Sorbose, L-Gulose, and L-Sorbosone by *Ketogulonicigenium vulgare* DSM 4025. *Biosci. Biotechnol. Biochem.* **2005**, *69*, 659–662. [CrossRef]
48. Zhang, J.; Zhou, J.; Liu, J.; Chen, K.; Liu, L.; Chen, J. Development of Chemically Defined Media Supporting High Cell Density Growth of *Ketogulonicigenium vulgare* and *Bacillus megaterium. Bioresour. Technol.* **2011**, *102*, 4807–4814. [CrossRef]
49. He, J.W. Detoxification of Deoxynivalenol by a Soil Bacterium *Devosia mutans* 17-2-E-8. Ph.D. Thesis, University of Guelph, Guelph, ON, Canada, 2015.
50. Yang, Y.; Yu, S.; Tan, Y.; Liu, N.; Wu, A. Individual and Combined Cytotoxic Effects of Co-Occurring Deoxyniva- lenol Family Mycotoxins on Human Gastric Epithelial Cells. *Toxins* **2017**, *9*, 96. [CrossRef]
51. Takeda, K.; Matsumura, H.; Ishida, T.; Samejima, M.; Ohno, H.; Yoshida, M.; Igarashi, K.; Nakamura, N. Charac- terization of a Novel PQQ-Dependent Quinohemoprotein Pyranose Dehydrogenase from *Coprinopsis cinerea* Classified into Auxiliary Activities Family 12 in Carbohydrate-Active Enzymes. *PLoS ONE* **2015**, *10*, e0115722. [CrossRef]

52. Good, N.M.; Vu, H.N.; Suriano, C.J.; Subuyuj, G.A.; Skovran, E.; Martinez-Gomez, N.C. Pyrroloquinoline Quinone Ethanol Dehydrogenase in *Methylobacterium extorquens* AM1 Extends Lanthanide-Dependent Metabolism to Multicarbon Substrates. *J. Bacteriol.* **2016**, *198*, 3109–3118. [CrossRef]
53. Chattopadhyay, A.; Förster-Fromme, K.; Jendrossek, D. PQQ-Dependent Alcohol Dehydrogenase (QEDH) of *Pseudomonas Aeruginosa* Is Involved in Catabolism of Acyclic Terpenes. *J. Basic Microb.* **2010**, *50*, 119–124. [CrossRef]
54. Sarmiento-Pavía, P.D.; Sosa-Torres, M.E. Bioinorganic Insights of the PQQ-Dependent Alcohol Dehydrogenases. *J. Biol. Inorg. Chem.* **2021**, *26*, 177–203. [CrossRef] [PubMed]
55. Toyama, H.; Mathews, F.S.; Adachi, O.; Matsushita, K. Quinohemoprotein Alcohol Dehydrogenases: Structure, Function, and Physiology. *Arch. Biochem. Biophys.* **2004**, *428*, 10–21. [CrossRef] [PubMed]
56. Yang, H.; Yan, R.; Li, Y.; Lu, Z.; Bie, X.; Zhao, H.; Lu, F.; Chen, M. Structure–Function Analysis of a Quinone-Dependent Dehydrogenase Capable of Deoxynivalenol Detoxification. *J. Agric. Food. Chem.* **2022**, *70*, 6764–6774. [CrossRef] [PubMed]
57. Keltjens, J.T.; Pol, A.; Reimann, J.; Op den Camp, H.J.M. PQQ-Dependent Methanol Dehydrogenases: Rare-Earth Elements Make a Difference. *Appl. Microbiol. Biotechnol.* **2014**, *98*, 6163–6183. [CrossRef]
58. Toyama, H.; Fujii, A.; Matsushita, K.; Shinagawa, E.; Ameyama, M.; Adachi, O. Three Distinct Quinoprotein Al- cohol Dehydrogenases Are Expressed When *Pseudomonas putida* Is Grown on Different Alcohols. *J. Bacteriol.* **1995**, *177*, 2442–2450. [CrossRef]
59. Studier, F.W. Protein Production by Auto-Induction in High-Density Shaking Cultures. *Protein Expr. Purif.* **2005**, *41*, 207–234. [CrossRef]
60. Tamura, K.; Stecher, G.; Kumar, S. MEGA11: Molecular Evolutionary Genetics Analysis Version 11. *Mol. Biol. Evol.* **2021**, *38*, 3022–3027. [CrossRef]
61. Di Veroli, G.Y.; Fornari, C.; Goldlust, I.; Mills, G.; Koh, S.B.; Bramhall, J.L.; Richards, F.M.; Jodrell, D.I. An Auto mated Fitting Procedure and Software for Dose-Response Curves with Multiphasic Features. *Sci. Rep.* **2015**, *5*, 14701. [CrossRef]
62. Megateli, S.; Dosnon-Olette, R.; Trotel-Aziz, P.; Geffard, A.; Semsari, S.; Couderchet, M. Simultaneous Effects of Two Fungicides (Copper and Dimethomorph) on Their Phytoremediation Using Lemna Minor. *Ecotoxicology* **2013**, *22*, 683–692. [CrossRef]
63. Wang, Y.; Wang, S. Complete Genome Sequence of *Paradevosia shaoguanensis* Type Strain J5-3, Obtained Using Nanopore and Illumina Sequencing Technologies. *Microbiol. Resour. Ann.* **2021**, *10*, e00099-21. [CrossRef] [PubMed]
64. Chen, C.; Chen, H.; Zhang, Y.; Thomas, H.R.; Frank, M.H.; He, Y.; Xia, R. TBtools: An Integrative Toolkit Devel- oped for Interactive Analyses of Big Biological Data. *Mol. Plant* **2020**, *13*, 1194–1202. [CrossRef] [PubMed]
65. Gough, J.; Karplus, K.; Hughey, R.; Chothia, C. Assignment of Homology to Genome Sequences Using a Library of Hidden Markov Models That Represent All Proteins of Known Structure. *J. Mol. Biol.* **2001**, *313*, 903–919. [CrossRef]
66. Pandurangan, A.P.; Stahlhacke, J.; Oates, M.E.; Smithers, B.; Gough, J. The SUPERFAMILY 2.0 Database: A Sig- nificant Proteome Update and a New Webserver. *Nucleic Acids Res.* **2019**, *47*, D490–D494. [CrossRef]
67. Almagro Armenteros, J.J.; Tsirigos, K.D.; Sønderby, C.K.; Petersen, T.N.; Winther, O.; Brunak, S.; Von Heijne, G.; Nielsen, H. SignalP 5.0 Improves Signal Peptide Predictions Using Deep Neural Networks. *Nat. Biotechnol.* **2019**, *37*, 420–423. [CrossRef]
68. Madeira, F.; Pearce, M.; Tivey, A.R.N.; Basutkar, P.; Lee, J.; Edbali, O.; Madhusoodanan, N.; Kolesnikov, A.; Lopez, R. Search and Sequence Analysis Tools Services from EMBL-EBI in 2022. *Nucleic Acids Res.* **2022**, *50*, W276–W279. [CrossRef] [PubMed]
69. Xia, Y.; Li, K.; Li, J.; Wang, T.; Gu, L.; Xun, L. T5 Exonuclease-Dependent Assembly Offers a Low-Cost Method for Efficient Cloning and Site-Directed Mutagenesis. *Nucleic Acids Res.* **2019**, *47*, e15. [CrossRef]

toxins

MDPI

Article

Dietary Catalase Supplementation Alleviates Deoxynivalenol-Induced Oxidative Stress and Gut Microbiota Dysbiosis in Broiler Chickens

Weiwei Wang, Jingqiang Zhu, Qingyun Cao, Changming Zhang, Zemin Dong, Dingyuan Feng, Hui Ye * and Jianjun Zuo *

Guangdong Provincial Key Laboratory of Animal Nutrition Control, College of Animal Science, South China Agricultural University, Guangzhou 510642, China
* Correspondence: magicsmall@scau.edu.cn (H.Y.); zuoj@scau.edu.cn (J.Z.)

Abstract: Catalase (CAT) can eliminate oxygen radicals, but it is unclear whether exogenous CAT can protect chickens against deoxynivalenol (DON)-induced oxidative stress. This study aimed to investigate the effects of supplemental CAT on antioxidant property and gut microbiota in DON-exposed broilers. A total of 144 one-day-old Lingnan yellow-feathered male broilers were randomly divided into three groups (six replicates/group): control, DON group, and DON + CAT (DONC) group. The control and DON group received a diet without and with DON contamination, respectively, while the DONC group received a DON-contaminated diet with 200 U/kg CAT added. Parameter analysis was performed on d 21. The results showed that DON-induced liver enlargement ($p < 0.05$) was blocked by CAT addition, which also normalized the increases ($p < 0.05$) in hepatic oxidative metabolites contents and caspase-9 expression. Additionally, CAT addition increased ($p < 0.05$) the jejunal CAT and GSH-Px activities coupled with T-AOC in DON-exposed broilers, as well as the normalized DON-induced reductions ($p < 0.05$) of jejunal villus height (VH) and its ratio for crypt depth. There was a difference ($p < 0.05$) in gut microbiota among groups. The DON group was enriched ($p < 0.05$) with some harmful bacteria (e.g., *Proteobacteria*, *Gammaproteobacteria*, *Enterobacteriales*, *Enterobacteriaceae*, and *Escherichia/Shigella*) that elicited negative correlations ($p < 0.05$) with jejunal CAT activity, and VH. DONC group was differentially enriched ($p < 0.05$) with certain beneficial bacteria (e.g., *Acidobacteriota*, *Anaerofustis*, and *Anaerotruncus*) that could benefit intestinal antioxidation and morphology. In conclusion, supplemental CAT alleviates DON-induced oxidative stress and intestinal damage in broilers, which can be associated with its ability to improve gut microbiota, aside from its direct oxygen radical-scavenging activity.

Keywords: antioxidant property; catalase; deoxynivalenol; gut microbiota; intestinal health

Key Contribution: Supplemental 200 U/kg catalase is beneficial in attenuating deoxynivalenol-induced oxidative stress and intestinal damage in broilers.

Citation: Wang, W.; Zhu, J.; Cao, Q.; Zhang, C.; Dong, Z.; Feng, D.; Ye, H.; Zuo, J. Dietary Catalase Supplementation Alleviates Deoxynivalenol-Induced Oxidative Stress and Gut Microbiota Dysbiosis in Broiler Chickens. *Toxins* **2022**, *14*, 830. https://doi.org/10.3390/toxins14120830

Received: 20 September 2022
Accepted: 25 November 2022
Published: 28 November 2022

1. Introduction

Deoxynivalenol (DON), also known as vomitoxin, is a secondary metabolite of *Fusarium graminearum*. As one of the most severe mycotoxins prevalent in multifarious crops, especially grains [1,2], DON can elicit serious detriments to animal growth and health when diets are prepared with DON-contaminated grains [3,4]. Animals exposed to DON may exhibit acute or chronic poisoning, as manifested by the structural and functional injuries of multiple organs (e.g., liver, kidney, and intestine), accompanied with a series of clinical symptoms, such as vomiting, anorexia, diarrhea, and intestinal bleeding [5,6]. Despite the lesser susceptibility to DON exposure than other monogastric animals (e.g., pigs), poultry exposed to DON can still display metabolic abnormalities and health disorders [5,6].

One of the toxicological mechanisms of DON for animals is to cause mitochondrial dysfunction, leading to increased generation of reactive oxygen radicals, concurrent with decreased production of antioxidants in the cells [6,7], which can break the redox balance inside the body [8]. Once the accumulation of oxygen radicals exceeds their elimination, there can be oxidative stress responses within the body with a peroxidation and destruction of certain biomacromolecules, especially DNA and proteins [8], subsequently impairing animal growth and health [6,8]. Thereby, suppression of the inducive oxidative stress is a potential strategy to attenuate the toxic effects of DON in chickens.

Antioxidant enzymes play crucial roles in contributing to the scavenging actions of antioxidant system on various free radicals generated in the body [9]. Among the antioxidant enzymes, catalase (CAT) is a key member capable of catalyzing the decomposition of hydrogen peroxide (H_2O_2), which is a kind of reactive oxygen radicals and may trigger oxidative stress of the body when it exceeds the physiologic concentration [10]. More importantly, CAT can prevent reactions of H_2O_2 with oxygen under the action of iron chelates from producing more toxic hydroxyl radicals (·OH) and some other radicals [11]. Additionally, certain hydrogen donors, such as methanol and ethanol, may also be scavenged during the catalytic reaction of CAT, which can benefit the mitigation of oxidative stress [11,12]. Previous reports have shown that dietary CAT addition resulted in an enhancement of antioxidant property, with simultaneous improvements of growth and health performance in both broilers and pigs [13,14]. Moreover, CAT addition was reported to alleviate oxidative stress-induced intestinal and hepatic damages in pigs [15–17]. However, few studies are available concerning the potentially beneficial roles of CAT in alleviating the detriments of DON to chickens.

Gut microbiota elicit profound roles in mediating the impacts of dietary treatments on chicken growth and health [18]. There is evidence that dietary mycotoxin, including DON contamination-induced toxicity and oxidative damages in animals, were at least partially realized by the inducive disturbance of gut microbiota [19–21]. Growing studies implied that CAT addition had an ability to enhance antioxidant capacity, as well as improve intestinal and hepatic health, through associating with an optimization of gut microbial composition in both broilers [14] and pigs [16,22]. Nevertheless, the protective effects of CAT on the antioxidant property and gut microbiota in chickens exposed to DON remain unclear. Accordingly, the present study aimed to investigate the potential roles of supplemental CAT in alleviating DON-induced oxidative stress and gut microbiota dysbiosis in broiler chickens.

2. Results

2.1. Growth Performance and Organ Indexes

As shown in Table 1, there were no differences ($p > 0.10$) in the initial body weight (IBW) and final body weight (FBW), average daily gain (ADG), and average daily feed intake (ADFI) coupled with feed conversion ratio (FCR) among groups. However, the DON group showed a decreasing trend ($p = 0.093$) of average daily feed intake (ADFI), relative to the control or DONC groups. Regarding the organ indexes (Table 2), the indexes of spleen and bursa of Fabricius showed no differences ($p > 0.10$) among groups. The liver index and kidney index in the DONC group were lower ($p < 0.05$) and tended to be higher ($p = 0.078$), respectively, than those of the DON group, but similar to ($p > 0.10$) the control group.

Table 1. Effect of catalase on growth performance [1] of broilers exposed to deoxynivalenol (DON).

	Treatments [2]			*p*-Value
	Control	DON	DONC	
IBW (g)	33.60 ± 0.21	33.50 ± 0.13	33.51 ± 0.17	0.593
FBW (g)	319.47 ± 22.41	309.83 ± 15.06	313.39 ± 13.59	0.635
ADG (g)	15.97 ± 1.12	15.49 ± 0.75	15.67 ± 0.68	0.636

Table 1. *Cont.*

	Treatments [2]			*p*-Value
	Control	DON	DONC	
ADFI (g)	31.53 ± 1.56	29.54 ± 1.38	30.77 ± 1.49	0.093
FCR	1.98 ± 0.12	1.91 ± 0.15	1.96 ± 0.13	0.676

[1] IBW, initial body weight; FBW, final body weight; ADG, average daily gain; ADFI, average daily feed intake; FCR, feed conversion ratio. [2] Broilers in control and DON group were fed a basal diet without and with DON contamination, respectively, while those in DONC group were fed a DON-contaminated basal diet supplemented with 200 U/kg catalase.

Table 2. Effect of catalase on organ indexes [1] of broilers exposed to DON.

	Treatments [2]			*p*-Value
	Control	DON	DONC	
Spleen (g/kg)	1.30 ± 0.33	1.52 ± 0.25	1.18 ± 0.45	0.419
Liver (g/kg)	23.66 ± 1.77 [b]	28.56 ± 1.81 [a]	25.16 ± 2.58 [b]	0.016
Bursa of Fabricius (g/kg)	3.28 ± 0.96	3.21 ± 0.31	3.16 ± 0.41	0.957
Kidney (g/kg)	9.56 ± 0.71	8.46 ± 0.77	9.19 ± 0.95	0.078

[a,b] Values within a row with unlike superscript letters differ significantly ($p < 0.05$). [1] Organ indexes were calculated as the ratio of organ weight (g) to body weight (kg). [2] Broilers in control and DON group were fed a basal diet without and with DON contamination, respectively, while those in DONC group were fed a DON-contaminated basal diet supplemented with 200 U/kg catalase.

2.2. Oxidative Status of the Intestine and Liver

As shown in Table 3, the contents of oxidative metabolites, including reactive oxide species (ROS), superoxide anion (O_2^-), hydroxyl radical (OH), 8-hydroxy-2′-deoxyguanosine (8-OHdG), and malondialdehyde (MDA) in the jejunum, did not differ ($p > 0.05$) among groups. However, the DON group showed increases ($p < 0.05$) in hepatic ROS, 8-OHdG, and MDA contents, relative to the control group. Supplemental CAT normalized hepatic ROS and 8-OHdG contents in DON-exposed birds to levels similar ($p > 0.05$) to the control group, and it abolished ($p < 0.05$) the DON-induced increase in hepatic MDA content. Regarding the antioxidant indicators, jejunal CAT and GSH-Px activities, along with hepatic T-AOC, were detected to be lower ($p < 0.05$) in the DON group versus control group. However, supplementing CAT to DON-exposed birds attenuated ($p < 0.05$) the decreased activities of jejunal CAT and GSH-Px and increased ($p < 0.05$) jejunal T-AOC.

Table 3. Effect of catalase on antioxidant parameters [1] of broilers exposed to DON.

		Treatments [2]			*p*-Value
		Control	DON	DONC	
Jejunum	ROS	62.12 ± 5.94	64.88 ± 3.94	64.83 ± 7.56	0.667
	O_2^-	111.62 ± 48.46	115.78 ± 78.19	101.99 ± 35.02	0.924
	OH	6.51 ± 2.67	6.96 ± 1.80	5.46 ± 1.21	0.466
	8-OHdG	27.49 ± 0.97	29.02 ± 2.59	28.09 ± 1.77	0.394
	MDA	7.49 ± 0.22	7.30 ± 0.40	7.34 ± 0.31	0.327
	CAT	545.59 ± 29.53 [a]	464.41 ± 8.66 [c]	497.50 ± 11.86 [b]	<0.001
	GSH-Px	114.79 ± 2.21 [a]	111.18 ± 2.29 [b]	117.75 ± 3.63 [a]	0.008
	T-AOC	3.68 ± 0.31 [b]	3.67 ± 0.23 [b]	4.62 ± 0.65 [a]	0.006

Table 3. *Cont.*

		Treatments [2]			*p*-Value
		Control	DON	DONC	
Liver	ROS	27.56 ± 0.52 [b]	29.75 ± 1.32 [a]	28.52 ± 1.34 [ab]	0.031
	O_2^-	421.58 ± 35.68	445.83 ± 53.19	443.60 ± 73.66	0.753
	OH	33.26 ± 2.77	34.10 ± 6.16	33.16 ± 5.50	0.949
	8-OHdG	27.56 ± 0.52 [b]	29.75 ± 1.32 [a]	28.52 ± 1.34 [ab]	0.031
	MDA	7.24 ± 0.086 [b]	7.76 ± 0.20 [a]	7.00 ± 0.26 [b]	0.012
	CAT	554.118 ± 19.88	565.15 ± 18.03	545.41 ± 30.67	0.764
	GSH-Px	118.92 ± 5.68	123.71 ± 5.04	121.78 ± 3.50	0.255
	T-AOC	5.74 ± 0.19 [a]	5.17 ± 0.42 [b]	5.27 ± 0.07 [b]	0.017

[a,b,c] Values within a row with unlike superscript letters differ significantly ($p < 0.05$). [1] ROS, reactive oxygen species (ng/mL); O_2^-, superoxide anion (nmol/mL); OH, hydroxyl radical (ng/mL); 8-OHdG, 8-hydroxy-2′-deoxyguanosine (ng/mL); MDA, malondialdehyde (nmol/mL); CAT, catalase (U/mL); GSH-Px, glutathione peroxidase (U/mL); T-AOC, total antioxidant capacity (U/mL). [2] Broilers in control and DON group were fed a basal diet without and with DON contamination, respectively, while those in DONC group were fed a DON-contaminated basal diet supplemented with 200 U/kg catalase.

2.3. Relative mRNA Expression of Antioxidation- and Apoptosis-Related Genes

As exhibited in Figure 1, the DON group displayed increases ($p < 0.05$) in the relative expression levels of nuclear factor erythroid-2 related factor (Nrf2) and Bcl-2 associated X protein (Bax) in the jejunum, together with heme oxygenase 1 (HO-1) and Caspase-9 in the liver, as compared with the control group. Supplementing CAT to DON-exposed birds did not change ($p > 0.05$) the relative expression of jejunal Nrf2 and Bax or hepatic HO-1. However, DON-exposed birds supplemented with CAT supported a similar ($p > 0.05$) expression of hepatic Caspase-9, relative to the control group.

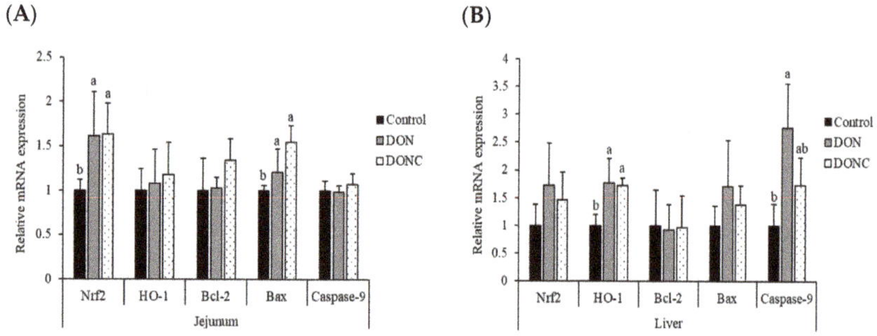

Figure 1. Effect of catalase on the relative mRNA expression of antioxidation-related genes (Nrf2 and HO-1) and apoptosis-related genes (Bcl-2, Bax and Caspase-9) in the jejunum (**A**) and liver (**B**) of broilers exposed to DON. Broilers in control and DON groups were fed a basal diet without and with DON contamination, respectively, while those in DONC group were fed a DON-contaminated basal diet supplemented with 200 U/kg catalase. [a,b] Values with unlike superscript letters differ significantly ($p < 0.05$). Nrf2, nuclear factor erythroid-2 related factor; HO-1, heme oxygenase 1; Bcl-2, B-cell lymphoma-2; Bax, Bcl-2 associated X protein.

2.4. Intestinal Morphological Structure

As presented in Figure 2A, jejunal villi in the control group were straight with a completed structure, while the DON group had a destruction of jejunal morphological structure, as evidenced by the breakage, shedding, and shortening of villi. However, the above phenomena were alleviated when birds were supplemented with CAT. Concretely, the broilers in the DON group had reduced ($p < 0.05$) villus height (VH) and villus height to crypt depth ratio (VCR), rather than crypt depth (CD) of the jejunum, compared with those in control group (Figure 2B–D). Nevertheless, the DON-induced reduction of jejunal VCR was reversed ($p < 0.05$) by CAT addition, which also rendered jejunal VH of DON-exposed birds, comparable ($p > 0.05$) to that in the control birds.

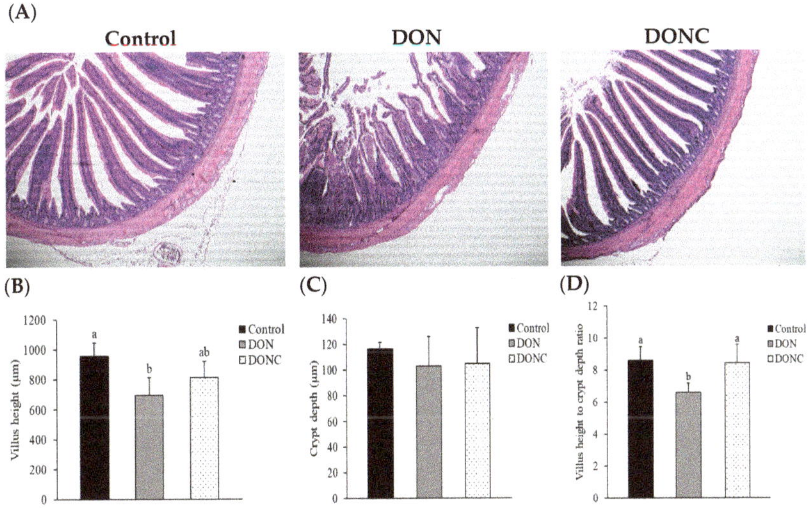

Figure 2. Effect of catalase on jejunal morphological structure of broilers exposed to DON. (**A**) Illustration (magnification 200×) of jejunal morphology of broilers from different groups. (**B**) Jejunal villus height of broilers from different groups. (**C**) Jejunal crypt depth of broilers from different groups. (**D**) Villus height to crypt depth ratio of the jejunum of broilers from different groups. [a,b] Values with unlike superscript letters differ significantly ($p < 0.05$). Broilers in control and DON groups were fed a basal diet without and with DON contamination, respectively, while those in DONC group were fed a DON-contaminated basal diet supplemented with 200 U/kg catalase.

2.5. Gut Microbiota

2.5.1. Diversity of Gut Microbiota

No difference ($p > 0.05$) was noted in the α-diversity indexes of gut microbiota among the groups (Figure S1). β-Diversity analysis manifested a difference ($p < 0.05$) in the similarity of gut microbiota among the groups (Figure 3A). This could be visualized by partial least squares discriminant analysis (PLS-DA) plot (Figure 3B), which revealed a distinct separation of microbiota among the groups.

2.5.2. Gut Microbial Composition

As shown in Figure S2, the predominant phylum of broiler gut was *Firmicutes*, followed by *Proteobacteria*. Within *Firmicutes*, the majority belonged to the classes *Clostridia* and *Bacilli*, while the majority within *Proteobacteria* were *Gammaproteobacteria*. Orders level analysis showed that the gut of the control group was mainly occupied by *Oscillospirales* and *Lachnospirales*, while the DON and DONC groups were dominated by *Oscillospirales*, *Lachnospirales*, and *Enterobacteriales*. At the family level, the major members in control group were *Lachnospiraceae* and *Ruminococcaceae*, while those in the DON and DONC groups were *Lachnospiraceae*, *Ruminococcaceae*, and *Enterobacteriaceae*. The dominating genera in the control group were *Faecalibacterium* and the unclassified *Lachnospiraceae*, while those in DON group were *Escherichia/Shigella* and the unclassified *Lachnospiraceae*. In comparison, DONC group was dominated by *Escherichia/Shigella* and *Ruminococcus torques group*.

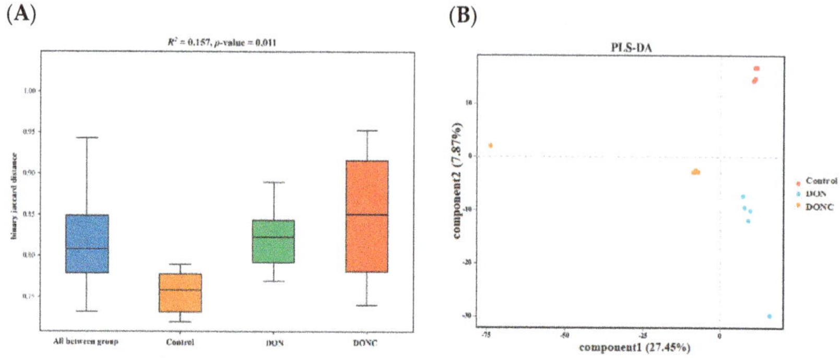

Figure 3. β-Diversity of broiler gut microbiota. (**A**) ANOSIM analysis (similarity analysis); (**B**) partial least squares discriminant analysis (PLS-DA). Broilers in control and DON groups were fed a basal diet without and with DON contamination, respectively, while those in DONC group were fed a DON-contaminated basal diet supplemented with 200 U/kg catalase.

2.5.3. Bacterial Richness among Groups

Bacterial richness ($p < 0.05$, linear discriminant analysis score (LDA) > 2.0) was identified by LDA combined effect size measurements (LEfSe) analysis. As illustrated in Figure 4, certain bacterial members, such as phylum Firmicutes, class Clostridia, orders Oscillospirales and Peptococcales, and families Ruminococcaceae and Peptococcaceae, together with genera Faecalibacterium, Flavonifractor, and Paludicola, were detected to be enriched in control group. Strikingly, the phylum Proteobacteria, class Gammaproteobacteria, order Enterobacteriales, family Enterobacteriaceae, and genus Escherichia/Shigella were enriched in the DON group. In comparison, the DONC group was differentially enriched with phylum Acidobacteriota, order Eubacteriales, and family Anaerofustaceae, along with genera Anaerofustis and Anaerotruncus.

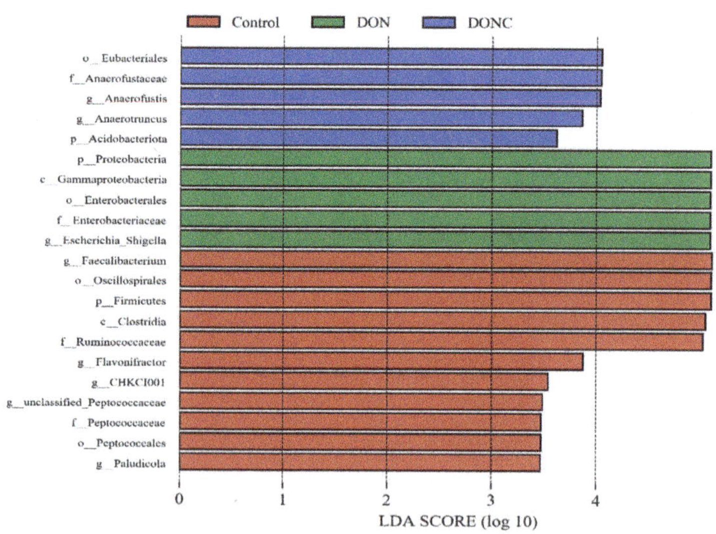

Figure 4. Linear discriminant analysis (LDA) combined effect size measurements (LEfSe) analysis of bacterial richness ($p < 0.05$, LDA > 3.0) in gut microbiota of broilers. Broilers in control and DON groups were fed a basal diet without and with DON contamination, respectively, while those in DONC group were fed a DON-contaminated basal diet supplemented with 200 U/kg catalase.

2.5.4. Correlations of Gut Microbiota with Other Intestinal Parameters

Spearman's correlation analysis was used to identify associations between gut microbiota and other intestinal parameters among groups. As shown in Figure S3, there were no correlations between gut microbiota and intestinal gene expression. However, the abundance of phylum Firmicutes showed a positive correlation ($p < 0.05$) with jejunal CAT activity and VH (Figure 5), whereas a contrasting pattern was found for the phylum Proteobacteria and its affiliate members, including class Gammaproteobacteria, order Enterobacteriales, family Enterobacteriaceae, and genus Escherichia/Shigella. The abundance of phylum Acidobacteriota had a positive correlation ($p < 0.05$) with jejunal T-AOC. The class Clostridia, order Peptococcales, and genus Flavonifractor were positively correlated ($p < 0.05$) with jejunal CAT activity, and the order Oscillospirales was positively correlated ($p < 0.05$) with jejunal CAT activity and VCR. Additionally, the genus Faecalibacterium elicited a positive correlation ($p < 0.05$) with jejunal VH.

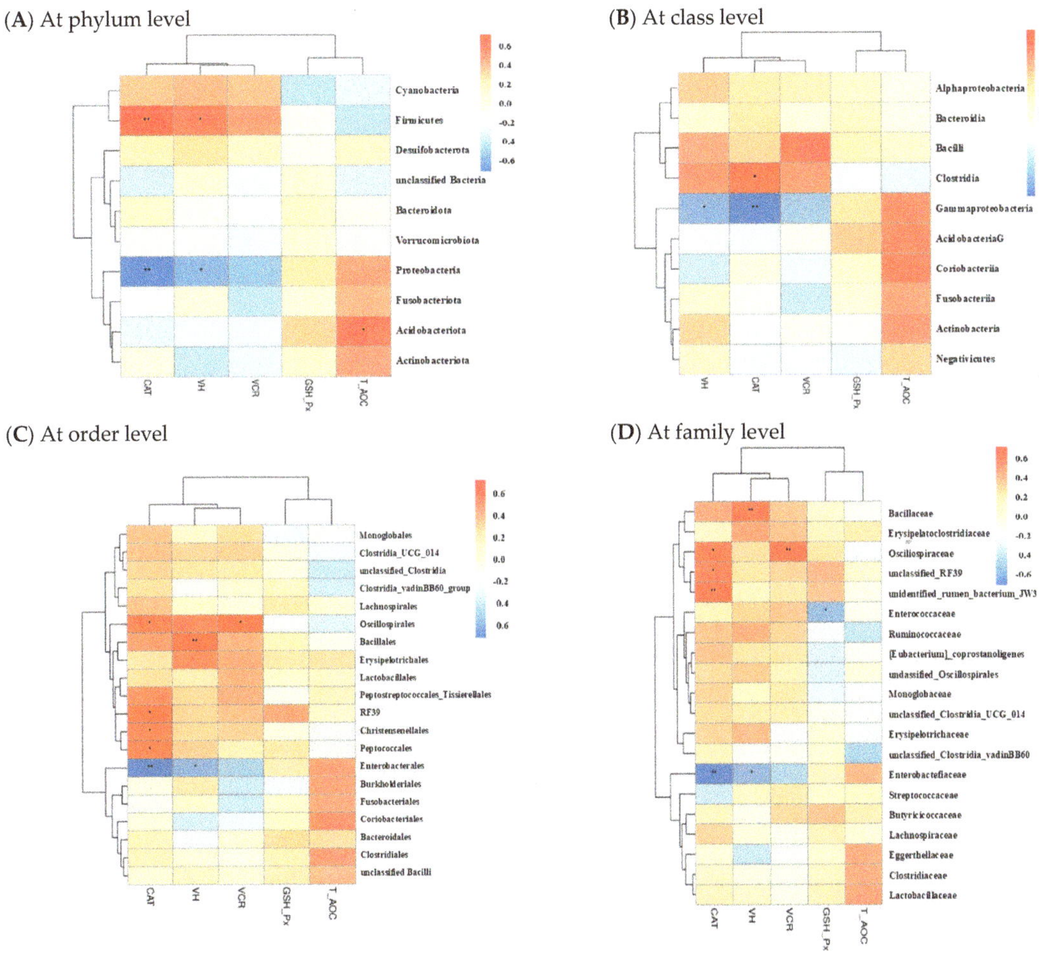

Figure 5. *Cont.*

(**E**) At genus level

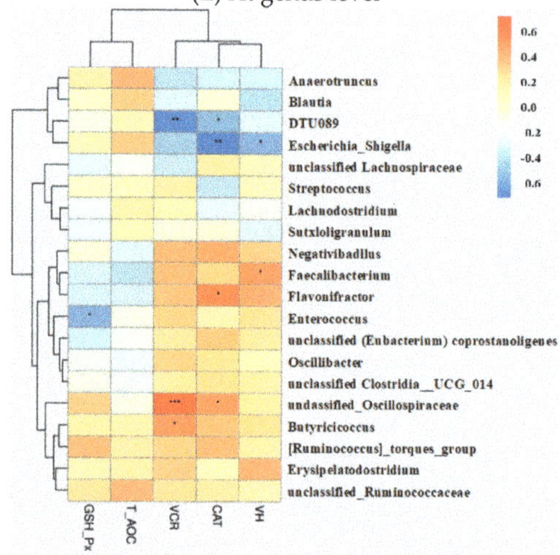

Figure 5. Correlation analysis of gut microbiota (**A**) at phyum level; (**B**) at class level; (**C**) at order level; (**D**) at family level; (**E**) at genus level) with intestinal antioxidant property and morphology in broilers. VH, villus height; CAT, catalase; VCR, villus height to crypt depth ratio; GSH-Px, glutathione peroxidase; T-AOC, total antioxidant capacity. The red and blue panes represent positive and negative correlations, respectively. Color intensity means the Spearman's r-value of correlations in each panel. The asterisks indicate significant correlations (* $p < 0.05$; ** $p < 0.01$; *** $p < 0.001$). Broilers in control and DON groups were fed a basal diet without and with DON contamination, respectively, while those in DONC group were fed a DON-contaminated basal diet supplemented with 200 U/kg catalase.

3. Discussion

The presence of a control + CAT group is not necessary for this study, based on the following two reasons: (1) this study was aimed to evaluate the efficacy of CAT in alleviating DON-induced detriments to broilers; (2) the beneficial effects of CAT addition on the growth and antioxidant capacity of both broilers and pigs under normal status had been confirmed in either our previous experiments or other studies [13,14].

DON contamination of feedstuffs, especially grains, has been an increasingly common problem, as it can cause growth retardation in animals [1,5]. There is evidence that poultry are far less sensitive to DON, relative to other monogastric animals (e.g., pigs), based on a previous study in which poultry showed no clinical response to a DON level of less than 20 mg/kg DON in diets, whereas 1–2 mg/kg DON caused toxicity in pigs [5]. Similarly, Dersjant-Li et al. [23] reported that dietary DON level below 16 mg/kg elicited marginal impact on broiler growth. Azizi et al. [24] recorded little impairment except for a reduction of feed intake during the early stage of growth of broilers when fortifying DON in diet at 10 mg/kg. In this study, dietary DON fortification at a subclinical level (an estimate of 7 mg/kg) caused little compromise of broiler growth performance except for a decreasing trend of ADFI. Supplemental CAT failed to improve growth performance of broilers exposed to DON, which did not agree with the study of Tang et al. [14] who found increases in weight gain, feed intake and feed efficiency of non-challenged broilers in response to CAT addition. The discrepancy might be due to the difference in raising condition of broilers.

Organ index namely the relative organ weight is a common indicator of health status in chickens. The effect of DON contamination of diets on broiler organ index revealed in previous studies is quite conflicting. For example, broilers fed a diet containing 5 mg/kg DON was reported to have negligible change in organ index [25], whereas it was also manifested to increase spleen index [26]. When fortifying DON at 10 mg/kg in diet, broilers were found to show an increase in jejunum index and a reduction of the index of bursa of Fabricius in the study of Awad et al. [27] and Wu et al. [28], respectively, with no changes in the indexes of other organs. The highly variable outcomes in these studies might be associated with the exposure time of DON and broiler breeds. In this study, we recorded an increase in the liver index, with a decreasing trend of kidney index in the broilers exposed to DON at a subclinical dosage (an estimate of 7 mg/kg). The increase in liver index might be ascribed to the fact that the increased free radicals induced by DON exposure perturbed protein synthesis via attack on ribosome and subsequently triggered a compensatory swelling of liver (the main site of protein synthesis) [8,29], while the decreasing trend of kidney index was presumably due to the inductive atrophy of kidney [30]. At present, no published study was available regarding the effects of CAT on organ index in animals. Herein, we noted that supplemental CAT abolished the increase in liver index and tended to reverse the decreasing trend of the kidney index of DON-exposed broilers. It was possible that the elimination of free radicals upon CAT addition resulted in less injuries of liver and kidney [22,31], thereby attenuating liver swelling and kidney atrophy.

The resultant redox imbalance of broiler organs, such as the intestine and liver, resulting from DON exposure have been well-established [5,32]. Analogously, this study manifested that broilers exposed to DON exhibited a disturbance of redox status, as exhibited by the increased contents of hepatic oxidative metabolites (ROS, MDA, and 8-OHdG), as well as impaired antioxidant property (decreases in hepatic T-AOC with jejunal CAT and GSH-Px activities). It is well-established that ROS represent the major kinds of free radicals with the ability to attack the unsaturated fatty acids in cell membrane, which cause lipid peroxidation chain reaction and, finally, generate a large amount of MDA [33]. The 8-OHdG, a biomarker of DNA oxidative damage, is also produced by ROS attacking the carbon atom of guanine base in DNA [34]. CAT and GSH-Px can prevent from redox imbalance by specifically scavenging hydrogen peroxide and lipid hydroperoxides, respectively [10]. In this study, the increased production of oxidative metabolites, together with impaired antioxidant property, due to DON exposure, was deduced to cause oxidative stress and the biomacromolecular damage of broilers [5,7]. However, supplementing CAT to DON-exposed birds normalized the hepatic ROS and 8-OHdG contents, alleviated the increase in hepatic MDA content with decreases in jejunal CAT and GSH-Px activities, and increased jejunal T-AOC. These results highlighted that exogenous CAT was beneficial for attenuating DON-induced oxidative stress in broilers. Likewise, some previous studies reported that dietary CAT addition improved redox status by enhancing the activities of antioxidant enzymes (SOD, CAT, and GSH-Px) and lowering oxidative metabolite MDA in the liver and intestine of both broilers [14] and pigs [16].

In order to cope with the tissue damage induced by free radicals, the body has evolved a complex mechanism responding to oxidative stress, among which Keapl/Nrf2-ARE is the most important endogenous pathway regulating the redox status inside the body [34]. Keap1, a cytosol binding protein of Nrf2, can bind to Nrf2 to prevent it from entering the nucleus under normal status, thereby avoiding the increase of cell sensitivity to stressors [34]. However, the overproduction of free radicals may activate Nrf2, which is then released from Keap1, enters the nucleus, and interacts with the antioxidant response element (ARE), subsequently promoting the expression of a series of downstream antioxidants and eliminating excess free radicals [35]. HO-1 is a crucial antioxidant enzyme with a binding site of a promoter similar to that of ARE, thus being targeted by Nrf2 [34]. It was emphasized that DON exposure-induced oxidative stress, which could be mediated by the Nrf2/HO-1 pathway [35]. Similarly, the current study revealed an upregulation of jejunal Nrf2 and hepatic HO-1 expression in broilers, due to DON exposure, which could be a feedback response of the host defense mechanism to the inducive oxidative stress. Strikingly, CAT addition did not affect either Nrf2 or HO-1 expression in DON-exposed broilers, demonstrating that exogenous CAT probably moderated DON-induced oxidative stress by directly eliminating radical accumulation and subsequently reducing the consumption of antioxidant enzymes, instead of promoting the expression of antioxidant enzyme genes via the Nrf2/HO-1 pathway. DON-induced oxidative stress was known to closely involve in cell apoptosis [8], as supported by the findings that DON exposure increased the expression of several key pro-apoptosis genes (e.g., Bax and Caspase family proteins) and anti-apoptosis gene Bcl-2, as well as their ratio (Bax/Bcl-2) in the intestine and liver [36,37]. Herein, we observed that broilers exposed to DON displayed an increased expression of jejunal Bax and hepatic caspase-9 (a critical mediator of cell apoptosis), highlighting an initiation of cell apoptosis in the jejunum and liver of broilers, in response to DON contamination in diet. When supplemented with CAT, the increased expression of hepatic caspase-9 in DON-exposed broilers was normalized, demonstrating a potential of CAT addition to mitigate DON-induced hepatic cell apoptosis in broilers to some extent. This was similar to some previous studies in which CAT addition decreased the mRNA expression of Bax and its ratio to Bcl-2 expression in the liver, as well as the mRNA and protein levels of hepatic and intestinal caspase family proteins (caspase-3 and -9) in pigs confronted with an oxidative stress induced by lipopolysaccharide [16,22].

Oxidative stress in broilers resulting from dietary DON contamination has been documented to trigger a destruction of the intestinal structure [24,28]. As expected, the present study showed that DON exposure led to an impairment of the jejunal morphology, as evidenced by the breakage, shedding, and incompleteness of the villi, together with reductions in VH and VCR. These were likely connected with the detected redox imbalance of jejunum. However, supplementing CAT to DON-exposed broilers normalized jejunal VH and VCR and might, in turn, favor maintenance of intestinal absorption and barrier function. This benefit was speculated to be responsible by the observed protective effect of CAT addition on jejunal antioxidant enzyme activities in DON-exposed broilers, because it has evidenced that exogenous CAT ameliorated intestinal morphological structure, probably through associating with the simultaneous increase in intestinal antioxidant capacity in both broilers [14] and pigs [16].

Gut microbiota are well-known for exerting essential roles in regulating host growth and health. Increasing studies verified that DON exposure-induced oxidative damage was involved in the gut microbiota disturbance of animals [18,28]. On the other hand, CAT addition was reported to promote intestinal and hepatic health via an association with improvement of gut microbiota in animals [14,16,22]. Similarly, the PLS-DA plot in this study disclosed a distinct shift of gut microbiota of broilers following DON exposure; however, this shift was alleviated by CAT addition. Bacterial richness analysis supported changes in gut microbial composition among groups. Thereinto, the control group was enriched with certain beneficial bacteria, such as *Firmicutes*, *Clostridia*, *Oscillospirales*, *Peptococcales*, *Ruminococcaceae*, *Faecalibacterium*, and *Flavonifractor*. In general, *Firmicutes* and *Clostridia* are the predominating commensal bacteria in animal gut and encompass plentiful potentially beneficial bacteria, therefore benefiting intestinal health of host [38,39]. *Oscillospirales*, *Peptococcales*, *Faecalibacterium*, *Ruminococcaceae*, and *Flavonifractor* were characterized as producers of butyric acid [40–42], which has a strong ability to enhance the antioxidant properties of broilers [43]. In the current study, the abundances of *Firmicutes*, *Clostridia*, *Oscillospirales*, *Peptococcales*, *Faecalibacterium*, and *Flavonifractor* were positively correlated with intestinal CAT activity and/or VH, suggesting that the enrichments of these bacterial members in gut could be conducive to intestinal redox homeostasis of broilers in control group. Comparatively, several potentially pathogenic or harmful bacterial members, including *Proteobacteria*, *Gammaproteobacteria*, *Enterobacteriales*, *Enterobacteriaceae*, and *Escherichia/Shigella* were enriched in the DON group. *Proteobacteria* includes a mass of typical pathogens, such as *Salmonella*, *Shigella*, *Klebsiella*, and pathogenic *Escherichia coli* that can generate considerable toxins, thereby serving as a momemtous indicator of gut microbiota disturbance and health disorders of animals [44,45]. It has been documented that the expansions of *Proteobacteria*, *Gammaproteobacteria*, *Enterobacteriales*, *Enterobacteriaceae*, and *Escherichia/Shigella* in the gut could cause accumulation of lipopolysaccharide, rendering the occurrence of oxidative stress in animals [19,46–49]. Herein, we detected negative correlations of the abundances of *Proteobacteria*, *Gammaproteobacteria*, *Enterobacteriales*, *Enterobacteriaceae*, and *Escherichia/Shigella* with both intestinal CAT activity and VH, demonstrating that the enrichments of these bacteria in the DON group conduced the simultaneous intestinal oxidative damage of broilers. In contrast to the DON group, the DONC group was differentially enriched with several potentially beneficial bacteria, such as the phylum *Acidobacteriota*, along with the genera *Anaerofustis* and *Anaerotruncus*. It was indicated that *Acidobacteriota* might favor the suppression of intestinal oxidative stress, due to its connection with intestinal anti-inflammation and the antioxidation of animals [50,51]. Likewise, this study disclosed a positive correlation of the abundance of *Acidobacteriota* with intestinal T-AOC of broilers. *Anaerofustis* and *Anaerotruncus* in the gut were indicated to prompt fiber digestion and the production of short-chain fatty acids that might allow for improvements of the intestinal antioxidant properties and morphological structures of animals [28,52–54]. Accordingly, the intestinal enrichments of *Acidobacteriota*, *Anaerofustis*, and *Anaerotruncus* due to CAT addition were probably associated with the observed alleviation of the intestinal oxidative damage of broilers exposed to DON. This was similar to a previous study in which dietary CAT addition improved the intestinal antioxidant capacity and gut microbial composition in broilers [14].

4. Conclusions

Supplemental CAT had a capacity to attenuate oxidative stress and intestinal injury of broilers exposed to DON. It is possible that the improved gut microbial composition (reflected by the enrichments of several beneficial bacteria) following CAT addition contributed to the observed protection against DON-induced oxidative damage in broilers. Our findings provided a strategy for limiting the detriments of dietary DON contamination to poultry.

5. Materials and Methods

5.1. Animals and Experimental Design

The experimental animal protocols for the present study were approved by the Animal Care and Use Committee of the South China Agricultural University. A total of 144 one-day-old Lingnan yellow-feathered male broilers were randomly assigned to 3 groups, with 6 replicates per group and 8 birds per replicate. The initial body weight of birds was similar across replicates. Broilers fed a basal diet were considered the control group, and the other two groups received a DON-contaminated basal diet added with 0 (DON group) or 200 U/kg CAT (DONC group). The content of DON (1.3 mg/kg) in basal diet was estimated using an AgraQuant® DON enzyme-linked immunosorbent assay (ELISA) kit (Romer Labs, Getzersdorf, Austria) with a Multiskan SkyHigh Microplate Reader (Thermo Fisher Scientific, Waltham, MA, USA). Dietary DON fortification was performed by supplementing DON-enriching rice to basal diet at the expense of corn. The final content of DON in diet was estimated at 7 mg/kg, which could be viewed as a subclinical dosage, as it was less than the clinical dosage (greater than 16 mg/kg) estimated in a previous study [23], but exceeded the maximum allowable dietary level (3 mg/kg), according to the Chinese Hygienical Standard for Feeds (GB13078-2017). CAT preparation (the theoretical value of enzyme activity was 200 U/g) was obtained from Vetland Bio-Technology Co., Ltd. (Shenyang, China). The actual activity of CAT in this supplement was determined to be 197 U/g. The supplemental CAT level was selected based on our preliminary experiment. Birds were housed in two-tier cages in an environmentally controlled room, in which the lighting program was 16 h per day, and room temperature was kept around 34 °C during the first three days and then gradually decreased to 24 °C on d 21. Birds had free access to the mash feed and water. The composition of basal diet is exhibited in Table S1. At 21 day (d) of age, birds were randomly selected from each replicate (6 birds/group) for determination of growth performance and sample collection. After sacrifice of these birds, visceral organs, including the liver, spleen, kidney, and bursa of Fabricius and intestines were separated. The midpoint of jejunal section was removed and cleaved into two segments, one of which was fixed in 4% paraformaldehyde solution, while the other one was quick-froze by liquid nitrogen and reserved at −80 °C. Moreover, cecal digesta was collected for sequencing analysis of gut microbiota.

5.2. Fabrication of DON-Enriching Rices

DON-enriching rices were fabricated according to the following procedures: (1) the *Fusarium graminearum* strain PH-1, kindly provided by Prof. Chenglan Liu (College of Plant Protection, South China Agricultural University), was revived, the resulting hyphae were harvested and aerobically plated on potato dextrose agar medium at 25 °C for 7 d; (2) fresh rices were placed into conical flasks and soaked with pure water overnight, followed by autoclaved sterilization; (3) the plating medium containing *Fusarium graminearum* hyphae was split into smaller portions and evenly scattered on the sterile rice, followed by static culture at 25 °C for 3 d with a subsequent shake culture at 25 °C for 18 d under aerobic condition. The contaminated rice were then collected, dried at 50 °C, and smashed, and the contents of DON (117.12 mg/kg) and its acetylated derivatives 3-acetyl-DON (2.39 mg/kg) and 15-acetyl-DON (8.03 mg/kg) in the powder were determined using an AgraQuant® DON ELISA kit (Romer Labs, Getzersdorf, Austria) and a liquid chromatography/mass spectrometry system (LCMS-8060, Shimadzu, Kyoto, Japan), respectively. The contents of some other mycotoxins (Table S2) were also quantified using the corresponding ELISA kits (Romer Labs, Getzersdorf, Austria).

5.3. Determination of Growth Performance and Organ Indexes

Body weight and feed intake were recorded for each replicate on d 21 for calculating the average body weight (ABW) of broilers at 21 d of age, along with average daily gain (ADG), average daily feed intake (ADFI), and feed conversion ratio (FCR) during 1–21 d of age. The collected organs, including liver, spleen, kidney, and bursa of Fabricius were weighed for determination of organ indexes, as calculated by the ratio of organ weight (g) to body weight (kg).

5.4. Measurement of Oxidative Status

The liver and jejunum samples were separately homogenized with 1:9 (w/v) cold saline, followed by centrifugation in a high-speed refrigerated centrifuge (6380R, Eppendorf, Hamburg, Germany) at 6000 rpm for 10 min at 4 °C to obtain the supernatants. The levels of oxidative metabolites, including reactive oxygen species (ROS), superoxide anion (O_2^-), hydroxyl radical (OH), 8-hydroxy-2′-deoxyguanosine (8-OHdG), and malondialdehyde (MDA), as well as the antioxidant indices, including the total antioxidant capacity (T-AOC), the levels of glutathione (GSH) coupled with the activities of CAT, total superoxide dismutase (T-SOD), and glutathione peroxidase (GSH-Px) in the supernatants were measured by micromethods using the corresponding kits provided by Meimian Bioengineering Institute (Nanjing, China) under the manufacturer's instructions.

5.5. Measurement of Gene Expression

Total RNA from the jejunum and liver samples was isolated and purified using the FastPure Cell/Tissue Total RNA Isolation Kit (Vazyme Biotech. Co., Ltd., Nanjing, China) under the corresponding instructions. The concentration of isolated RNA was measured with a NanoDrop-2000 spectrophotometer (Thermo Fisher Scientific, Waltham, MA, USA). RNA purity was estimated by detecting the absorbance ratio at 260:280 nm. RNA integrity was verified via detection of the 18S and 28S bands in 1% agarose gel electrophoresis. Thereafter, RNA samples were reverse transcribed to cDNA samples by using the HiScript II qRT SuperMix (Vazyme Biotech. Co., Ltd., Nanjing, China). Real-time PCR for examining gene expression was implemented in a CFX96Touch Real-Time PCR system (Bio-Rad Laboratories, Hercules, CA, USA) using the 2×Taq Master Mix (Vazyme Biotech. Co., Ltd., Nanjing, China). Reduced glyceraldehyde-phosphate dehydrogenase (GAPDH) acted as a reference gene. Information for the primers of GAPDH and target genes, including nuclear factor erythroid-2 related factor (Nrf2), heme oxygenase 1 (HO-1), B-cell lymphoma-2 (Bcl-2), Bcl-2 associated X protein (Bax), and caspase-9, are displayed in Table 4. The relative mRNA expressions of genes were calculated using the $2^{-\Delta\Delta Ct}$ method [55].

Table 4. Sequences for real-time PCR primers.

Genes [1]	Primer Sequence [2] (5′-3′)	Accession No.
GAPDH	F: GAGGGTAGTGAAGGCTGCTG R: CATCAAAGGTGGAGGAATGG	NM_204305.1
Nrf2	F: GATGTCACCCTGCCCTTAG R: CTGCCACCATGTTATTCC	NM_205117.1
HO-1	F: GCTGAAGAAAATCGCCCAA R: ATCTCAAGGGCATTCATTCGG	NM_205344.1
Bax	F: CAACAGGAAGAACACGCTGA R: TCAGTCTCGGCCCACTATCT	XM_015290060.1
Bcl-2	F: GACAACGGAGGATGGGATG R: CAGGCTCAGGATGGTCTTCA	NM_205339.2
Caspase-9	F: TCGAGCTGGCTCTGACATAGACTG R: AGGATGACCACGAGGCAGCAG	XM_424580.5

[1] *GAPDH*: reduced glyceraldehyde-phosphate dehydrogenase, *Nrf2*: nuclear factor erythroid 2-related factor 2, *HO-1*: heme oxygenase 1, *Bax*: B-cell lymphoma-2, *Bcl-2*: Bcl-2 associated X protein. [2] F, forward; R, reverse.

5.6. Analysis of Intestinal Morphological Structure

The fixed jejunal tissues were embedded in paraffin and stained with hematoxylin-eosin to obtain cross-sections. The intact and representative villi selected from each section were used for determining intestinal morphological structure with a light microscope. Villus height (VH) and crypt depth (CD) were defined as the height from villous tip to villus-crypt joint, respectively, based on which villus height to crypt depth ratio (VCR) was then calculated.

5.7. High-Throughput Sequencing of Gut Microbiota

Bacterial genomic DNA was extracted from cecal digesta using NucleoSpin® DNA Stool kit (Macherey-Nagel company, Düren, Germany). The concentration and quality of extracted DNA were checked using Nanodrop 2000 (Thermo Fisher Scientific, Waltham, MA, USA) and gel electrophoresis. Bacterial 16S rDNA sequences spanning the variable regions (V3-V4) were amplified using primers 338F (5′-ACT CCT ACG GGA GGC AGC A-3′) and 806R (5′-GGA CTA CHV GGG TWT CTA AT-3′). The amplified products were paired-end sequenced on an Illumina Novaseq platform (Illumina, San Diego, CA, USA) at Biomarker BioTech. Inc. (Beijing, China). The effective reads were clustered into operational taxonomic units and classified at various taxonomic levels based on a 97% sequence similarity. Bacterial α-diversity was analyzed using the QIIME2 software, and bacterial β-diversity was assessed by the partial least squares discriminant analysis (PLS-DA). The differences in bacterial abundances among groups were detected using the linear discriminant analysis (LDA) combined effect size measurements (LEfSe) analysis. Spearman correlation analysis was used for detecting the correlations between bacterial composition and other parameters.

5.8. Statistical Analysis

Data are expressed as the mean ± standard deviation and analyzed by one-way ANOVA using the general linear model procedure of SPSS 20.0. Differences among groups were examined using Duncan's multiple comparisons. Statistical significance was set at $p < 0.05$, and $0.05 \leq p < 0.10$ was thought as a tendency towards significance.

Supplementary Materials: The following supporting information can be downloaded at: https://www.mdpi.com/article/10.3390/toxins14120830/s1. Tables S1 and S2 and Figures S1–S3. Table S1: Composition of the basal diet (air-dry basis). Table S2: The contents of some mycotoxins aside from deoxynivalenol and its acetylated precursors in DON-enriching rice powder. Figure S1: Alpha-diversity analysis of gut microbiota among groups. (A) Shannon index; (B) Simpson index; (C) Chao1 index; (D) ACE index. Control, birds were fed a basal diet; DON, birds were fed a basal diet contaminated with DON; DONC, birds were fed a DON-contaminated basal diet supplemented with 200 U/kg CAT. Figure S2: Gut microbial composition of broilers at phylum (A), class (B), order (C), family (D), and genus (E) levels. Control, birds were fed a basal diet; DON, birds were fed a basal diet contaminated with DON; DONC, birds were fed a DON-contaminated basal diet supplemented with 200 U/kg CAT. Figure S3: Correlation analysis between gut microbiota (A) at phyum level; (B) at class level; (C) at order level; (D) at family level; (E) at genus level) and intestinal gene expression in broilers. Bax, Bcl-2 associated X protein; Nrf2, nuclear factor erythroid-2 related factor. The red and blue panes represent positive and negative correlations, respectively. Color intensity means the Spearman's r-value of correlations in each panel. The asterisks indicate significant correlations (* $p < 0.05$; ** $p < 0.01$). Control, birds were fed a basal diet; DON, birds were fed a basal diet contaminated with DON; DONC, birds were fed a DON-contaminated basal diet supplemented with 200 U/kg CAT.

Author Contributions: W.W. contributed to the conceptualization of this work and wrote the manuscript. J.Z. (Jingqiang Zhu) conducted the investigation. Q.C. contributed to parameter determination. C.Z. assisted with bioinformatic analysis. Z.D. contributed to data curation. D.F. administrated this project. H.Y. supervised this work. J.Z. (Jianjun Zuo) revised the manuscript and obtained the funding. All authors have read and agreed to the published version of the manuscript.

Funding: This study was financially supported by the Science and Technology Planning Project of Guangzhou, the National Natural Science Foundation of China (No. 32102584) and the Modern Feed Industry Innovation Team Project of Guangdong Province (No. 2021KJ115).

Institutional Review Board Statement: This study was conducted according to the guidelines of the Declaration of Helsinki and approved by the Animal Care and Use Committee of the South China Agricultural University (No. SCAU20210612, date of approval 12 June 2021). The welfare of all chickens was guaranteed in accordance with the national standard Laboratory Animal Requirements of Environment and Housing Facilities (GB 14925–2001).

Informed Consent Statement: Not applicable.

Data Availability Statement: The datasets used in this study are available from the corresponding author upon reasonable request.

Acknowledgments: We gratefully thank Yiliang Chen, Zheng Fan, Chong Ling, and Shuming Zhang (South China Agricultural University) for their assistance with sample collection.

Conflicts of Interest: The authors declare no conflict of interest.

References

1. Bennett, J.W.; Klich, M.A. Mycotoxins. *Clin. Microbiol. Rev.* **2003**, *16*, 497–516. [CrossRef] [PubMed]
2. Li, X.Y.; Zhao, L.H.; Fan, Y.; Jia, Y.X.; Sun, L.; Ma, S.S.; Ji, C.; Ma, Q.G.; Zhang, J.Y. Occurrence of mycotoxins in feed ingredients and complete feeds obtained from the Beijing region of China. *J. Anim. Sci. Biotechno.* **2014**, *5*, 37. [CrossRef]
3. Li, X.Y.; Guo, Y.P.; Zhao, L.H.; Fan, Y.; Ji, C.; Zhang, J.Y.; Ma, Q.G. Protective effects of Devosia sp ANSB714 on growth performance, immunity function, antioxidant capacity and tissue residues in growing-finishing pigs fed with deoxynivalenol contaminated diets. *Food Chem. Toxicol.* **2018**, *121*, 246–251. [CrossRef] [PubMed]
4. Zhao, L.H.; Li, X.Y.; Ji, C.; Rong, X.P.; Liu, S.J.; Zhang, J.Y.; Ma, Q.G. Protective effect of Devosia sp ANSB714 on growth performance, serum chemistry, immunity function and residues in kidneys of mice exposed to deoxynivalenol. *Food Chem. Toxicol.* **2016**, *92*, 143–149. [CrossRef] [PubMed]
5. Pestka, J.J. Deoxynivalenol: Toxicity, mechanisms and animal health risks. *Anim. Feed Sci. Tech.* **2007**, *137*, 283–298. [CrossRef]
6. Paraskeuas, V.; Griela, E.; Bouziotis, D.; Fegeros, K.; Antonissen, G.; Mountzouris, K.C. Effects of deoxynivalenol and fumonisins on broiler gut cytoprotective capacity. *Toxins* **2021**, *13*, 729. [CrossRef]
7. Wu, Q.H.; Wang, X.; Yang, W.; Niissler, A.K.; Xiong, L.Y.; Kuca, K.; Dohnal, V.; Zhang, X.J.; Yuan, Z.H. Oxidative stress-mediated cytotoxicity and metabolism of T-2 toxin and deoxynivalenol in animals and humans: An update. *Arch. Toxicol.* **2014**, *88*, 1309–1326. [CrossRef]

8. You, L.; Zhao, Y.Y.; Kuca, K.; Wang, X.; Oleksak, P.; Chrienova, Z.; Nepovimova, E.; Jacevic, V.; Wu, Q.H.; Wu, W.D. Hypoxia, oxidative stress, and immune evasion: A trinity of the trichothecenes T-2 toxin and deoxynivalenol (DON). *Arch. Toxicol.* **2021**, *95*, 1899–1915. [CrossRef] [PubMed]
9. Mates, J.M. Effects of antioxidant enzymes in the molecular control of reactive oxygen species toxicology. *Toxicology* **2000**, *153*, 83–104. [CrossRef]
10. Nemoto, T.; Kawakami, S.; Yamashita, F.; Hashida, M. Efficient protection by cationized catalase against H_2O_2 injury in primary cultured alveolar epithelial cells. *J. Control. Release* **2007**, *121*, 74–80. [CrossRef]
11. Switala, J.; Loewen, P.C. Diversity of properties among catalases. *Arch. Biochem. Biophys.* **2002**, *401*, 145–154. [CrossRef] [PubMed]
12. Sies, H. Hydrogen peroxide as a central redox signaling molecule in physiological oxidative stress: Oxidative eustress. *Red. Biol.* **2017**, *11*, 613–619. [CrossRef]
13. Guo, G.L.; Zhou, T.T.; Ren, F.Y.; Sun, J.Z.; Deng, D.; Huang, X.G.; Wassie, T.; Qazi, I.H.; Wu, X. Effect of maternal catalase supplementation on reproductive performance, antioxidant activity and mineral transport in sows and piglets. *Animals* **2022**, *12*, 828. [CrossRef]
14. Tang, M.H.; Fang, R.J.; Xue, J.J.; Yang, K.L.; Lu, Y. Effects of catalase on growth performance, antioxidant capacity, intestinal morphology, and microbial composition in yellow broilers. *Front. Vet. Sci.* **2022**, *9*, 802051. [CrossRef] [PubMed]
15. Li, Y.; Zhao, X.L.; Jiang, X.M.; Chen, L.; Hong, L.; Zhuo, Y.; Lin, Y.; Fang, Z.Z.; Che, L.Q.; Feng, B.; et al. Effects of dietary supplementation with exogenous catalase on growth performance, oxidative stress, and hepatic apoptosis in weaned piglets challenged with lipopolysaccharide. *J. Anim. Sci.* **2020**, *98*, 1–10. [CrossRef]
16. Chen, J.L.; Li, F.C.; Yang, W.R.; Jiang, S.Z.; Li, Y. Supplementation with exogenous catalase from *Penicillium notatum* in the diet ameliorates lipopolysaccharide-induced intestinal oxidative damage through affecting intestinal antioxidant capacity and microbiota in weaned pigs. *Microbiol. Spectr.* **2021**, *9*, e0065421. [CrossRef]
17. Sun, X.J.; Piao, L.G.; Jin, H.F.; Nogoy, K.M.C.; Zhang, J.F.; Sun, B.; Jin, Y.; Lee, D.H.; Choi, S.; Li, X.Z. Dietary glucose oxidase and/or catalase supplementation alleviates intestinal oxidative stress induced by diquat in weaned piglets. *Anim. Sci. J.* **2021**, *92*, e13634. [CrossRef]
18. Bindari, Y.R.; Gerber, P.F. Centennial review: Factors affecting the chicken gastrointestinal microbial composition and their association with gut health and productive performance. *Poult. Sci.* **2022**, *101*, 101612. [CrossRef]
19. Qiu, Y.Q.; Yang, J.; Wang, L.; Yang, X.F.; Gao, K.G.; Zhu, C.; Jiang, Z.Y. Dietary resveratrol attenuation of intestinal inflammation and oxidative damage is linked to the alteration of gut microbiota and butyrate in piglets challenged with deoxynivalenol. *J. Anim. Sci. Biotechnol.* **2021**, *12*, 71. [CrossRef]
20. Wang, S.; Yang, J.C.; Zhang, B.Y.; Zhang, L.; Wu, K.T.; Yang, A.; Li, C.; Wang, Y.N.; Zhang, J.C.; Qi, D.S. Potential link between gut microbiota and deoxynivalenol-induced feed refusal in weaned piglets. *J. Agric. Food Chem.* **2019**, *67*, 4976–4986. [CrossRef]
21. Qing, H.R.; Huang, S.M.; Zhan, K.; Zhao, L.H.; Zhang, J.Y.; Ji, C.; Ma, Q.G. Combined toxicity evaluation of ochratoxin A and aflatoxin B1 on kidney and liver injury, immune inflammation, and gut microbiota alteration through pair-feeding pullet model. *Front. Immunol.* **2022**, *13*, 920147. [CrossRef] [PubMed]
22. Li, Y.; Zhao, X.L.; Zhang, L.J.; Zhan, X.Y.; Liu, Z.G.; Zhuo, Y.; Lin, Y.; Fang, Z.Z.; Che, L.Q.; Feng, B.; et al. Effects of a diet supplemented with exogenous catalase from *Penicillium notatum* on intestinal development and microbiota in weaned piglets. *Microorganisms* **2020**, *8*, 391. [CrossRef] [PubMed]
23. Dersjant-Li, Y.M.; Verstegen, M.W.A.; Gerrits, W.J.J. The impact of low concentrations of aflatoxin, deoxynivalenol or fumonisin in diets on growing pigs and poultry. *Nutr. Res. Rev.* **2003**, *16*, 223–239. [CrossRef] [PubMed]
24. Azizi, T.; Daneshyar, M.; Allymehr, M.; Jalali, A.S.; Behroozyar, H.K.; Tukmechi, A. The impact of deoxynivalenol contaminated diet on performance, immune response, intestine morphology and jejunal gene expression in broiler chicken. *Toxicon* **2021**, *199*, 72–78. [CrossRef] [PubMed]
25. Awad, W.A.; Hess, M.; Twarużek, M.; Grajewski, J.; Kosicki, R.; Böhm, J.; Zentek, J. The impact of the *Fusarium* mycotoxin deoxynivalenol on the health and performance of broiler chickens. *Int. J. Mol. Sci.* **2011**, *12*, 7996–8012. [CrossRef]
26. Chen, S.S.; Li, Y.H.; Lin, M.F. Chronic exposure to the *Fusarium* mycotoxin deoxynivalenol: Impact on performance, immune organ, and intestinal integrity of slow-growing chickens. *Toxins* **2017**, *9*, 334–354. [CrossRef]
27. Awad, W.A.; Bohm, J.; Razzazi-Fazeli, E.; Ghareeb, K.; Zentek, J. Effect of addition of a probiotic microorganism to broiler diets contaminated with deoxynivalenol on performance and histological alterations of intestinal villi of broiler chickens. *Poult. Sci.* **2006**, *85*, 974–979. [CrossRef] [PubMed]
28. Wu, S.R.; Liu, Y.L.; Duan, Y.L.; Wang, F.Y.; Guo, F.S.; Yan, F.; Yang, X.J.; Yang, X. Intestinal toxicity of deoxynivalenol is limited by supplementation with *Lactobacillus plantarum* JM113 and consequentially altered gut microbiota in broiler chickens. *J. Anim. Sci. Biotechnol.* **2018**, *9*, 74. [CrossRef] [PubMed]
29. Chen, Y.P.; Cheng, Y.F.; Wen, C.; Wang, W.B.; Kang, Y.R.; Wang, A.Q.; Zhou, Y.M. The protective effects of modified palygorskite on the broilers fed a purified zearalenone-contaminated diet. *Poult. Sci.* **2019**, *98*, 3802–3810. [CrossRef]
30. Reddy, K.E.; Song, J.; Lee, H.J.; Kim, M.; Kim, D.W.; Jung, H.J.; Kim, B.; Lee, Y.; Yu, D.; Kim, D.W.; et al. Effects of high levels of deoxynivalenol and zearalenone on growth performance, and hematological and immunological parameters in pigs. *Toxins* **2018**, *10*, 114. [CrossRef]
31. Kobayashi, M.; Sugiyama, H.; Makino, H. Catalase deficiency renders remnant kidneys more susceptible to oxidant tissue injury and renal fibrosis in mice. *Kidney Int.* **2005**, *68*, 1018–1031. [CrossRef] [PubMed]

32. de Souza, M.; Baptista, A.A.S.; Valdiviezo, M.J.J.; Justino, L.; Menck-Costa, M.F.; Ferraz, C.R.; da Gloria, E.M.; Verri, W.A.; Bracarense, A.P.F.R.L. *Lactobacillus* spp. reduces morphological changes and oxidative stress induced by deoxynivalenol on the intestine and liver of broilers. *Toxicon* **2020**, *185*, 203–212. [CrossRef]
33. Jelic, M.D.; Mandic, A.D.; Maricic, S.M.; Srdjenovic, B.U. Oxidative stress and its role in cancer. *J. Cancer Res. Ther.* **2021**, *17*, 22–28. [CrossRef]
34. Magesh, S.; Chen, Y.; Hu, L.Q. Small molecule modulators of Keap1-Nrf2-ARE pathway as potential preventive and therapeutic agents. *Med. Res. Rev.* **2012**, *32*, 687–726. [CrossRef] [PubMed]
35. Yu, M.; Chen, L.K.; Peng, Z.; Wang, D.; Song, Y.D.; Wang, H.Y.; Yao, P.; Yan, H.; Nüssler, A.K.; Liu, L.G.; et al. Embryotoxicity caused by DON-induced oxidative stress mediated by Nrf2/HO-1 pathway. *Toxins* **2017**, *9*, 188. [CrossRef]
36. Yang, X.; Liang, S.S.; Guo, F.S.; Ren, Z.Z.; Yang, X.J.; Long, F.Y. Gut microbiota mediates the protective role of *Lactobacillus plantarum* in ameliorating deoxynivalenol-induced apoptosis and intestinal inflammation of broiler chickens. *Poult. Sci.* **2020**, *99*, 2395–2406. [CrossRef]
37. Li, J.B.; Bai, Y.S.; Ma, K.D.; Ren, Z.S.; Li, J.P.; Zhang, J.; Shan, A.S. Dihydroartemisinin alleviates deoxynivalenol induced liver apoptosis and inflammation in piglets. *Ecotoxicol. Environ. Saf.* **2022**, *241*, 113811. [CrossRef]
38. Davey, L.; Halperin, S.A.; Lee, S.F. Thiol-disulfide exchange in gram-positive *Firmicutes*. *Trends Microbiol.* **2016**, *24*, 902–915. [CrossRef]
39. Lopetuso, L.R.; Scaldaferri, F.; Petito, V.; Gasbarrini, A. Commensal *Clostridia*: Leading players in the maintenance of gut homeostasis. *Gut Pathog.* **2013**, *5*, 23. [CrossRef]
40. Ahrens, A.P.; Culpepper, T.; Saldivar, B.; Anton, S.; Stoll, S.; Handberg, E.M.; Xu, K.; Pepine, C.; Triplett, E.W.; Aggarwal, M. A six-day, lifestyle-based immersion program mitigates cardiovascular risk factors and induces shifts in gut microbiota, specifically *Lachnospiraceae, Ruminococcaceae, Faecalibacterium prausnitzii*: A pilot study. *Nutrients* **2021**, *13*, 3459. [CrossRef]
41. Khorraminezhad, L.; Leclercq, M.; O'Connor, S.; Julien, P.; Weisnagel, S.J.; Gagnon, C.; Droit, A.; Rudkowska, I. Dairy product intake modifies gut microbiota composition among hyperinsulinemic individuals. *Eur. J. Nutr.* **2021**, *60*, 159–167. [CrossRef]
42. Scott, K.P.; Martin, J.C.; Duncan, S.H.; Flint, H.J. Prebiotic stimulation of human colonic butyrate-producing bacteria and *bifidobacteria*, In Vitro. *FEMS Microbiol. Ecol.* **2014**, *87*, 30–40. [CrossRef]
43. Zhang, W.H.; Gao, F.; Zhu, Q.F.; Li, C.; Jiang, Y.; Dai, S.F.; Zhou, G.H. Dietary sodium butyrate alleviates the oxidative stress induced by corticosterone exposure and improves meat quality in broiler chickens. *Poult. Sci.* **2011**, *90*, 2592–2599. [CrossRef]
44. Litvak, Y.; Byndloss, M.X.; Tsolis, R.M.; Bäumler, A.J. Dysbiotic *Proteobacteria* expansion: A microbial signature of epithelial dysfunction. *Curr. Opin. Microbiol.* **2017**, *39*, 1–6. [CrossRef] [PubMed]
45. Guo, X.; Huang, C.; Xu, J.; Xu, H.M.; Liu, L.; Zhao, H.L.; Wang, J.Q.; Huang, W.Q.; Peng, W.; Chen, Y.; et al. Gut microbiota is a potential biomarker in inflammatory bowel disease. *Front. Nutr.* **2022**, *8*, 818902. [CrossRef] [PubMed]
46. Liu, B.Y.; Chen, B.W.; Chen, B.W.; Yi, J.; Long, H.P.; Wen, H.Q.; Tian, F.M.; Liu, Y.F.; Xiao, L.; Li, L.S. Liuwei Dihuang Decoction alleviates cognitive dysfunction in mice with D-galactose-induced aging by regulating lipid metabolism and oxidative stress via the microbiota-gut-brain axis. *Front. Neurosci.* **2022**, *16*, 949298. [CrossRef] [PubMed]
47. Cai, C.X.; Zhang, Z.X.; Morales, M.; Wang, Y.N.; Khafipour, E.; Friel, J. Feeding practice influences gut microbiome composition in very low birth weight preterm infants and the association with oxidative stress: A prospective cohort study. *Free Radic. Biol. Med.* **2019**, *142*, 146–154. [CrossRef] [PubMed]
48. Zhang, X.G.; Zhang, B.; Li, L.; Li, X.B.; Zhang, J.Q.; Chen, G.Y. Fermented noni (*Morinda citrifolia* L.) fruit juice improved oxidative stress and insulin resistance under the synergistic effect of Nrf2/ARE pathway and gut flora in db/db mice and HepG2 cells. *Food Funct.* **2022**, *13*, 8254–8273. [CrossRef]
49. Wu, Y.P.; Wang, B.K.; Tang, L.; Zhou, Y.H.; Wang, Q.; Gong, L.; Ni, J.J.; Li, W.F. Probiotic *Bacillus* alleviates oxidative stress-induced liver injury by modulating gut-liver axis in a rat model. *Antioxidants* **2022**, *11*, 291. [CrossRef]
50. Wang, W.W.; Wang, Y.; Duan, Y.X.; Meng, Z.Q.; An, X.P.; Qi, J.W. Regulation of wheat bran feruloyl oligosaccharides in the intestinal antioxidative capacity of rats associated with the p38/JNK-Nrf2 signaling pathway and gut microbiota. *J. Sci. Food. Agric.* **2022**, *102*, 6992–7002. [CrossRef]
51. Liu, S.L.; Wang, S.D.; Liu, X.W.; Wen, L.F.; Zou, J.X. Effects of dietary antimicrobial peptides on intestinal morphology, antioxidant status, immune responses, microbiota and pathogen disease resistance in grass carp *Ctenopharyngodon idellus*. *Microb. Pathogenesis* **2022**, *165*, 105386. [CrossRef] [PubMed]
52. Niu, Q.; Li, P.H.; Hao, S.S.; Kim, S.W.; Du, T.R.; Hua, J.D.; Huang, R.H. Characteristics of gut microbiota in sows and their relationship with apparent nutrient digestibility. *Int. J. Mol. Sci.* **2019**, *20*, 870. [CrossRef] [PubMed]
53. Polansky, O.; Sekelova, Z.; Faldynova, M.; Sebkova, A.; Sisak, F.; Rychlik, I. Important metabolic pathways and biological processes expressed by chicken cecal microbiota. *Appl. Environ. Microbiol.* **2015**, *82*, 1569–1576. [CrossRef] [PubMed]
54. Zhang, Q.Y.; Zhang, S.; Wu, S.; Madsen, M.H.; Shi, S.R. Supplementing the early diet of broilers with soy protein concentrate can improve intestinal development and enhance short-chain fatty acid-producing microbes and short-chain fatty acids, especially butyric acid. *J. Anim. Sci. Biotechnol.* **2022**, *13*, 97. [CrossRef] [PubMed]
55. Livak, K.J.; Schmittgen, T.D. Analysis of relative gene expression data using real-time quantitative PCR and the 2(T)(-Delta Delta C) method. *Methods* **2001**, *25*, 402–408. [CrossRef]

toxins

MDPI

Article

Transcriptomics Reveals the Effect of Thymol on the Growth and Toxin Production of *Fusarium graminearum*

Lian-Qun Wang [1,2], Kun-Tan Wu [1], Ping Yang [1], Fang Hou [2], Shahid Ali Rajput [3], De-Sheng Qi [1,*] and Shuai Wang [1,*]

1 Department of Animal Nutrition and Feed Science, College of Animal Science and Technology, Huazhong Agricultural University, Wuhan 430070, China; wlq2005199@126.com (L.-Q.W.); kuntanwu@webmail.hzau.edu.cn (K.-T.W.); yp923418@webmail.hzau.edu.cn (P.Y.)
2 Department of Animal Science, College of Animal Science and Technology, Tarim University, Aral 843300, China; liuyy221@126.com
3 Faculty of Veterinary and Animal Science, Muhammad Nawaz Shareef University of Agriculture, Multan 60000, Punjab, Pakistan; shahid.ali@mnsuam.edu.pk
* Correspondence: qds@mail.hzau.edu.cn (D.-S.Q.); wangshuai@mail.hzau.edu.cn (S.W.)

Abstract: *Fusarium graminearum* is a harmful pathogen causing head blight in cereals such as wheat and barley, and thymol has been proven to inhibit the growth of many pathogens. This study aims to explore the fungistatic effect of thymol on *F. graminearum* and its mechanism. Different concentrations of thymol were used to treat *F. graminearum*. The results showed that the EC_{50} concentration of thymol against *F. graminearum* was 40 μg/mL. Compared with the control group, 40 μg/mL of thymol reduced the production of Deoxynivalenol (DON) and 3-Ac-DON by 70.1% and 78.2%, respectively. Our results indicate that thymol can effectively inhibit the growth and toxin production of *F. graminearum* and cause an extensive transcriptome response. Transcriptome identified 16,727 non-redundant unigenes and 1653 unigenes that COG did not annotate. The correlation coefficients between samples were all >0.941. When FC was 2.0 times, a total of 3230 differential unigenes were identified, of which 1223 were up-regulated, and 2007 were down-regulated. Through the transcriptome, we confirmed that the expression of many genes involved in *F. graminearum* growth and synthesis of DON and other secondary metabolites were also changed. The gluconeogenesis/glycolysis pathway may be a potential and important way for thymol to affect the growth of *F. graminearum* hyphae and the production of DON simultaneously.

Keywords: thymol; *Fusarium graminearum*; deoxynivalenol; mycelial growth; toxin production; gluconeogenesis/glycolysis

Key Contribution: This study revealed the genetic mechanisms for the responses of mycelial growth and toxin production of *Fusarium graminearum* to thymol.

Citation: Wang, L.-Q.; Wu, K.-T.; Yang, P.; Hou, F.; Rajput, S.A.; Qi, D.-S.; Wang, S. Transcriptomics Reveals the Effect of Thymol on the Growth and Toxin Production of *Fusarium graminearum*. *Toxins* **2022**, *14*, 142. https://doi.org/10.3390/toxins14020142

Received: 28 December 2021
Accepted: 11 February 2022
Published: 15 February 2022

1. Introduction

Crops and food are often contaminated by molds and mycotoxins [1]. *Fusarium* head blight (FHB) is the most common disease of wheat in the world [2]. Some species of *Fusarium* can cause FHB, like *F. graminearum*, *F. culmorum*, and *F. avenaceum* [3]. The infection starts from the flowering period and progresses to the harvest period, causing the crops to be unharvested. A large amount of feed and food are rendered unfit for consumption due to mycotoxins contamination around the globe [4]. Finished feeds have high contamination rates and are often co-contaminated with multiple mycotoxins [5]. Common mycotoxins include aflatoxin B1 (AFB1) [6,7] and zearalenone (ZEN) [8], but deoxynivalenol (DON) and its acetyl-derivatives (3-Ac-DON and 15-Ac-DON) produced by *F. graminearum* are particularly concerning [9]. DON often contaminates wheat, barley, and oats. It can affect animal growth, immunity, and intestinal barrier function. In severe cases, it can induce vomiting, refusal to feed, and gastrointestinal bleeding in pigs [10]. Therefore, great attention is paid to controlling mycotoxins in feed to reduce economic losses. For compound feed for pigs, China's Hygienical standard for feeds (GB 13078-2017) stipulates that the maximum limit is 1 mg/kg, and in the United States, it is 5 mg/kg. Due to the urgency of controlling FHB and the toxicity of trichothecenes, *F. graminearum* is listed as one of the top 10 fungal plant pathogens. Benzimidazole (such as carbendazim) and triazole (such as tebuconazole) have been widely used to control contamination by *F. graminearum* [11]. However, chemical fungicides are toxic and induce *F. graminearum* to synthesize DON [11]. Its residues on crops pose potential risks to environmental, animal, and human health. Therefore, people are paying more and more attention to finding biological antifungal agents alternative to synthetic pesticides. Researchers are particularly interested in essential oils (EO) extracted from plants, among natural fungistatic products. Due to its great potential to inhibit pathogens and their medicinal properties, it is considered one of the most promising biological antifungal agents.

EO is considered to be a secondary metabolite of aromatic flowers and plants. Because of their antiviral, fungistatic, and insecticidal properties in plants, they have been included in the category of natural preservatives [12]. Thymol (2-isopropyl-5-methylphenol) is a natural phenolic monoterpene compound, mainly found in *thymus Vulgaris, orange peel*, and *origanum heracleoticum* [13]. Thymol also has an excellent inhibitory effect on other toxin-producing molds, such as *Aspergillus flavus* [14], *Rhizoctonia solani*, *Pyricularia oryzae* [15], *Rhizopus stolonifera* [16], and *Fusarium solani* [17]. At the same time, it can also reduce the production of DON [18] and ZEN [19], so using thymol to inhibit *F. graminearum* has good potential and application prospects.

Some physical adsorbents can adsorb trichothecenes [20], but the effect is still not satisfactory. Adding thymol to feed can inhibit the mycelial growth of *F. graminearum* and reduce the production of trichothecenes from the source. It will be more conducive to the control of mycotoxins in the feed. Thymol is natural and degradable, and it also has the advantage of enhancing the antioxidant capacity of animals. Therefore, this experiment aims to study the effect of thymol on the growth and toxin production of *F. graminearum*. At present, there are also some reports that thymol inhibits the growth of *F. graminearum*, such as accumulating reactive oxygen species (ROS), destroying the integrity of cell walls and cell membranes [21], through ergosterol biosynthesis [22], and that it can block the overproduction of ROS [23]. However, there is a lack of research on other approaches. The fungistatic mechanism of EO is usually not a single pathway. Many studies have reported that EO can also change the membrane potential and destroy the integrity of cell membranes [24], inhibit DNA repair and transcription processes [25], or form chimeras with DNA and other pathways [26,27]. Therefore, we also adopted transcriptome sequencing technology (RNA-Seq), a high-throughput and high-resolution tool widely used to study fungi [28,29]. It can provide a comprehensive view of the *F. graminearum* transcriptome to comprehensively understand and explore other mechanisms by which thymol inhibits the growth of *F. graminearum*.

2. Results

2.1. The Effect of Thymol on the Growth of F. graminearum Hypha

First, we added different concentrations of thymol to the medium to determine its inhibitory effect on *F. graminearum* (Figure 1). The results show that thymol had a good fungistatic effect, and the 10 μg/mL thymol treatment group significantly reduced the colony diameter on the fourth day (Figures 1C and 2B). The inhibitory effect of thymol on the growth of *F. graminearum* had an obvious dose effect and time effect (Figures 1 and 2A). At the same time, the inhibition rate of different concentrations of thymol on mycelial growth was calculated after the fourth day of culture (Figure 2B). The mycelial growth inhibition rate (MGIR) of *F. graminearum* reached 100%, and the thymol concentration was 160 μg/mL. The EC_{50} and EC_{90} calculated by the regression equation were 40.15 μg/mL and 139.12 μg/mL, respectively.

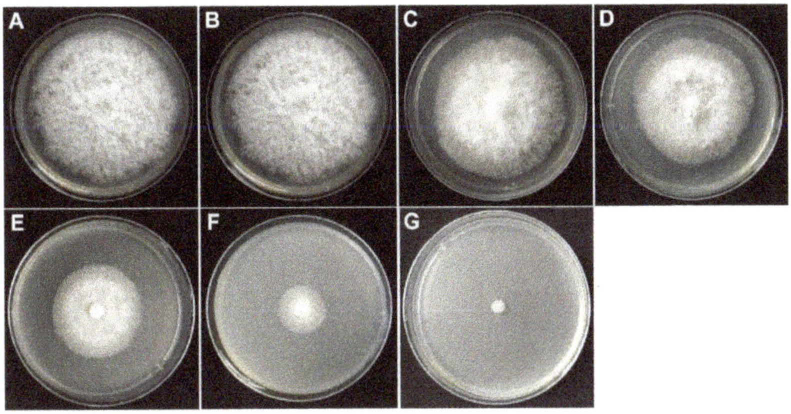

Figure 1. The effect of thymol on the growth of *F. graminearum* hyphae. (**A**) Control group; (**B**) 5 μg/mL thymol treatment group; (**C**) 10 μg/mL thymol treatment group; (**D**) 20 μg/mL thymol treatment group; (**E**) 40 μg/mL thymol treatment group; (**F**) 80 μg/mL thymol treatment group; (**G**) 160 μg/mL thymol treatment group.

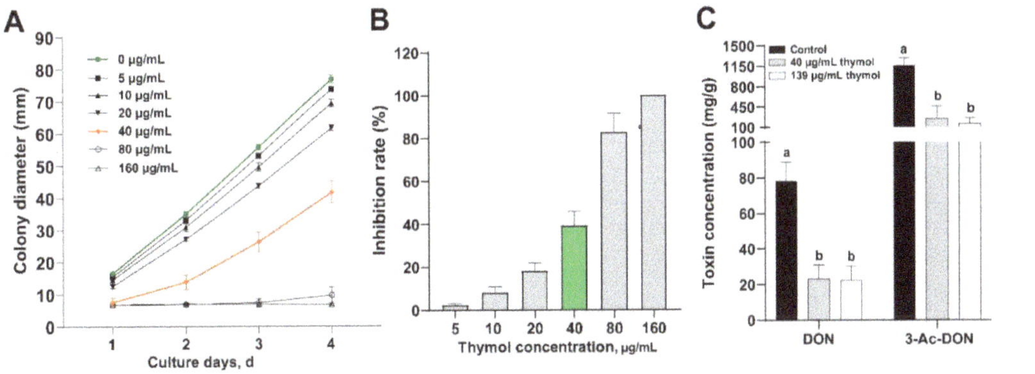

Figure 2. Thymol inhibits the growth of *F. graminearum* and DON synthesis. (**A**) Effect of different concentrations of thymol on the mycelial diameter; (**B**) Inhibition rate of different concentrations of thymol on the growth of *F. graminearum*; (**C**) The effect of thymol on the synthesis of DON and 3-Ac-DON at 40 μg/mL (EC_{50}) or 139 μg/mL(EC_{90}) concentrations. [a,b] Columns with different lowercase letters indicated significant differences between the compared groups ($p < 0.05$).

2.2. The Effect of Thymol on DON Production by F. graminearum

According to the previous growth inhibition test results, 40 μg/mL or 139 μg/mL of thymol was used to treat *F. graminearum* to evaluate its effect on DON. The results are shown in Figure 2C. The contents of DON and 3-Ac-DON in the control group were 78.0 ± 10.8 mg/g and 1160 ± 130.5 mg/g, respectively. The DON and 3-Ac-DON in the EC_{50} thymol treatment group were 23.3 ± 7.5 mg/g and 255.0 ± 209.3, respectively. Compared with the control group, the DON and 3-Ac-DON of the thymol treatment group decreased by 70.1% and 78.2%, respectively. The DON and 3-Ac-DON of the EC_{90} thymol treatment group were 22.3 ± 8.0 mg/g and 166.4.0 ± 91.6 mg/g, respectively, which decreased by 71.4% and 85.7%, respectively. The results show that 40 μg/mL thymol could significantly inhibit the production of DON and 3-Ac-DON by *F. graminearum*.

2.3. The Effect of Thymol on the Transcriptome of F. graminearum

To evaluate the quality of RNA-Seq data, we conducted some quality control analyses. Illumina sequencing produced 47,244,179 reads (control group) and 49,121,259 reads (thymol group). Strict data cleaning and quality inspection of the Illumina platform sequencing results, the error rate, GC percentage, and Q20 percentage were 0.02%, 52.2%, and 98.3%, respectively. Using Trinity to assemble all clean data de novo, a total of 16,727 non-redundant unigenes were identified, and the proportion of the expected length of the sequence to the total BUSCO score was 95.9%. The Mapped ratio between the sequencing data and the assembly results was 89.27%, indicating that this study's assembled data were high quality.

To better understand the functions of these non-redundant unigenes, all unigenes were compared with the NCBI non-redundant protein database sequences, and the e-value threshold was 10^{-5}. The comparison analysis showed that a total of 12,597 unigenes matched the known proteins in the NR database. The matching percentages of *F. graminearum*, *Fusarium pseudograminearum*, and *Fusarium culmorum* were 73.05%, 5.61%, and 4.22%, respectively. All predicted unigenes were classified by functional annotation and classification through the Gene Ontology (GO) and Cluster of Orthologous Groups of Proteins (COG) database (Figure 3). The Top3 of the Biological Process (BP) were Cellular process, Metabolic process, and Localization; the Top3 of the Cellular Component (CC) were Cell part, Membrane part, and Organelle; the Top3 of the Molecular Function (MF) were Catalytic activity, Binding, and Transcription regulator activity. COG annotated 281, 230, and 356 unigenes with known functional classifications on Cellular processes and signaling, Information storage and processing, and Metabolism, respectively. In addition, 1653 (65.6%) unigenes were annotated by COG as Function unknown, and many unigenes were not matched to the database. These unigenes have the potential to be translated into functional proteins. This study's RNA-seq data helps enrich the annotations of the unigenes group of *F. graminearum*. The qRT-PCR results of the candidate genes were compared with the corresponding RNA-seq data, and the results were the same (Figure S1 and Table S1), which confirmed th e accuracy of the expression profile based on the RNA-seq data.

RSEM quantitatively analyzed the expression level of the unigenes, and the quantitative index was TPM. The overall distribution diagram of the expression level is shown in Figure 4D. At the same time, the correlation of the expression levels between samples was analyzed, and the heat map results are shown in Figure 4C. The correlation coefficient between samples was >0.986 and >0.941 in the control and treatment groups, respectively. The results of the PCA are shown in Figure 4A. PC1 was 53.02%, and PC2 was 15.94%, indicating a high similarity between samples. These results suggest that the correlation between biological replicates was high and that the experimental design was reasonable. On this basis, the unigenes with a differential expression caused by thymol treatment were further screened out (Figure 4B), and p-adjust was <0.05. When Fold change (FC) = 1.5, a total of 4417 differential unigenes were identified, of which 1989 were up-regulated, and 2428 were down-regulated. When FC was 2.0 times, a total of 3230 differential unigenes were identified, of which 1223 were up-regulated, and 2007 were down-regulated. When FC was 3.0 times, a total of 1944 differential unigenes were identified, of which 529 were up-regulated and 1415 were down-regulated. When FC was 6.0 times, 884 differential unigenes were identified, of which 165 were up-regulated, and 719 were down-regulated.

First, we used a functional enrichment analysis to understand the differentially expressed unigenes' functional pathways. The functional enrichment analysis of GO is shown in Figure 5A. The Top3 of BP were Catalytic activity, Oxidoreductase activity, and Cofactor binding. The Top3 of CC were Membrane part, Intrinsic component of membrane, and Integral component of membrane; the Top3 of MF were tRNA aminoacylation, Amino acid activation, and Polysaccharide catabolic process. The functional enrichment analysis of the Kyoto Encyclopedia of Genes and Genomes (KEGG) can be seen in Figure 5B. The Top3 of MF were Ribosome, Protein processing in endoplasmic reticulum, and Glycolysis/Gluconeogenesis.

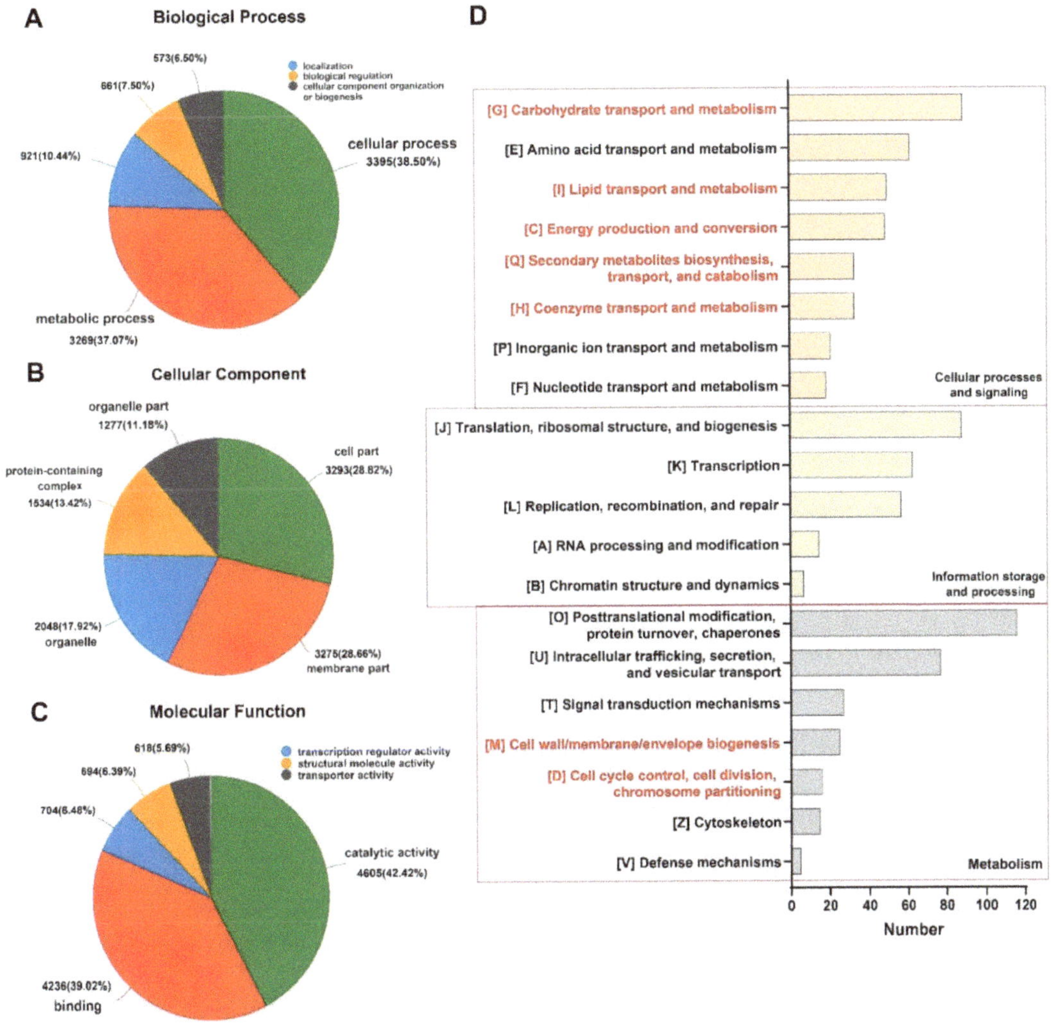

Figure 3. Function annotation analysis. (**A–C**) GO function annotation analysis; (**D**) COG function annotation analysis. Red markers represent annotations related to energy metabolism processes.

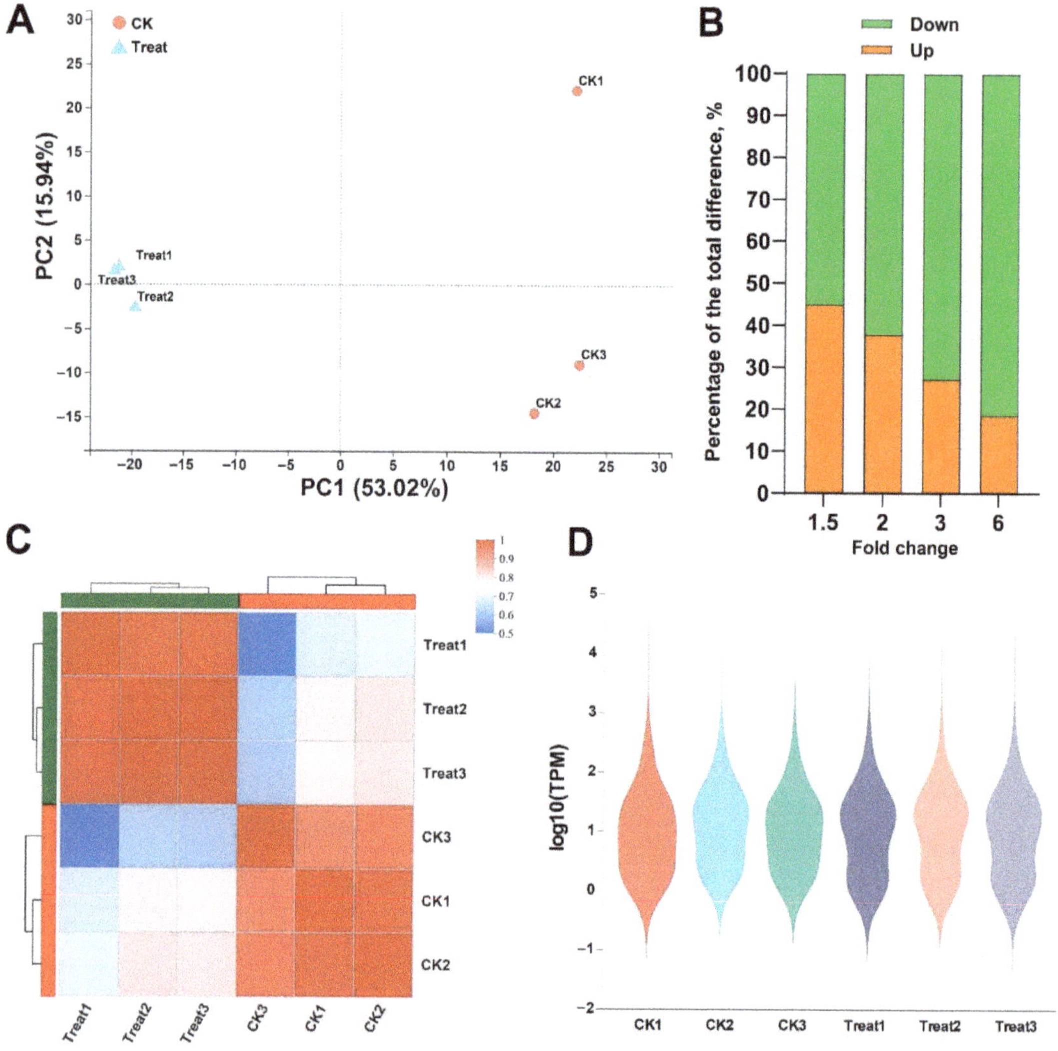

Figure 4. Analysis of the relationship between samples. (**A**) PCA analysis between samples; (**B**) Percentage of up-regulated/down-regulated unigenes identified by different multiples of difference; (**C**) Heat map of correlation analysis between samples; (**D**) Distribution of expression levels. CK is the control group, and Treat is the thymol treatment group.

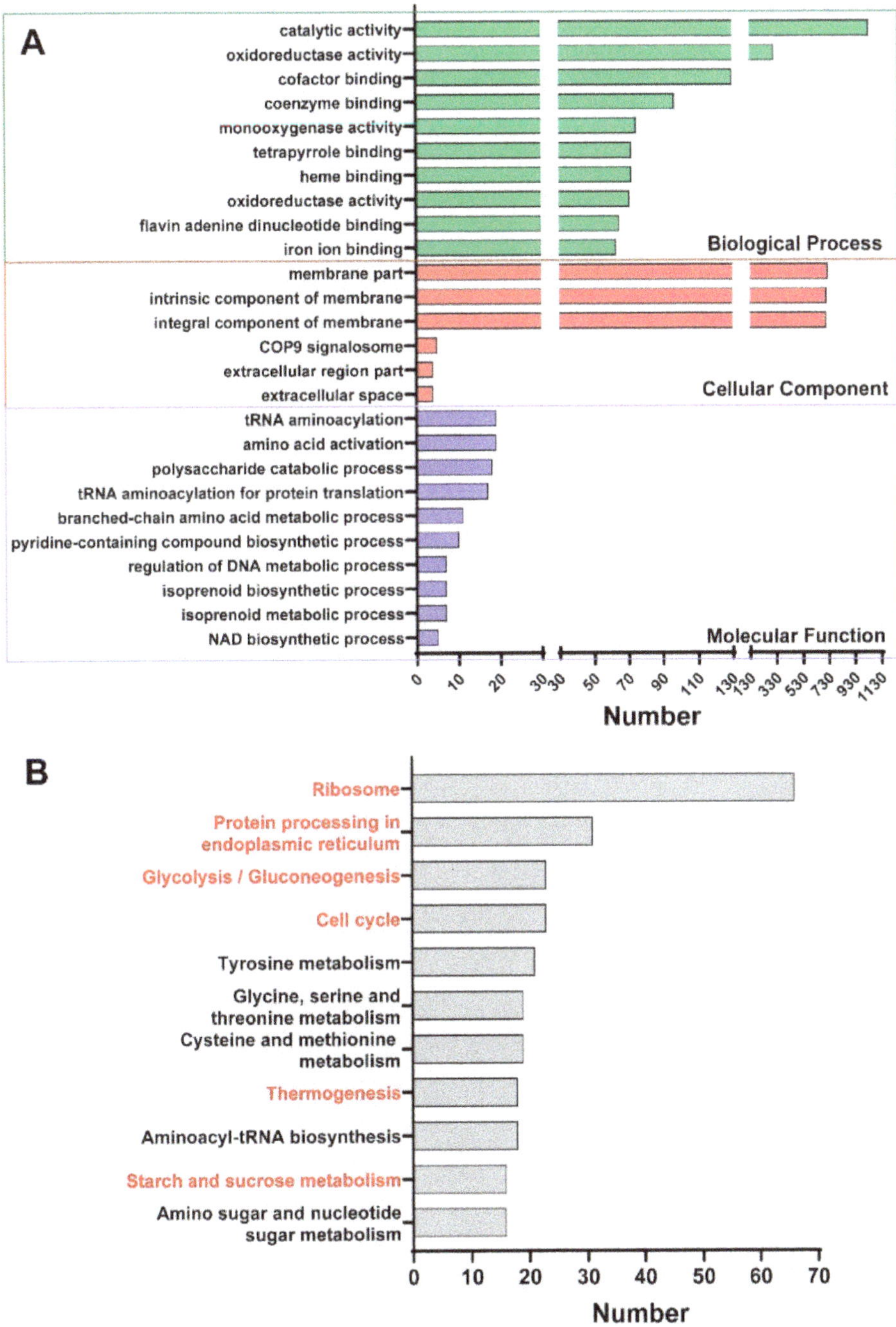

Figure 5. Functional enrichment analysis (**A**) GO function enrichment analysis; (**B**) KEGG function enrichment analysis.

2.4. Unigenes Related to Mycelial Growth and DON Production

The differentially expressed unigenes were functionally annotated and summarized in each NR, EggNOG, GO, KEGG, and Swiss-Prot database, and then the unigenes of interest were examined. Many unigenes related to mycelial growth (Table 1), DON synthesis (Table 2), secondary metabolites (Supplementary Materials Table S2), and glycolysis process (Supplementary Materials Table S3) were identified.

Table 1. Unigenes related to mycelial growth.

Name	COG_ID	FC	Swiss-Prot_Description	Functional Categories
CND2	COG5229	2.37	Condensin complex subunit 2	Chromatin structure and dynamics
MPIP	COG5105	3.05	M-phase inducer phosphatase	Cell cycle control, cell division, chromosome partitioning
HSK1	ENOG410YKCM	3.25	Cell cycle serine/threonine-protein kinase hsk1	
CDC45	ENOG410XR8E	2.00	Cell division control protein 45 homolog	
APC10	COG5156	3.50	Anaphase-promoting complex subunit 10	
CALM	COG5126	2.98	Calmodulin	
HXT2	ENOG410XNQK	0.46	High-affinity glucose transporter HXT2	Carbohydrate transport and metabolism
PHSG	COG0058	2.41	Glycogen phosphorylase	
HSP80	COG0326	0.06	Heat shock cognate protein 80	Coenzyme transport and metabolism
FSR4	ENOG4111F9D	2.44	Trans-enoyl reductase fsr4	Lipid transport and metabolism
PARP2	ENOG410XP18	13.54	ADP-ribose polymerase 2	Translation, ribosomal structure, and biogenesis
SMC2	COG1196	2.79	Structural maintenance of chromosomes protein 2	Transcription
SAK1	ENOG410XSHE	2.00	Protein sak1	
MCM1	COG5068	2.02	Transcription factor of morphogenesis MCM1	
ORC4	ENOG410XSK0	2.70	Origin recognition complex subunit 4	Replication, recombination, and repair
RAD1	ENOG410YHQU	2.26	DNA damage checkpoint control protein rad1	
SKP1	COG5201	2.42	E3 ubiquitin ligase complex SCF subunit scon-3	Posttranslational modification, protein turnover
CDK1	ENOG410XPP3	2.26	Cyclin-dependent kinase 1	Signal transduction mechanisms
PPZ	COG0639	0.49	Serine/threonine-protein phosphatase PP-Z	
PHLN	COG3511	0.08	Non-hemolytic phospholipase C	Cell wall/membrane/envelope biogenesis
GAL10	COG1087	3.03	Bifunctional protein GAL10	

Table 2. Unigenes related to the synthesis of DON.

Name	COG_ID	FC	Swiss-Prot_Description	Functional Categories
TRI4	COG2124	0.01	Cytochrome P450 monooxygenase	Cytochrome P450
TRI13	COG2124	0.02	Cytochrome P450 monooxygenase	
TRI11	COG2124	0.02	Trichothecene C-15 hydroxylase	
TRI1	COG2124	0.01	Cytochrome P450 monooxygenase	
TRI8	ENOG410YF7H	0.01	Trichothecene C-3 esterase	Trichothecene C-3 esterase
TRI6	ENOG410YRUA	0.01	Trichothecene biosynthesis transcription regulator 6	Trichothecene biosynthesis transcription regulator
TRI5	ENOG410YBAU	0.01	Trichodiene synthase	Trichodiene synthase
CLM1	ENOG410YEIS	0.01	Longiborneol synthase CLM1	
TRI3	ENOG4112AIP	0.01	Trichothecene 15-O-acetyltransferase	Trichothecene 15-O-acetyltransferase
TRI14	ENOG41118DK	0.03	Core trichothecene cluster	Core trichothecene cluster
TR101	ENOG410YJA8	0.09	Trichothecene 3-O-acetyltransferase	Transferase family
TRI10	ENOG410YFZP	0.06	Trichothecene biosynthesis transcription regulator 10	
TRI12	ENOG410XNQK	0.01	Trichothecene efflux pump TRI12	Fungal trichothecene efflux pump
ACLA	ENOG410XNQK	0.01	MFS efflux transporter aclA	
STR3	ENOG41109Q9	0.06	Siderophore iron transporter 3	
LACP	ENOG410XNQK	0.08	Lactose permease	
NAG3	COG0477	0.00	Major facilitator superfamily multidrug transporter NAG3	
GSFJ	ENOG410XNQK	0.08	Probable efflux pump gsfJ	
YJ94	ENOG410ZNSM	0.00	Uncharacterized membrane protein	
MF227	ENOG4111EZJ	0.02	Probable efflux pump mfs2	
ATB_A	COG0477	0.07	Efflux pump atB	
FUS6	ENOG410XNQK	0.06	Efflux pump FUS6	
FUB11	COG0477	0.06	Efflux pump FUBT	
MIRB	ENOG410XNQK	0.01	Siderophore iron transporter mirB	
ROQT	ENOG410XNQK	0.03	Efflux pump roqT	
SIT1	ENOG410ZYHG	0.15	Siderophore iron transporter 1	

3. Discussion

3.1. The Effect of Thymol on the Growth of Mycelium

The fungistatic activity of thymol against pathogenic microorganisms has been widely reported [17,30]. For example, the EC_{50} of *Staphylococcus aureus*, *Staphylococcus luteus*, *Escherichia coli*, and *Bacillus cereus* is 27.64–128.58 μM [31]. Studies have found that seven kinds of thymol have suitable antifungal activities, one of which has an activity similar to the commercial fungicide thiabendazole [32]. Thymol inhibits the growth of *Fusarium oxysporum* with an EC_{50} of 80 μg/mL [21], and *Allium tuberosum R.* inhibits the growth of *Fusarium oxysporum* with an EC_{50} of 400 mg/mL [33]. In contrast, thymol has a better fungistatic effect. Thymol shows extensive antifungal activity against various isolates of *F. graminearum*, inhibiting the production of conidia, the growth of hyphae [20], and the production of DON [34]. By detecting the sensitivity of 59 *F. graminearum* strains to thymol, the EC_{50} values of thymol of these strains were 22.53–51.76 μg/mL, and the average value was 26.3 μg/mL [11]. In this study, the EC_{50} of thymol against *F. graminearum* was 40 μg/mL, compared with the EC_{50} of eugenol and its derivatives of 395.7–1163.9 μM, indicating that the sensitivity of *F. graminearum* to eugenol is lower than thymol [35]. Thymol can also be used in combination, such as cinnamaldehyde and carvacrol, as a synergist to enhance antifungal activity [36] and reduce the production of trichothecenes by 95–99% [24]. These studies have shown that thymol can very effectively inhibit the growth of *F. graminearum*, and it is a very potent plant and antifungal agent.

COG annotation results annotated many transcripts related to fungal growth, such as "Cell cycle control, Cell division, Chromosome partitioning" (16 unigenes) and "Cell wall/membrane/envelope biogenesis" (25 unigenes). From the Top10 of KEGG functional enrichment analysis, many transcripts were significantly enriched in "Protein processing in endoplasmic reticulum" (31 unigenes), and "Cell cycle" (23 unigenes) related to fungal growth was also screened (Figure 5B). Based on the functional annotations of related databases and previous research reports, we screened out unigenes associated with the effect of thymol on mycelial growth (Table 1). From the table, we can see that many related to cell growth processes, such as cycle control, the translation process, and ribosome and cell wall/membrane; cell division and replication processes, such as classification, chromosomes, and transcription processes; and material metabolism processes, such as carbohydrates, coenzymes, and lipids. The expression of unigenes during transportation and metabolism changed significantly.

Interestingly, COG annotated many unigenes related to "Ribosome" (66 unigenes). DON can interact with the peptidyl-transferase region of the 60 S ribosomal subunit to induce "ribosomal stress toxicity" [10]. It suggests that thymol may alleviate toxicity by alleviating the ribosomal stress caused by DON to animals [37]. However, the mechanism of action of thymol on the ribosomes of *F. graminearum* itself is still unclear, and further research is needed. The results show that thymol inhibits the growth of *F. graminearum* by affecting the expression of related unigenes in the various processes of mycelial growth.

3.2. The Effect of Thymol on DON Production by F. graminearum

During the process of *F. graminearum* infecting plants, it can produce a variety of secondary metabolites, and one of the most concerning products is DON [38]. Therefore, we also summarized the table where thymol affects mycelial DON (Table 2) and the synthesis of secondary metabolites (Supplementary Materials Table S2). *Tri* gene refers to a gene cluster related to the biosynthetic pathway of trichothecenes. *Tri1*, *Tri4*, *Tri13*, and *Tri11* are the more important CYPs in fungi. *Tri4* encodes a multifunctional oxygenase that converts trichodiene to isotrichotriol [39]. *Tri1* and *Tri11* encode 3-acetyltrich-othecen C-8 hydroxylase and isotrichodermin C-15 hydroxylase, respectively. *Tri13*, as the 3-acetyl trichothecenes C-4 hydroxylase, is responsible for the hydroxylation of C-4 [40]. *Tri4* is involved in the synthesis of the trichothecenes framework [34]. Compared with plants and animals, few fungal CYPs have been thoroughly studied for their functions. They may be the key enzymes fungi use to metabolize phenolic compounds and aromatic hydrocarbon compounds [41]. *Tri5* is the first gene in DON biosynthesis. *Tri6* and *Tri10* are unigenes that regulate the synthesis of DON [42]. Fusarium's self-protection mechanism pumps the toxin out of the cell through *Tri12* and reduces the toxicity of intermediates in the biosynthesis of trichothecenes through *Tri101* [42]. Thymol can reduce the expression of *Tri5* [43] and inhibit the function of the toxin efflux pump, thereby enhancing the sensitivity of the fungus to tetracycline and benzalkonium chloride [44]. This is consistent with our results; 15 unigenes related to the fungal trichothecene efflux pump, such as *Tri10*, *Tri12*, *FUS6*, *FUB11*, and *ROQT*, were significantly down-regulated after thymol treatment. Studies have found that plant essential oils reduce the production of DON, 3-Ac-DON, and 15-Ac-DON by 96.33–100% [18], consistent with our results. The results indicate that thymol may inhibit the expression of unigenes clusters related to the trichothecene biosynthetic pathway and inhibit the Fungal trichothecene efflux pump, thereby inhibiting the synthesis of DON.

3.3. *The Effect of Thymol on Glycolysis in F. graminearum*

In addition to the synthesis of DON, thymol also affects the synthesis of many secondary metabolites [12]. For example, histone acetyltransferases, such as *Elp3*, *Sas3*, and *Gcn5*, are related to the regulatory effect induced by DON [45–47]. Earlier studies reported that thymol might cause cell membrane damage by inducing lipid peroxidation and inhibiting ergosterol biosynthesis, thereby inhibiting the growth of pathogenic fungus [22]. Interestingly, we found that thymol may also inhibit toxins' growth and production by inhibiting the fungus's glycolysis process through a further analysis of transcriptomics data. Supplementary Materials Table S3 found that the expression levels of many unigenes related to carbohydrates and protein methylases, acetylases, oxidoreductases, and hydrolases have undergone significant changes. *ADH* is responsible for catalyzing the last methanol synthesis step [48]. *ALDOC* participates in the aldol condensation reaction in glycolysis [49]. *NAGA* encodes α-N-acetylgalactosaminidase, which is mainly involved in regulating the metabolism of glycoproteins and glycolipids in lysosomes [50]. *PME* catalyzes the hydrolysis of pectin with pectinic acid and methanol. *DAK1* catalyzes the production of dihydroxyacetone phosphate and enters the glycolysis pathway. *ENOA* is the gene encoding enolase, the metallocenes responsible for catalyzing the ninth step of glycolysis, converting 2-phosphoglycerate to phosphoenolpyruvate. *CHI1* catalyzes the hydrolysis of chitin to N-acetylglucosamine. As a key enzyme in the glycolysis process, the phosphoglycerate kinase encoded by *PGK* can catalyze ATP production.

It shows that the synthesis of secondary metabolites is closely associated with gluconeogenesis/glycolysis. Cinnamaldehyde can regulate intracellular glucose metabolism through α-enolase [51]. *Chuzhou chrysanthemum* can inhibit the growth of *E. coli* through the hexose monophosphate pathway [12,52]. It proves that EO can hinder the growth of a fungus by affecting energy metabolism. Glycolysis is an important metabolic process of *F. graminearum*, so we infer that thymol should also be able to exert fungistatic effects through glycometabolism and energy utilization pathways (Supplementary Materials Table S3). The COG annotation results (Figure 3D) annotate that many unigenes are related to energy metabolism processes, such as "Coenzyme transport and metabolism" (33 unigenes), "Secondary metabolites biosynthesis, transport and catabolism" (33 unigenes), "Energy Production and conversion and Lipid transport and metabolism" (49 unigenes), "Amino acid transport and metabolism" (62 unigenes), and "Carbohydrate transport and metabolism" (89 unigenes). From the Top10 of the KEGG functional enrichment analysis, many unigenes were also screened to be significantly enriched in pathways related to the energy metabolism process (Figure 5B), such as "Starch and sucrose metabolism" (16 unigenes), "Thermogenesis" (18 unigenes), and "Glycolysis/Gluconeogenesis" (23 unigenes).

Studies have found that EO does not completely inhibit the production of AFB1 by inhibiting the growth of fungi. It may also interfere with the process of carbohydrate decomposition and metabolism, resulting in an insufficient supply of acetyl-CoA, thereby reducing the ability of fungi to produce aflatoxin [53]. This is because acetyl-CoA is a key component in the glycolysis process and a crucial substrate and raw material in the production of DON. Thymol can inhibit the expression of acetyl-CoA carboxylase and fatty acid synthase [54] and affect the utilization of farnesyl pyrophosphate FPP (the precursor of DON synthesis) [43]. Acetyl-CoA and Tri5 work together. Many studies have shown that thymol can affect many intermediate products in the tricarboxylic acid (TCA) cycle [17], such as citrate, fumarate, succinate, and α-ketoglutarate [21,55]. Thymol mediates its bactericidal activity against Staphylococcus aureus by targeting aldehyde-ketoreductase to consume NADPH [56]. On Caenorhabditis elegans, thymol accelerates glucose metabolism by regulating multiple targets in the glycolytic pathway and participates in the degradation of fatty acids [57]. In summary, thymol can affect the energy homeostasis in cells [58]. It may interfere with the glycolysis process and the formation of the DON toxin via acetyl-CoA or other common substances. This is a new idea to study the effect of EO on fungi. Acetyl-CoA is used as the raw material for DON synthesis. If the acetyl-CoA produced by glycolysis is not enough to supply its DON synthesis, it will eventually reduce the production of DON by *F. graminearum*. As far as we know, this is the first article combining thymol on the growth inhibition and toxin production of *F. graminearum* and RNA-Seq to understand the effect of thymol on *F. graminearum* fully.

4. Conclusions

Thymol can effectively inhibit the growth of *F. graminearum* and the production of DON. These results prove that after thymol treatment, many genes related to growth, DON, and the secondary metabolite synthesis process of *F. graminearum* undergo significant changes, which ultimately affect the growth and toxin production of *F. graminearum*. The study has enriched the data about thymol's influence on the genes of *F. graminearum* from the transcriptomics level. In addition, since the acetyl-CoA produced by the gluconeogenesis/glycolysis process can simultaneously participate in growth and toxin production, we believe that gluconeogenesis/glycolysis can be a breakthrough point for future research on the regulation of other plant essential oils in *F. graminearum*.

5. Materials and Methods

5.1. Fungal Strain, Media and Culture Condition

The *F. graminearum* strain W3008 was kindly provided by the College of Plant Science and Technology of Huazhong Agricultural University, China [59]. The strain was routinely cultured at 25 °C on potato dextrose agar (PDA, Hopebio, Qingdao, China) plates and was preserved in 20% disinfected glycerol at −80 °C for long-term storage [60]. Thymol (HPLC grade standard, purity > 98%, B21153, Shanghai yuanye Bio-Technology Co., Ltd. Shanghai, China) was dissolved in acetone into a 100 mg/mL stock solution, protected from light, and stored at 4 °C.

5.2. Determination of the Sensitivity of Mycelial Growth to Thymol

According to the results of our previous experiments, thymol was diluted by a certain multiple and then added to the PDA medium. The control group only added an equal volume of acetone (0.5%); the final concentrations of the thymol treatment group were 0, 5, 10, 20, 40, 80, and 160 μg/mL, and the acetone concentration in all groups was 5 μL/mL (0.5% of acetone used). A 6 mm diameter bacterial cake was taken from the edge of a 3-day-old colony with a sterile puncher, placed in the culture medium's center, and cultured for 4 days. The colony diameter was measured by the cross method every 24 h to evaluate the sensitivity of mycelial growth to thymol. The experiment was repeated 3 times, with 3 repetitions for each concentration. The percentage of mycelial radial growth inhibition on the 4th day of the culture after inoculation was calculated by the MGIR formula, MGIR (%) = $[(C-N)/(C-6)] \times 100$. Where C and N are the average diameter values of the control and treatment groups, respectively. The thymol concentration of 50% (EC_{50}) and 90% (EC_{90}) of mycelial growth inhibition rate were calculated by the regression equation (See Supplementary Material "regression equation" and Figure S2).

5.3. Changes in DON and 3-Ac-DON

The preparation method of the conidia can be seen the supplementary material "Conidiation Assays" [46]. We added 1 mL of the spore suspension (5×10^5 CFU/mL) to a flask containing 100 mL of GYEP (glucose yeast extract peptone) medium, and we incubated it with shaking (180 r/min) at 25 °C for 24 h [60]. Thymol was then added to the culture, and the same amount of acetone (0.5%) was added to the control culture. The final concentration of the thymol treatment group was 40 μg/mL (this concentration was close to the EC_{50}) or 139 μg/mL (this concentration was close to the EC_{90}). The acetone concentration in each culture was 5 μL/mL (0.5% of acetone used), and each treatment had 3 replicates. After 7 days of continuous cultivation, the mycelium was collected and dried at 60 °C for 3 h [61].

The filtrate was used for DON and 3-Ac-DON quantification. DON production in vitro was expressed as a ratio of DON content to dry mycelia weight (μg/g) [62]. The extraction and purification methods (see Supplementary Material "Changes in DON and 3-Ac-DON" for details) of DON and 3-Ac-DON were improved from Stroka [63], and the quantitative method was based on Diao [64]. The pure products of DON and 3-Ac-DON were from FERMENTEK, with a purity of >99.6%.

5.4. Transcriptome Analysis

A total of 100 mL of GYEP medium containing 1 mL of conidia suspension (5×10^5 conidia/mL) was incubated with shaking (180 rpm/min) at 25 °C; after 24 h, thymol was added to the culture; the concentration of the thymol treatment group was 40 μg/mL, and the amount of acetone (0.5%) in groups control and treatment was same. Then, we continued incubating for 24 h, filtered, and collected hyphae [60]. The quality of RNA was evaluated after total RNA isolation using TRIzol reagent (Invitrogen, Shanghai, China). After the mRNA was isolated and fragmented, the library was sequenced using IlluminaHiSeq^{TM}2000. Quality control of the original reading was performed, and we removed the linker and other low-quality base sequences to obtain the clean data reading.

Trinity was used to assemble all samples from scratch. TransRate and CD-HIT were optimized and filtered to remove common errors and redundancies. Then, BUSCO was used for assembly evaluation, and finally, the clean reads of each sample were compared with the reference sequence obtained by the Trinity assembly to obtain the mapping result of each sample. Unigene was the longest transcript in the transcript cluster, and unigenes were used for functional database annotation analyses (NR, Swiss-prot, Pfam, COG, GO, and KEGG). The RPKM method was used to calculate the expression value of unigenes (Reads Perkbper Millionread). RSEM compared the quality-controlled sequencing data with the assembled transcriptome sequence through comparison software, such as the bowtie, and then it estimated the expression abundance of unigenes/transcripts based on the comparison results. The quantitative expression index was TPM for homogenization so that the total expression in the sample was consistent for a more intuitive comparison. We used DESeq2 to analyze the differences between groups based on the quantitative expression results. The screening threshold was $|\log2FC| \geq 1.585$ and p-adjust < 0.05 to obtain unigenes with a differential expression between the two groups. Finally, the functional enrichment analysis of GO and KEGG was performed on the differentially expressed unigenes.

5.5. qRT-PCR

The total RNA was extracted from the sample as described above for real-time quantitative qRT-PCR analysis. Reverse transcription was performed using ABScriptIII Reverse Transcriptase kit (RK20408, ABclonal Technology Co., Ltd. Wuhan, China) with gDNA Eraser. Then, cDNA and 2×Universal SYBR Green Fast qPCR Mix reagent (RK21203, ABclonal Technology Co., Ltd.) were added to the 384 plates, respectively. We used a CFX384 real-time PCR system (Bio-Rad, Hercules, CA, USA) to complete qRT-PCR detection. The PCR program was as follows: 95 °C for 1 min, 40 cycles of 95 °C for 10 s, 60 °C for 5 s, and 72 °C for 10 s. The melting curve analysis was performed between 60 °C and 95 °C. The primers were synthesized by TSINGKE (Beijing, China). The qRT-PCR experiment was repeated 3 times, and each sample was repeated 3 times for analysis. *EF1α* was used as the reference gene for normalized expression data, and the relative gene expression level was calculated based on $2^{-\Delta\Delta Ct}$. Detailed information about gene-specific primers and alignment results are listed in the Supplementary Materials.

5.6. Statistical Analysis

The data were shown as mean ± standard deviation, comparing colony diameter, inhibition rate, DON, and 3-Ac-DON concentration. The values were analyzed by a one-way analysis of variance (ANOVA), followed by Duncan's multiple range test, and different letters indicated significant differences at $p < 0.05$, compared with different concentrations of thymol. All analyses were performed with the GraphPad Prism 8.0 software (GraphPad Software, Inc., San Diego, CA, USA).

5.7. Availability of Supporting Data

The original Illumina sequencing dataset was submitted to the NCBI Sequence Read Archive with the accession number PRJNA792342. CK1-CK3 is the control group, and Treat1-Treat3 is the 40 μg/mL thymol treatment group.

Supplementary Materials: The following supporting information can be downloaded at: https://www.mdpi.com/article/10.3390/toxins14020142/s1, Table S1: Primer sequence of quantitative fluorescence PCR; Table S2: Unigenes related to the synthesis of secondary metabolites; Table S3: Unigenes related to glycolysis; Figure S1: Comparison of gene expression (*Tri5, Tri6, Tri8, Tri14, Tri101, LEU1, 6PGD1, ERG6*) levels based on RNA-seq and qRT-PCR; Figure S2: Sensitivity regression equation.

Author Contributions: Supervision and funding acquisition, D.-S.Q. and S.W.; investigation and writing—original draft preparation, L.-Q.W. and K.-T.W.; writing—review and editing, L.-Q.W., K.-T.W., P.Y., and F.H.; visualization and formal analysis, P.Y., F.H., and S.A.R. All authors have read and agreed to the published version of the manuscript.

Funding: This research was funded by the National Key Research and Development Program of China (no. 2016YFD0501207).

Institutional Review Board: Not applicable.

Data Availability Statement: Not applicable.

Conflicts of Interest: The authors declare no conflict of interest.

References

1. Vandicke, J.; De Visschere, K.; Ameye, M.; Croubels, S.; De Saeger, S.; Audenaert, K.; Haesaert, G. Multi-Mycotoxin Contamination of Maize Silages in Flanders, Belgium: Monitoring Mycotoxin Levels from Seed to Feed. *Toxins* **2021**, *13*, 202. [CrossRef] [PubMed]
2. Goswami, R.S.; Kistler, H.C. Heading for disaster: *Fusarium graminearum* on cereal crops. *Mol. Plant Pathol.* **2004**, *5*, 515–525. [CrossRef] [PubMed]
3. Mielniczuk, E.; Skwaryło-Bednarz, B. Fusarium Head Blight, mycotoxins and strategies for their reduction. *Agronomy* **2020**, *10*, 509. [CrossRef]
4. Bhat, R.; Rai, R.V.; Karim, A.A. Mycotoxins in Food and Feed: Present Status and Future Concerns. *Compr. Rev. Food Sci. Food Saf.* **2010**, *9*, 57–81. [CrossRef]
5. Gruber-Dorninger, C.; Jenkins, T.; Schatzmayr, G. Global mycotoxin occurrence in feed: A ten-year survey. *Toxins* **2019**, *11*, 375. [CrossRef]
6. Wu, K.; Jia, S.; Zhang, J.; Zhang, C.; Wang, S.; Rajput, S.A.; Sun, L.; Qi, D. Transcriptomics and flow cytometry reveals the cytotoxicity of aflatoxin B1 and aflatoxin M1 in bovine mammary epithelial cells. *Ecotoxicol. Environ. Saf.* **2021**, *209*, 111823. [CrossRef]
7. Wu, K.; Liu, M.; Wang, H.; Rajput, S.A.; Shan, Y.; Qi, D.; Wang, S. The Mechanism Underlying the Extreme Sensitivity of Duck to Aflatoxin B1. *Oxid. Med. Cell. Longev.* **2021**, *2021*, 9996503. [CrossRef]
8. Wu, K.; Ren, C.; Gong, Y.; Gao, X.; Rajput, S.A.; Qi, D.; Wang, S. The insensitive mechanism of poultry to zearalenone: A review. *Anim. Nutr.* **2021**, *7*, 587–594. [CrossRef]
9. Wang, S.; Wu, K.; Xue, D.; Zhang, C.; Rajput, S.A.; Qi, D. Mechanism of deoxynivalenol mediated gastrointestinal toxicity: Insights from mitochondrial dysfunction. *Food Chem. Toxicol.* **2021**, *153*, 112214. [CrossRef]
10. Khoshal, A.K.; Novak, B.; Martin, P.; Jenkins, T.; Neves, M.; Schatzmayr, G.; Oswald, I.P.; Pinton, P. Co-occurrence of DON and Emerging Mycotoxins in Worldwide Finished Pig Feed and Their Combined Toxicity in Intestinal Cells. *Toxins* **2019**, *11*, 727. [CrossRef]
11. Audenaert, K.; Callewaert, E.; Hofte, M.; De Saeger, S.; Haesaert, G. Hydrogen peroxide induced by the fungicide prothioconazole triggers deoxynivalenol (DON) production by *Fusarium graminearum*. *BMC Microbiol.* **2010**, *10*, 112. [CrossRef] [PubMed]
12. Ni, Z.; Wang, X.; Shen, Y.; Thakur, K.; Han, J.; Zhang, J.; Hu, F.; Wei, Z. Recent updates on the chemistry, bioactivities, mode of action, and industrial applications of plant essential oils. *Trends Food Sci. Technol.* **2021**, *110*, 78–89. [CrossRef]
13. Prakash, B.; Kedia, A.; Mishra, P.K.; Dubey, N.K. Plant essential oils as food preservatives to control moulds, mycotoxin contamination and oxidative deterioration of agri-food commodities—Potentials and challenges. *Food Control* **2015**, *47*, 381–391. [CrossRef]

14. Hu, L.B.; Ban, F.F.; Li, H.B.; Qian, P.P.; Shen, Q.S.; Zhao, Y.Y.; Mo, H.Z.; Zhou, X. Thymol Induces Conidial Apoptosis in *Aspergillus flavus* via Stimulating K(+) Eruption. *J. Agric. Food Chem.* **2018**, *66*, 8530–8536. [CrossRef]
15. Wang, K.; Jiang, S.; Yang, Y.; Fan, L.; Su, F.; Ye, M. Synthesis and antifungal activity of carvacrol and thymol esters with heteroaromatic carboxylic acids. *Nat. Prod. Res.* **2019**, *33*, 1924–1930. [CrossRef]
16. Kong, J.; Zhang, Y.; Ju, J.; Xie, Y.; Guo, Y.; Cheng, Y.; Qian, H.; Quek, S.Y.; Yao, W. Antifungal effects of thymol and salicylic acid on cell membrane and mitochondria of *Rhizopus stolonifer* and their application in postharvest preservation of tomatoes. *Food Chem.* **2019**, *285*, 380–388. [CrossRef]
17. Kong, J.; Xie, Y.; Yu, H.; Guo, Y.; Cheng, Y.; Qian, H.; Yao, W. Synergistic antifungal mechanism of thymol and salicylic acid on *Fusarium solani*. *LWT* **2021**, *140*, 110787. [CrossRef]
18. Perczak, A.; Gwiazdowska, D.; Marchwińska, K.; Juś, K.; Gwiazdowski, R.; Waśkiewicz, A. Antifungal activity of selected essential oils against *Fusarium culmorum* and *F. graminearum* and their secondary metabolites in wheat seeds. *Arch. Microbiol.* **2019**, *201*, 1085–1097. [CrossRef]
19. Velluti, A.; Sanchis, V.; Ramos, A.J.; Turon, C.; Marin, S. Impact of essential oils on growth rate, zearalenone and deoxynivalenol production by *Fusarium graminearum* under different temperature and water activity conditions in maize grain. *J. Appl. Microbiol.* **2004**, *96*, 716–724. [CrossRef]
20. Wei, J.T.; Wu, K.T.; Sun, H.; Khalil, M.M.; Dai, J.F.; Liu, Y.; Liu, Q.; Zhang, N.Y.; Qi, D.S.; Sun, L.H. A Novel Modified Hydrated Sodium Calcium Aluminosilicate (HSCAS) Adsorbent Can Effectively Reduce T-2 Toxin-Induced Toxicity in Growth Performance, Nutrient Digestibility, Serum Biochemistry, and Small Intestinal Morphology in Chicks. *Toxins* **2019**, *11*, 199. [CrossRef]
21. Zhang, M.; Ge, J.; Yu, X. Transcriptome Analysis Reveals the Mechanism of Fungicidal of Thymol Against *Fusarium oxysporum f. sp. niveum*. *Curr. Microbiol.* **2018**, *75*, 410–419. [CrossRef] [PubMed]
22. Gao, T.; Zhou, H.; Zhou, W.; Hu, L.; Chen, J.; Shi, Z. The Fungicidal Activity of Thymol against *Fusarium graminearum* via Inducing Lipid Peroxidation and Disrupting Ergosterol Biosynthesis. *Molecules* **2016**, *21*, 770. [CrossRef] [PubMed]
23. Ye, X.; Ling, T.; Xue, Y.; Xu, C.; Zhou, W.; Hu, L.; Chen, J.; Shi, Z. Thymol Mitigates Cadmium Stress by Regulating Glutathione Levels and Reactive Oxygen Species Homeostasis in Tobacco Seedlings. *Molecules* **2016**, *21*, 1339. [CrossRef] [PubMed]
24. Zhang, Y.; Liu, X.; Wang, Y.; Jiang, P.; Quek, S. Antibacterial activity and mechanism of cinnamon essential oil against *Escherichia coli* and *Staphylococcus aureus*. *Food Control* **2016**, *59*, 282–289. [CrossRef]
25. Mith, H.; Dure, R.; Delcenserie, V.; Zhiri, A.; Daube, G.; Clinquart, A. Antimicrobial activities of commercial essential oils and their components against food-borne pathogens and food spoilage bacteria. *Food Sci. Nutr.* **2014**, *2*, 403–416. [CrossRef] [PubMed]
26. Wang, L.; Hu, W.; Deng, J.; Liu, X.; Zhou, J.; Li, X. Antibacterial activity of Litsea cubeba essential oil and its mechanism against *Botrytis cinerea*. *Rsc Adv.* **2019**, *9*, 28987–28995. [CrossRef]
27. Wang, L.H.; Zhang, Z.H.; Zeng, X.A.; Gong, D.M.; Wang, M.S. Combination of microbiological, spectroscopic and molecular docking techniques to study the antibacterial mechanism of thymol against *Staphylococcus aureus*: Membrane damage and genomic DNA binding. *Anal. Bioanal. Chem.* **2017**, *409*, 1615–1625. [CrossRef]
28. Zhang, F.; Guo, Z.; Zhong, H.; Wang, S.; Yang, W.; Liu, Y.; Wang, S. RNA-Seq-based transcriptome analysis of aflatoxigenic *Aspergillus flavus* in response to water activity. *Toxins* **2014**, *6*, 3187–3207. [CrossRef]
29. Lin, J.Q.; Zhao, X.X.; Zhi, Q.Q.; Zhao, M.; He, Z.M. Transcriptomic profiling of *Aspergillus flavus* in response to 5-azacytidine. *Fungal Genet. Biol.* **2013**, *56*, 78–86. [CrossRef]
30. Kachur, K.; Suntres, Z. The antibacterial properties of phenolic isomers, carvacrol and thymol. *Crit. Rev. Food Sci. Nutr.* **2020**, *60*, 3042–3053. [CrossRef]
31. Marchese, A.; Orhan, I.E.; Daglia, M.; Barbieri, R.; Di Lorenzo, A.; Nabavi, S.F.; Gortzi, O.; Izadi, M.; Nabavi, S.M. Antibacterial and antifungal activities of thymol: A brief review of the literature. *Food Chem.* **2016**, *210*, 402–414. [CrossRef] [PubMed]
32. Alves, E.M.; Ribeiro, R.; Martins, M.L.; Antonio, D.S.C.T.; Santana, F.C.; Lirian, J.C.; Borges, W.S.; Costa, A.V.; Queiroz, V.T.; Scherer, R.; et al. Thymol as an Interesting Building Block for Promising Fungicides against *Fusarium solani*. *J. Agric. Food Chem.* **2021**, *69*, 6958–6967. [CrossRef] [PubMed]
33. Zhang, X.; Wang, H.; Zhu, W.; Li, W.; Wang, F. Transcriptome Analysis Reveals the Effects of Chinese Chive (*Allium tuberosum R.*) Extract on *Fusarium oxysporum* f. sp. radicis-lycopersici Spore Germination. *Curr. Microbiol.* **2020**, *77*, 855–864. [CrossRef] [PubMed]
34. Oufensou, S.; Balmas, V.; Azara, E.; Fabbri, D.; Dettori, M.A.; Schuller, C.; Zehetbauer, F.; Strauss, J.; Delogu, G.; Migheli, Q. Naturally Occurring Phenols Modulate Vegetative Growth and Deoxynivalenol Biosynthesis in *Fusarium graminearum*. *ACS Omega* **2020**, *5*, 29407–29415. [CrossRef]
35. Cui, W.; Du, K.; Ling, Y.; Yang, C. Activity of eugenol derivatives against *Fusarium graminearum* Q1 strain and screening of isoeugenol mixtures. *J. Plant Pathol.* **2021**, *103*, 915–921. [CrossRef]
36. Venturini, T.P.; Rossato, L.; Chassot, F.; De Azevedo, M.I.; Al-Hatmi, A.M.S.; Santurio, J.M.; Alves, S.H. Activity of cinnamaldehyde, carvacrol and thymol combined with antifungal agents against *Fusarium* spp. *J. Essent. Oil Res.* **2021**, *33*, 502–508. [CrossRef]
37. Khmelnitskiy, G.O.; Korzunenko, V.D. Effect of combined sorbent preparation on growth, performance and haematobiochemical alterations in broiler chickens under the influence of T-2 toxin and deoxynivalenol mixed toxicosis. *Bangladesh J. Vet. Med.* **2013**, *11*, 7–11. [CrossRef]
38. Chen, Y.; Kistler, H.C.; Ma, Z. *Fusarium graminearum* Trichothecene Mycotoxins: Biosynthesis, Regulation, and Management. *Annu. Rev. Phytopathol.* **2019**, *57*, 15–39. [CrossRef]

39. Adnan, M.; Fang, W.; Sun, P.; Zheng, Y.; Abubakar, Y.S.; Zhang, J.; Lou, Y.; Zheng, W.; Lu, G.D. R-SNARE FgSec22 is essential for growth, pathogenicity and DON production of *Fusarium graminearum*. *Curr. Genet.* **2020**, *66*, 421–435. [CrossRef]
40. Shin, J.; Kim, J.E.; Lee, Y.W.; Son, H. Fungal Cytochrome P450s and the P450 Complement (CYPome) of *Fusarium graminearum*. *Toxins* **2018**, *10*, 112. [CrossRef]
41. Bombassaro, A.; Schneider, G.X.; Costa, F.F.; Leao, A.; Soley, B.S.; Medeiros, F.; Da, S.N.; Lima, B.; Castro, R.; Bocca, A.L.; et al. Genomics and Virulence of *Fonsecaea pugnacius*, Agent of Disseminated Chromoblastomycosis. *Front. Genet.* **2020**, *11*, 822. [CrossRef] [PubMed]
42. Hao, G.; McCormick, S.; Tiley, H.; Usgaard, T. Detoxification and Excretion of Trichothecenes in Transgenic Arabidopsisthaliana Expressing *Fusarium graminearum* Trichothecene 3-O-acetyltransferase. *Toxins* **2021**, *13*, 320. [CrossRef] [PubMed]
43. Oufensou, S.; Dessì, A.; Dallocchio, R.; Balmas, V.; Azara, E.; Carta, P.; Migheli, Q.; Delogu, G. Molecular Docking and Comparative Inhibitory Efficacy of Naturally Occurring Compounds on Vegetative Growth and Deoxynivalenol Biosynthesis in Fusarium culmorum. *Toxins* **2021**, *13*, 759. [CrossRef] [PubMed]
44. Miladi, H.; Zmantar, T.; Chaabouni, Y.; Fedhila, K.; Bakhrouf, A.; Mahdouani, K.; Chaieb, K. Antibacterial and efflux pump inhibitors of thymol and carvacrol against food-borne pathogens. *Microb. Pathog.* **2016**, *99*, 95–100. [CrossRef]
45. Chen, Y.; Wang, J.; Yang, N.; Wen, Z.; Sun, X.; Chai, Y.; Ma, Z. Wheat microbiome bacteria can reduce virulence of a plant pathogenic fungus by altering histone acetylation. *Nat. Commun.* **2018**, *9*, 3429. [CrossRef]
46. Kong, X.; van Diepeningen, A.D.; van der Lee, T.; Waalwijk, C.; Xu, J.; Xu, J.; Zhang, H.; Chen, W.; Feng, J. The *Fusarium graminearum* Histone Acetyltransferases Are Important for Morphogenesis, DON Biosynthesis, and Pathogenicity. *Front. Microbiol.* **2018**, *9*, 654. [CrossRef]
47. Lee, Y.; Min, K.; Son, H.; Park, A.R.; Kim, J.C.; Choi, G.J.; Lee, Y.W. ELP3 is involved in sexual and asexual development, virulence, and the oxidative stress response in *Fusarium graminearum*. *Mol. Plant Microbe Interact.* **2014**, *27*, 1344–1355. [CrossRef]
48. Wen, N.; Liu, W.; Hou, Y.; Zhao, Z. The kinetics behavior of the reduction of formaldehyde catalyzed by Alcohol Dehydrogenase (ADH) and partial uncompetitive substrate inhibition by NADH. *Appl. Biochem. Biotechnol.* **2013**, *170*, 370–380. [CrossRef]
49. Fujita, H.; Aoki, H.; Ajioka, I.; Yamazaki, M.; Abe, M.; Oh-Nishi, A.; Sakimura, K.; Sugihara, I. Detailed expression pattern of aldolase C (Aldoc) in the cerebellum, retina and other areas of the CNS studied in Aldoc-Venus knock-in mice. *PLoS ONE* **2014**, *9*, e86679. [CrossRef]
50. Li, Y.; Ma, C.; Li, W.; Yang, Y.; Li, X.; Liu, J.; Wang, J.; Li, S.; Liu, Y.; Li, K.; et al. A missense variant in NDUFA6 confers schizophrenia risk by affecting YY1 binding and NAGA expression. *Mol. Psychiatry* **2021**. [CrossRef]
51. Zhang, W.; Gao, J.; Shen, F.; Ma, X.; Wang, Z.; Hou, X.; Hao, E.; Hou, Y.; Bai, G. Cinnamaldehyde changes the dynamic balance of glucose metabolism by targeting ENO1. *Life Sci.* **2020**, *258*, 118151. [CrossRef] [PubMed]
52. Cui, H.; Bai, M.; Sun, Y.; Abdel-Samie, M.A.; Lin, L. Antibacterial activity and mechanism of *Chuzhou chrysanthemum* essential oil. *J. Funct. Foods* **2018**, *48*, 159–166. [CrossRef]
53. Tian, J.; Ban, X.; Zeng, H.; He, J.; Huang, B.; Wang, Y. Chemical composition and antifungal activity of essential oil from Cicuta virosa L. var. latisecta Celak. *Int. J. Food Microbiol.* **2011**, *145*, 464–470. [CrossRef] [PubMed]
54. Kang, D.; Lee, Y.; Oh, S.M.; Yoon, D.; Choi, D.J.; Kwon, D.; Kang, O.; Lee, D.Y. Inhibitory Effects of Thymol Isolated from *Curcuma longa* L. on Adipogenesis in HepG2 Cells. *Processes* **2020**, *8*, 1191. [CrossRef]
55. Araniti, F.; Miras-Moreno, B.; Lucini, L.; Landi, M.; Abenavoli, M.R. Metabolomic, proteomic and physiological insights into the potential mode of action of thymol, a phytotoxic natural monoterpenoid phenol. *Plant Physiol. Biochem.* **2020**, *153*, 141–153. [CrossRef]
56. Zhou, W.; Wang, Z.; Mo, H.; Zhao, Y.; Li, H.; Zhang, H.; Hu, L.; Zhou, X. Thymol Mediates Bactericidal Activity against *Staphylococcus aureus* by Targeting an Aldo–Keto Reductase and Consequent Depletion of NADPH. *J. Agric. Food Chem.* **2019**, *67*, 8382–8392. [CrossRef]
57. Ban, F.; Hu, L.; Zhou, X.H.; Zhao, Y.; Mo, H.; Li, H.; Zhou, W. Inverse molecular docking reveals a novel function of thymol: Inhibition of fat deposition induced by high-dose glucose in *Caenorhabditis elegans*. *Food Sci. Nutr.* **2021**, *9*, 4243–4253. [CrossRef]
58. Baldissera, M.D.; Souza, C.F.; De Matos, A.; Baldisserotto, B.; Da, S.A.; Monteiro, S.G. Tissue oxidative damage mediates impairment on phosphotransfer network during thymol intake: Effects on hepatic and renal bioenergetics. *Chem. Biol. Interact.* **2018**, *296*, 83–88. [CrossRef]
59. Wang, S.; Yang, J.; Zhang, B.; Wu, K.; Yang, A.; Li, C.; Zhang, J.; Zhang, C.; Rajput, S.A.; Zhang, N.; et al. Deoxynivalenol Impairs Porcine Intestinal Host Defense Peptide Expression in Weaned Piglets and IPEC-J2 Cells. *Toxins* **2018**, *10*, 541. [CrossRef]
60. Hellin, P.; King, R.; Urban, M.; Hammond-Kosack, K.E.; Legreve, A. The adaptation of *Fusarium culmorum* to DMI Fungicides Is Mediated by Major Transcriptome Modifications in Response to Azole Fungicide, Including the Overexpression of a PDR Transporter (FcABC1). *Front. Microbiol.* **2018**, *9*, 1385. [CrossRef]
61. Liu, X.; Yu, F.; Schnabel, G.; Wu, J.; Wang, Z.; Ma, Z. Paralogous cyp51 genes in *Fusarium graminearum* mediate differential sensitivity to sterol demethylation inhibitors. *Fungal Genet. Biol.* **2011**, *48*, 113–123. [CrossRef] [PubMed]
62. Li, J.; Duan, Y.; Bian, C.; Pan, X.; Yao, C.; Wang, J.; Zhou, M. Effects of validamycin in controlling *Fusarium* head blight caused by *Fusarium graminearum*: Inhibition of DON biosynthesis and induction of host resistance. *Pestic. Biochem. Physiol.* **2019**, *153*, 152–160. [CrossRef] [PubMed]

63. Stroka, J.; Derbyshire, M.; Mischke, C.; Ambrosio, M.; Kroeger, K.; Arranz, I.; Sizoo, E.; van Egmond, H. Liquid chromatographic determination of deoxynivalenol in baby food and animal feed: Interlaboratory study. *J. AOAC Int.* **2006**, *89*, 1012–1020. [CrossRef] [PubMed]
64. Diao, X.; Han, Y.; Liu, C. The Fungicidal Activity of Tebuconazole Enantiomers against *Fusarium graminearum* and Its Selective Effect on DON Production under Different Conditions. *J. Agric. Food Chem.* **2018**, *66*, 3637–3643. [CrossRef]

toxins

MDPI

Article

A Newly Isolated *Alcaligenes faecalis* ANSA176 with the Capability of Alleviating Immune Injury and Inflammation through Efficiently Degrading Ochratoxin A

Rui Zheng [†], Hanrui Qing [†], Qiugang Ma [*], Xueting Huo, Shimeng Huang, Lihong Zhao, Jianyun Zhang and Cheng Ji

State Key Laboratory of Animal Nutrition, College of Animal Science and Technology, China Agricultural University, Beijing 100193, China
* Correspondence: maqiugang@cau.edu.cn
† These authors contributed equally to this work.

Abstract: Ochratoxin A (OTA) is one of the most prevalent mycotoxins that threatens food and feed safety. Biodegradation of OTA has gained much attention. In this study, an *Alcaligenes faecalis* strain named ANSA176, with a strong OTA-detoxifying ability, was isolated from donkey intestinal chyme and characterized. The strain ANSA176 could degrade 97.43% of 1 mg/mL OTA into OTα within 12 h, at 37 °C. The optimal levels for bacterial growth were 22–37 °C and pH 6.0–9.0. The effects of ANSA176 on laying hens with an OTA-contaminated diet were further investigated. A total of 36 laying hens were assigned to three dietary treatments: control group, OTA (250 μg/kg) group, and OTA + ANSA176 (6.2 $\times$ 10^8 CFU/kg diet) group. The results showed that OTA decreased the average daily feed intake (ADFI) and egg weight (EW); meanwhile, it increased serum alanine aminopeptidase (AAP), leucine aminopeptidase (LAP), β2-microglobulin (β2-MG), immunoglobulin G (IgG), tumor necrosis factor-α (TNF-α), and glutathione reductase (GR). However, the ANSA176 supplementation inhibited or attenuated the OTA-induced damages. Taken together, OTA-degrading strain *A. faecalis* ANSA176 was able to alleviate the immune injury and inflammation induced by OTA.

Keywords: ochratoxin A; biodegradation; *Alcaligenes faecalis* ANSA176; immune injury; inflammation; layers

Key Contribution: We isolated an *A. faecalis* strain ANSA176 from donkey intestinal chyme which showed strong OTA degradation ability. In addition, ANSA176 could alleviate OTA-induced immune response and inflammation, indicating that ANSA176 might be used as a potential bioproduct for improving animal health and food safety.

Citation: Zheng, R.; Qing, H.; Ma, Q.; Huo, X.; Huang, S.; Zhao, L.; Zhang, J.; Ji, C. A Newly Isolated *Alcaligenes faecalis* ANSA176 with the Capability of Alleviating Immune Injury and Inflammation through Efficiently Degrading Ochratoxin A. *Toxins* **2022**, *14*, 569. https://doi.org/10.3390/toxins14080569

Received: 21 July 2022
Accepted: 17 August 2022
Published: 20 August 2022

1. Introduction

Ochratoxin A (OTA) is one of the most prevalent mycotoxins produced by fungi of the genus *Aspergillus* and *Penicillium* [1]. The natural contamination of OTA widely occurs in plant- and animal-derived food commodities, including cereal, fruit, coffee, meat, milk, and other agricultural food and feed [2,3]. Studies have indicated that OTA possesses carcinogenic, teratogenic, genotoxic, and immunotoxic properties [2,4,5]. The International Agency for Research on Cancer (IARC) has classified OTA as a probable carcinogen for humans (group 2B) [6].

In animals, OTA has a broad range of harmful effects on livestock health and productivity, resulting in food-safety incidents and economic losses [7]. Due to the high plasma-protein binding affinity and low metabolism rate of OTA, its bioaccumulation and carryover effects in edible tissues and animal-derived products further endanger human health [2,8]. For poultry's complete feed, the European Commission Recommendation 2006/576/EC limits the maximum tolerable level for OTA to 100 μg/kg [9]. Residues of

OTA are detectable in the liver, kidney, and eggs in OTA-challenged laying hens [10]. Feeding OTA contaminated diets to poultry typically causes decreased production performance, altered serum biochemical profiles, and impaired immune response [10,11]. The detrimental effects of OTA on the poultry industry have been indicated to be time- and dose-dependent, with the liver and kidney being the main target organs [4,12]. Exposure of 0.5–3 mg/kg OTA to poultry typically results in renal lesions, followed by decreases in feed consumption, grow rate, feed-conversion efficiency, egg production, and egg quality [7,12,13]. The effects of 1–5 mg/kg OTA on poultry serum biochemistry include decreased total protein (TP), albumin (ALB), and globulin levels and increased alkaline phosphatase (ALP) and gamma glutamine transpeptidase (GGT) levels [12]. Studies have been carried out for decades to elucidate the critical role of oxidative stress in OTA-induced toxicity [14]. The investigation of oxidative stress or antioxidant potential in OTA-fed (0–6.4 mg/kg) broiler chicks reveals a significant dose-dependent decrease in the superoxide dismutase (SOD), glutathione peroxidase (GPx), and total antioxidant status (TAS) in plasma and tissues [15].

To date, numerous physical, chemical, and biological methods have been proposed to eliminate OTA [16]. Among them, biodetoxification of OTA is a highly promising approach due to its specific, efficient, and environmentally friendly advantages [3]. Biological detoxification methods could be classified into adsorption and degradation. Many yeasts have been reported to remove OTA via high adsorption, in which cell viability is not a prerequisite and the most important influence factor is cell wall components [3,17]. *Lactobacillus rhamnosus* GG has also been found to decrease OTA during the first 15 h of culture growth [18]. However, the amount of toxin elimination could be partially reversed, indicating both a limitation and risk. The degradation of OTA to ochratoxin α (OTα: less toxic or almost nontoxic) and L-β-phenylalanine (Phe) by breaking the amide bond is considered the most important biodegradation mechanism [3,16]. A good deal of microorganisms have been identified as being able to degrade OTA, including fungi [19–21], bacteria [22,23], and yeast [24]. For example, *Aspergillus niger* M00120 isolated from the soil has the strong ability to detoxify 99% of OTA in 2 days [25]. Furthermore, the product has been identified as OTα and assessed for cytotoxic response, indicating that it does not induce cellular oxidative damage [25]. As for the bacteria, the strain *Acinetobacter* sp. Neg1 that was isolated from the OTA-contaminated vineyard is capable of converting 91% of OTA into OTα in six days, at 24 °C [26]. The biodegradation production has been confirmed as OTα by liquid chromatography with high-resolution mass spectrometer (LC–HRMS) [26]. However, the degradation efficiencies of some microbes might be limited, and those isolated from the gastrointestinal tract of healthy animal might be worth investigating. Notably, the application of microorganism to both the food and feed industries must be cautious of its safety.

The objective of this study was to obtain a high-efficient OTA-degrading strain, elucidate the detoxification mechanism, and examine its protective efficacy in OTA-fed poultry. The central hypothesis is that supplementation of this novel OTA biodegrading strain could alleviate the OTA-induced immune injury and inflammation in laying hens.

2. Results

2.1. Isolation and Characterization of OTA-Degrading Strain

Colonies grown on the Luria-Bertani (LB) agar plate were screened from the donkey intestinal chyme. One pure strain, ANSA176, exhibiting the highest OTA removal ability was isolated and stored in our laboratory (Figure 1). The degradation test showed that ANSA176 was able to degrade 97.43% of OTA and produce OTα within 12 h (Figure 2), indicating the cleavage of the amide bond in OTA. The constructed phylogenetic tree (Figure 3) was consistent with the phylogeny of some *Alcaligenes faecalis* (*A. faecalis* ANSA176). Considering the microscopic observations, biochemical characteristics (Table 1), and the 16S rRNA gene sequence, the strain used in this study was confirmed and named *A. Faecalis* ANSA176.

Table 1. Biochemical and physiological characteristics of *A. faecalis* ANSA176.

Substrate/Test	Result	Substrate/Test	Result	Substrate/Test	Result	Substrate/Test	Result
Cell shape	Rod	Gram Staining	–	Catalase	+	Oxidase	+
BIOLOG GEN III (carbon source utilization assays)							
Negative control	–	α-D-glucose	–	Gelatin	–	D-raffinose	–
3-methyl-D-glucose	–	D-mannose	–	Gentiobiose	–	α-D-lactose	–
D-glucose-6-PO_4	–	D-fructose	–	L-alanine	+	D-melibiose	–
Glucuronamide	+	D-galactose	–	L-arginine	–	L-lactic acid	+
D-cellobiose	–	Dextrin	–	L-glutamic acid	+	Citric acid	+
Glycyl-L-proline	–	L-rhamnose	–	L-histidine	+	α-ketoglutaric acid	+
L-aspartic acid	+	Inosine	–	L-pyroglutamic acid	+	D-malic acid	–
p-hydroxyphenyl acetic acid	+	D-sorbitol	–	L-serine	–	L-malic acid	+
Methyl pyruvate	+	D-mannitol	–	Pectin	–	Bromo succinic acid	+
D-lactic acid methyl ester	–	D-arabitol	–	Sucrose	–	γ-aminobutyric acid	–
β-methyl-D-glucoside	–	Myo-inositol	–	Turanose	–	α-hydroxybutyric acid	+
β-hydroxy-D,L-butyric acid	+	Glycerol	–	D-gluconic acid	–	D-salicin	–
N-acetyl-D-glucosamine	–	D-maltose	–	Stachyose	–	α-keto-butyric acid	+
N-acetyl-β-D-mannosamine	–	D-trehalose	–	Mucic acid	–	Propionic	+
N-acetyl-D-galactosamine	–	D-aspartic acid	–	Quinic acid	–	Acetic acid	+
N-neuraminic acid	–	D-serine	+	Saccharic acid	–	Formic acid	+
BIOLOG GEN III (chemical sensitivity assays; +, not sensitive; −, sensitive)							
Positive control	+	Tetrazolium blue	+	pH 6.0	+	Nalidixic acid	–
Na bromate	–	Fusidic acid	+	pH 5.0	+	LiCl	+
Rifamycin SV	+	D-serine	+	1% Na-lactase	+	K-tellurite	–
Minocycline	+	Lincomycin	+	1% NaCl	+	Aztreonam	+
Vancomycin	+	Guanidine HCl	+	4% NaCl	+	Na-butyrate	+
Niaproof 4	+	Tetrazolium violet	+	8% NaCl	+		

"+" and "−" represent the positive and negative response, respectively.

Figure 1. Isolation of the OTA-degrading strain. (**a**) Colony morphology. (**b**) Cell morphology. (**c**) Degradation of OTA by ANSA176 during a 24 h period. [a–b] The different letters in 1c mean significant difference ($p < 0.05$).

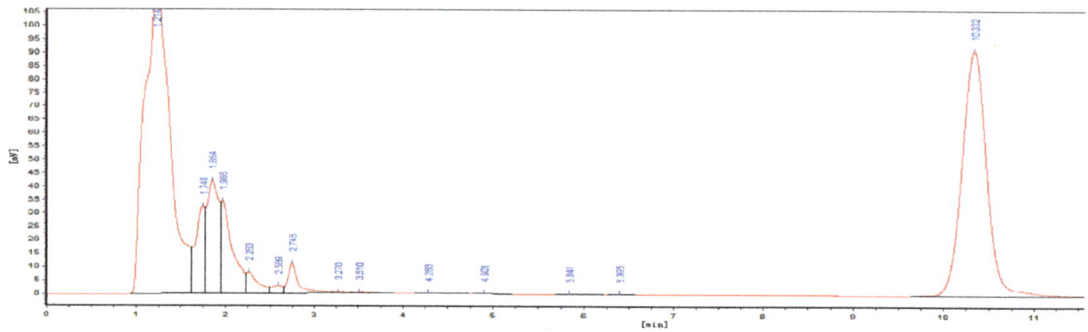

(**a**)

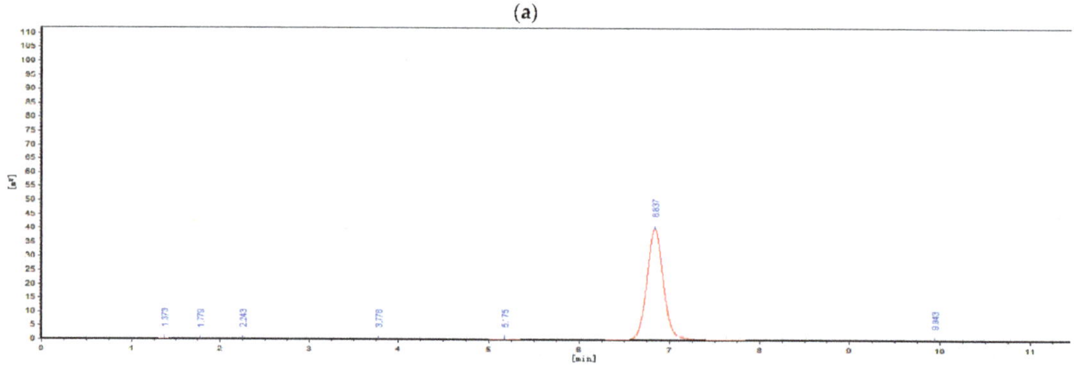

(**b**)

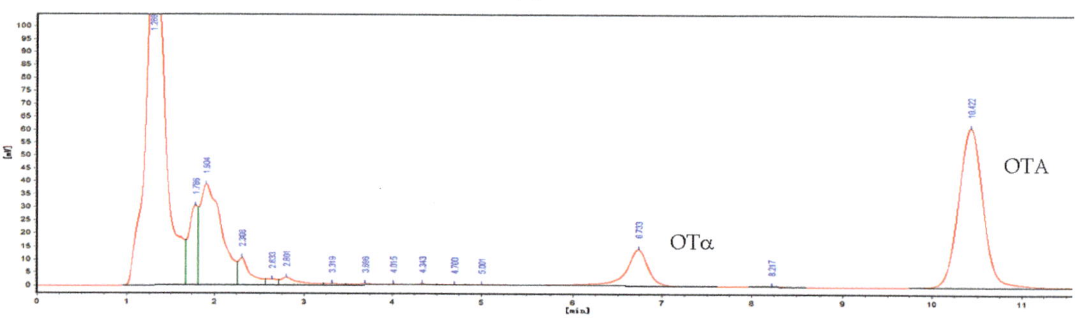

(**c**)

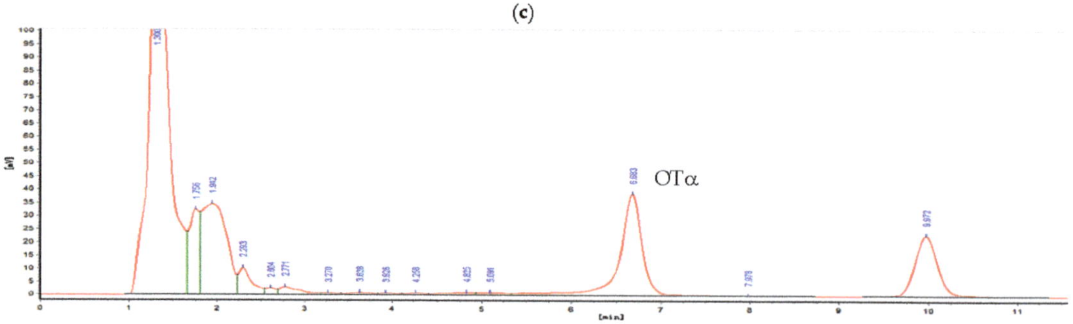

(**d**)

Figure 2. *Cont.*

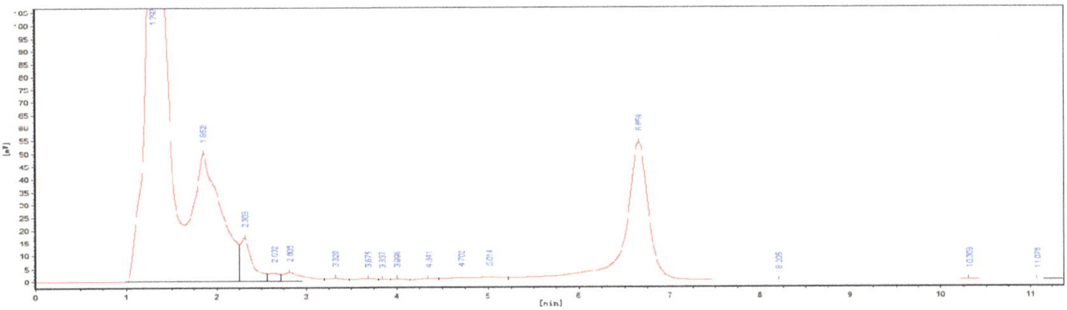

(e)

Figure 2. Determination of OTA degradation by HPLC. (**a**) LB culture medium with OTA standard. (**b**) OTα standard. (**c**) 30 % of OTA was degraded by ANSA176. (**d**) 75 % of OTA was degraded by ANSA176. (**e**) OTA was completely degraded by ANSA176.

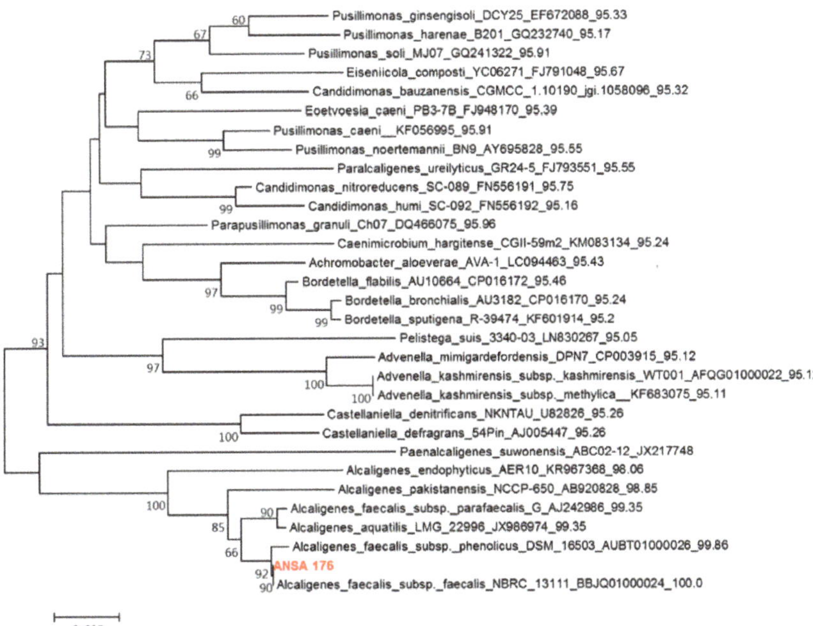

Figure 3. Phylogenetic tree of the isolated ANSA176 and related taxa.

Furthermore, we determined the growth of ANSA176 by OD_{600} measurement. As shown in Figure 4, the optimal temperatures for growth were observed to be between 22 and 37 °C (OD_{600} = 1.73–1.87). The lower (17 °C) and upper (42 °C) temperatures limited growth to some extent. In addition, the approximate pH levels for growth were 6.0–9.0 (OD_{600} = 1.54–1.79) at 37 °C incubation. At pH 10.0, the growth was reduced. Bacterial species subjected to the tested pH ranging from 2.5 to 5.0 were unable to grow.

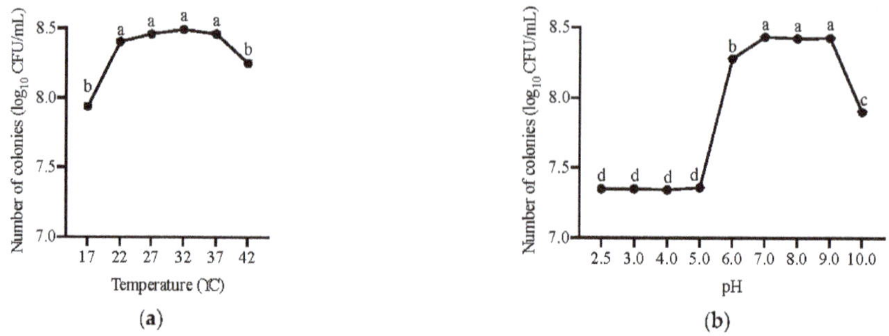

Figure 4. Growth characteristics of ANSA176. (**a**) Growth at different temperatures during a 24 h period. (**b**) Growth at different pH during a 24 h period. Data are presented as means ± SEM. [a–d] The different letters mean significant difference ($p < 0.05$).

2.2. Protection of A. faecalis ANSA176 against OTA-Induced Damages in Laying Hens

2.2.1. Production Performance in Laying Hens

In Figure 5a, we can see that the average daily feed intake (ADFI) of the OTA-fed group (2.02%, $p < 0.05$) and the OTA + ANSA176 group (6.75%, $p < 0.05$) was significantly lower than that of the control group. In the OTA + ANSA176 group, the ADFI was even lower (4.83%, $p < 0.05$) than that of the OTA-fed group. However, no statistically significant differences ($p > 0.10$) were found in the feed/egg ratio (FER, Figure 5b) when compared the OTA-fed group (an increase of 3.63%) or the OTA + ANSA176 group (a decrease of 5.16%) to the control group. Moreover, the FER was significantly decreased (8.48%, $p < 0.05$) in the OTA + ANSA176 group when compared to the OTA-fed group. During the experiment period, feeding layers with OTA at the concentration of 250 μg/kg decreased the egg production ratio (EPR; see Figure 5c; 4.61%, $p > 0.10$) and average daily egg mass (EM; see Figure 5d; 5.44%, $p > 0.10$) when compared with the control. Meanwhile, the supplementation of ANSA176 to the OTA-fed group ameliorated the negative effects by increasing EPR and EM at 0.94% ($p > 0.10$) and 2.46% ($p > 0.10$), respectively.

The effects OTA and ANSA176 on egg weight (EW), shell percentage, yolk percentage, and albumen percentages were significantly different ($p < 0.05$) during the experimental period, while there were no statistically significant differences ($p > 0.10$) in shell color, shell thickness, shell strength, Haugh unit, and yolk color (Table 2). As shown in Table 2, the EW in the OTA group was significantly lower than the control and OTA + ANSA176 groups ($p < 0.05$). The higher shell and yolk proportions and lower albumen proportion were found in the OTA group compared to those of the control group ($p < 0.05$). Meanwhile, the supplementation of ANSA176 in the OTA-containing diet ameliorated these changes. In addition, residues of OTA an OTα were not detected (limit of detection = 0.1 μg/kg) in the eggs of all groups.

Table 2. Effect of OTA and ANSA176 on the egg quality of layer hens.

Treatment	Control	OTA	OTA + ANSA176	SEM [1]	*p*-Value
Egg weight (g)	54.95 ^{a}	46.64 ^{b}	55.08 ^{a}	1.20	<0.01
Shell percentage (%)	9.92 ^{b}	12.17 ^{a}	10.23 ^{b}	0.30	<0.01
Yolk percentage (%)	25.90 ^{b}	31.81 ^{a}	25.35 ^{b}	1.07	<0.01
Albumen percentage (%)	64.18 ^{a}	56.01 ^{b}	64.43 ^{a}	1.31	<0.01
Shell color	58.03	58.01	57.69	1.31	0.98
Shell thickness (mm)	0.41	0.41	0.41	0.00	0.37
Shell strength (N)	40.75	38.77	39.90	1.12	0.46
Haugh unit	83.87	81.14	82.93	1.37	0.36
Yolk color	5.37	5.33	4.93	0.16	0.12

[1] Pooled standard error of the mean. [a,b] Means with different letters within a row mean significant difference ($p < 0.05$).

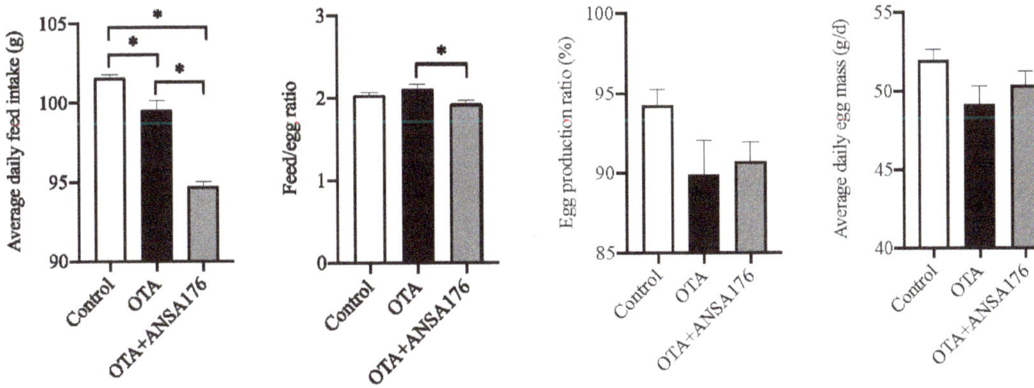

(a) (b) (c) (d)

Figure 5. Effects of OTA and ANSA176 on production performance of layers. (**a**) Average daily feed intake. (**b**) Feed/egg ratio. (**c**) Egg production ratio. (**d**) Average daily egg mass. Data are presented as means ± SEM. The differences are defined as * $p < 0.05$.

2.2.2. Kidney and Liver Damage Related Parameters in Laying Hens

The administration of OTA resulted in mild pathomorphological changes in kidney and liver. Histopathological examination revealed that OTA caused kidney tubulonephrosis with degeneration of the epithelial cells in proximal tubules and glomerulonephrosis with enlarged glomeruli and swollen capillary endothelial cells (Figure 6a). In addition, OTA induced inflammatory infiltration and hyaline degeneration in the liver, which were ameliorated in the OTA + ANSA176 group (Figure 6b).

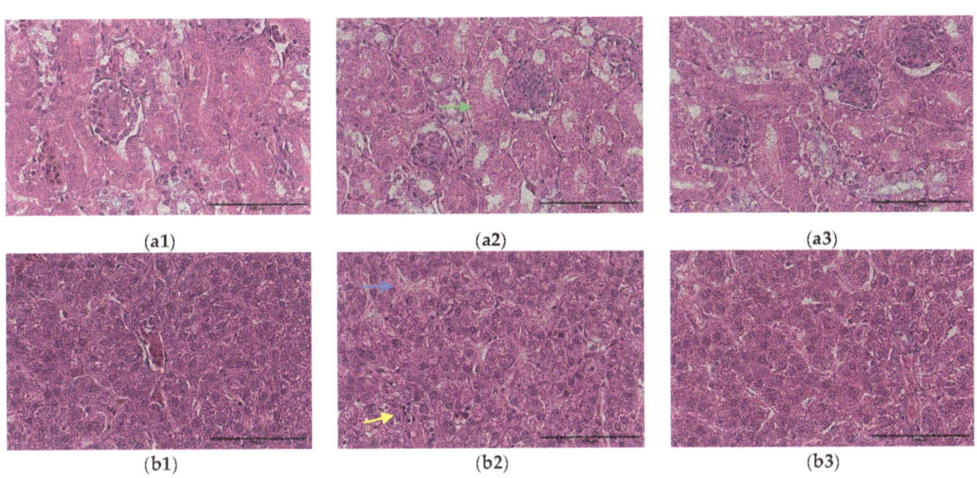

Figure 6. *Cont.*

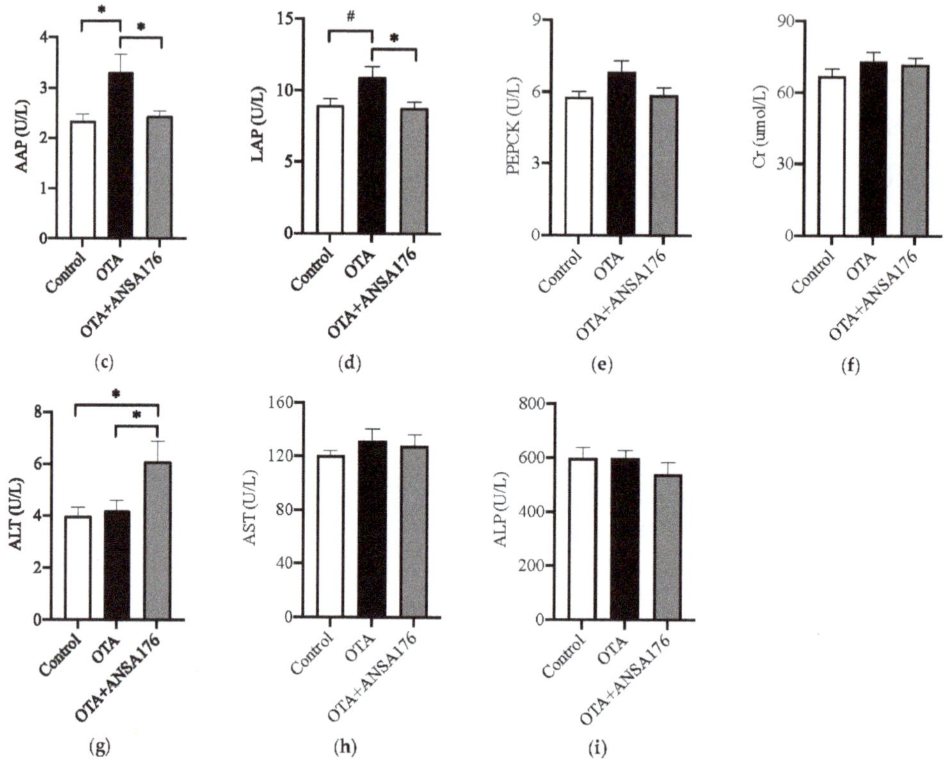

Figure 6. Effects of OTA and ANSA176 on kidney and liver damage related parameters in layers. (**a**,**b**) Histopathological examination of kidney (**a1**, control; **a2**, OTA; **a3**, OTA + ANSA176) and liver (**b1**, control; **b2**, OTA; **b3**, OTA + ANSA176) tissues by H&E staining. Magnification times, 400 × (scale bars = 100 μm). Lesions such as epithelial cells degeneration (green arrow), inflammatory infiltration (yellow arrow), and hyaline degeneration (blue arrow) are shown with arrows. (**c**–**i**) Serum biochemical parameters. Data are presented as means ± SEM. The differences are defined as # $0.05 \leq p < 0.10$ and * $p < 0.05$.

As shown in Figure 6c,d, the serum alanine aminopeptidase (AAP) ($p < 0.05$) and leucine aminopeptidase (LAP) ($p = 0.05$) levels increased in the OTA group compared to the control group, whereas the supplementation of ANSA176 obviously decreased ($p < 0.05$) them. The concentration of serum alanine aminotransferase (ALT) was significantly higher ($p < 0.05$) in the OTA + ANSA176 group than in the control and OTA groups. No differences ($p > 0.10$) were observed in phosphoenolpyruvate carboxykinase (PEPCK), creatinine (Cr), aspartate aminotransferase (AST), and alkaline phosphatase (ALP) among the three groups (Figure 6e,f,h,i).

2.2.3. Immune and Inflammatory Response in Laying Hens

The immune and inflammatory response of OTA and ANSA176-fed layers are presented in Figure 7. In the serum, the levels of β2-microglobulin (β2-MG) ($p < 0.05$), immunoglobulin G (IgG) ($p = 0.06$), and tumor necrosis factor-α (TNF-α) ($p = 0.05$) after OTA feeding showed increases compared with the control group. Meanwhile, the concentrations of IgG ($p < 0.05$), lysozyme (LZM) ($p < 0.05$), interleukin-2 (IL-2) ($p < 0.05$), and TNF-α ($p = 0.09$) in the OTA + ANSA176 groups were lower than the OTA-fed group. There were no obvious differences ($p > 0.10$) in total protein (TP), albumin (ALB), IgA, IgM, and IL-10 among all groups.

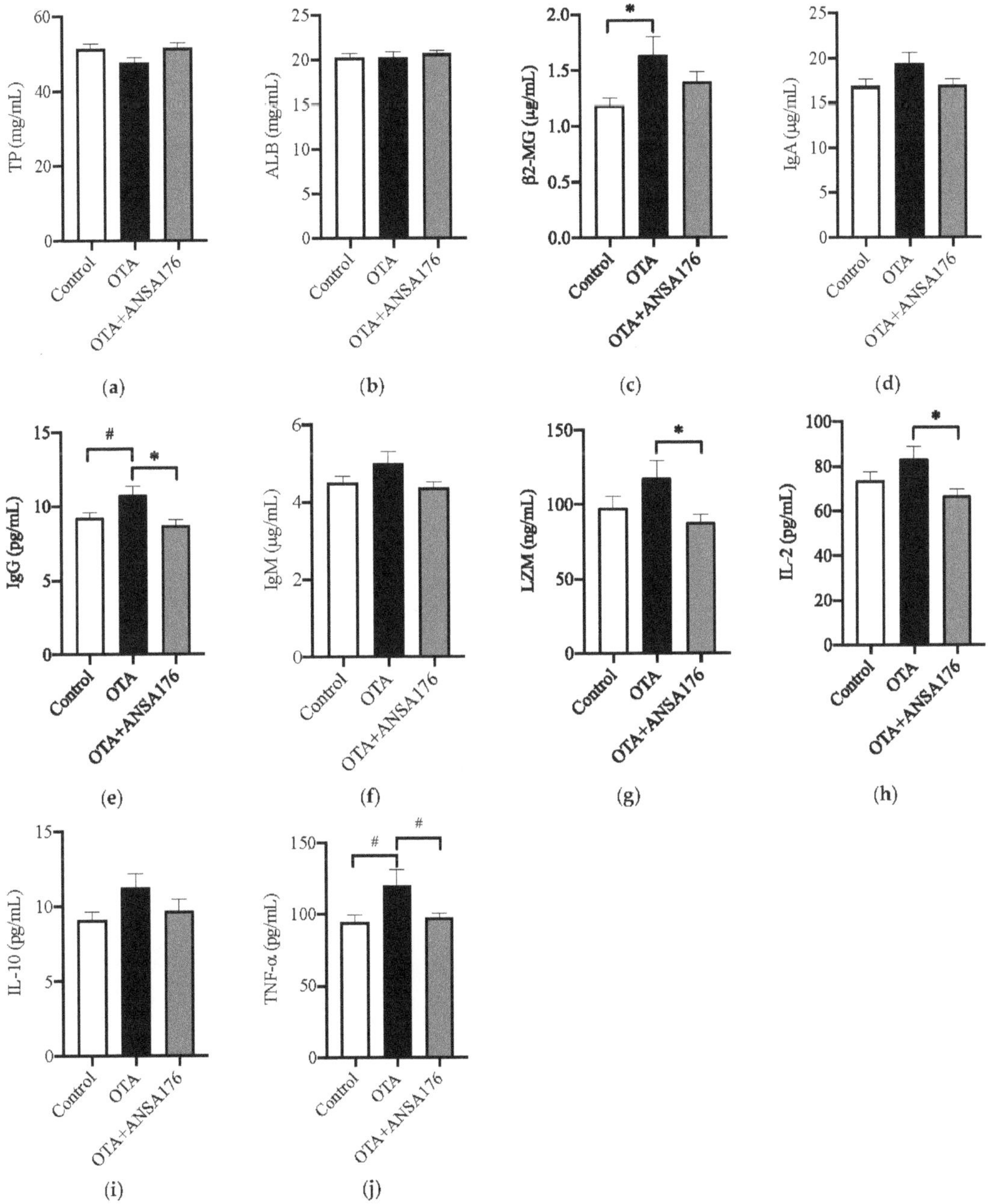

Figure 7. Effects of OTA and ANSA176 on immune and inflammatory response of layers. (**a**–**j**) Levels of TP, ALB, β2-MG, IgA, IgG, IgM, LZM, IL-2, IL-10, and TNF-α. Data are presented as means ± SEM. The differences are defined as # $0.05 \leq p < 0.10$ and * $p < 0.05$.

2.2.4. Oxidative Stress and Antioxidant Status in Laying Hens

Figure 8 exhibits the oxidative stress and antioxidant status in laying hens caused by the OTA and ANSA176 diet. The serum total antioxidant capacity (T-AOC) and superoxide dismutase (SOD) levels were significantly higher ($p < 0.05$) in the OTA + ANSA176 group compared with other groups. The glutathione reductase (GR) level was significantly higher ($p < 0.05$) in the OTA-fed group than in the control group. The levels of malonaldehyde (MDA), total glutathione (T-GSH), and glutathione peroxidase (GSH-Px) differed non-significantly ($p > 0.10$) among all groups.

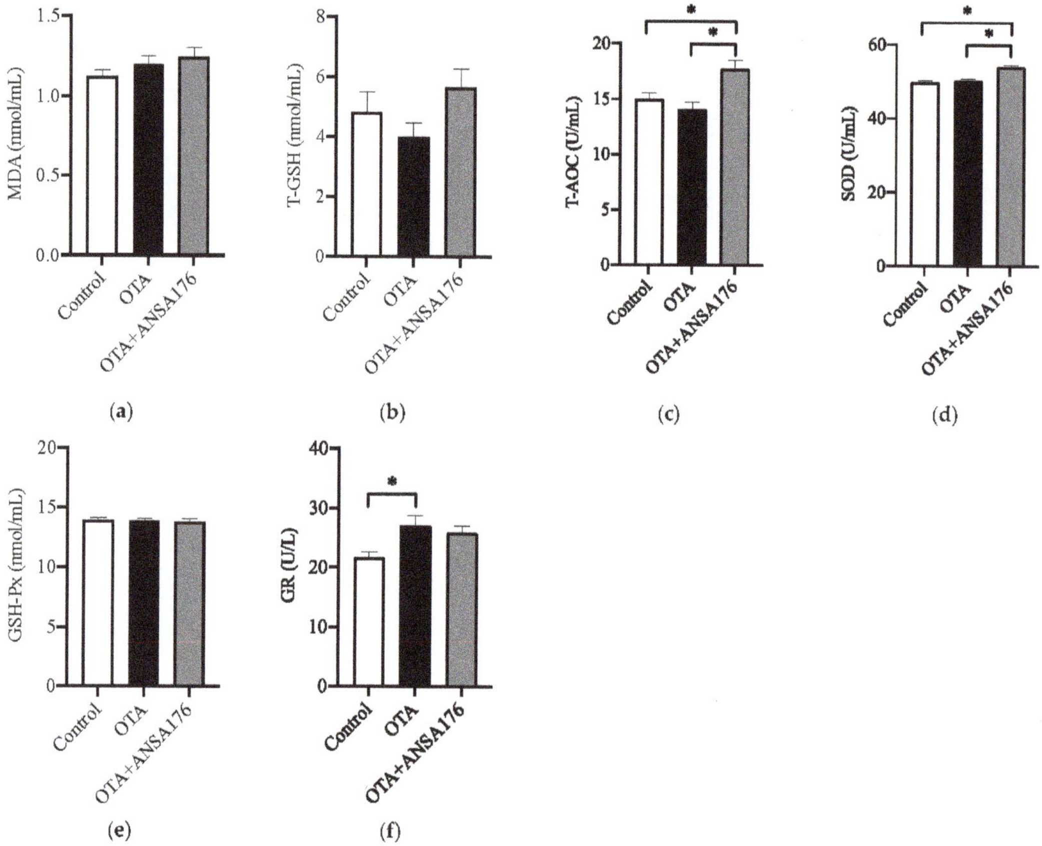

Figure 8. Effects of OTA and ANSA176 on oxidative stress and antioxidant status of layers. (**a–f**) Levels of MDA, T-GSH, T-AOC, SOD, GSH-Px, and GR. Data are presented as means ± SEM. The differences are defined as * $p < 0.05$.

3. Discussion

Ochratoxin A is a toxic secondary fungal metabolite that widely contaminates agriculture products, thereby threatening animal and human health. Microorganisms with an OTA-biodegradation property have great administration prospects in the food and feed industries due to their advantages in high specificity and effectivity [3]. In this study, we isolated and identified the *A. faecalis* ANSA176 strain from donkey intestinal chyme and found that it is capable of degrading 97.43% of OTA to OTα within 12 h in vitro. In addition, ANSA176 showed optimal growth at 22–37 °C and pH 6.0–9.0. Taken together, these results indicated a potential application of ANSA176 for the OTA biodegradation use due to its easy culture and high degradation efficiency. Microorganisms with a confirmed ability to transform OTA into OTα have been previously studied [3,27]. It is noteworthy that an *A. faecalis* ASAGF 0D-1 strain isolated from soil and a novel N-acyl-L-amino acid amidohydrolase cloned from *A. faecalis* DSM 16503 have been reported to degrade OTA into OTα [23,28]. The biodegradation of OTA into barely or non-toxic metabolites (OTα and Phe) by microorganisms and their intracellular or extracellular enzymes is a preferred method to eliminate the toxic and carcinogenic potential. In many cases, biodegradation entailed the use of enzymes, including crude and purified ones that were able to cleave OTA [3,16]. In recent years, carboxypeptidases from *Bacillus amyloliquefaciens* ASAG1 [29], *Acinetobacter* sp. neg1 ITEM 17016 [30], and *Bacillus subtilis* ANSB168 [31] have been cloned and expressed to hydrolyze OTA. Except for the well-studied carboxypeptidase, several other enzymes were also indicated to biodegrade OTA, including protease A [32], amidohydrolase [33], lipase A [34], and hydrolase [35]. However, the degradation efficiencies of some microbes might be limited. Isolating strains from the gastrointestinal tract of healthy animals might be worth investigating. Our assay may help shed some light on investigating highly efficient enzymes. Further purification and identification of the degradation enzymes involved in OTA degradation by the strain ANSA176 is still ongoing.

The exposure of OTA could cause safety issues and enormous economic losses. Meanwhile, the efficacy and melioration effects of microorganism on OTA challenged poultry are still limited [3]. A study of varying levels of OTA showed that 2 mg/kg OTA decreased EPR and increased FER, while 4 mg/kg OTA decreased shell thickness [36]. Denli et al. found that the 2 mg/kg OTA-contaminated diet significantly decreased EM compared to the control [11]. Moreover, OTA at a level of 0.25 mg/kg did not influence the number of daily eggs produced, but 1 mg/kg of OTA reduced that, indicating the dose-dependent effect [13]. In order to provide a theoretical basis for the practical application of *A. faecalis* ANSA176, we further analyzed its alleviation effects on the laying performance of OTA-fed hens. In this experiment, decreases in ADFI and EW were recorded in the OTA group, which were in line with other studies [11,36]. Notably, administration of ANSA176 obviously ameliorated OTA-induced effects on layer's EM and FER. In addition, the OTA changed shell, yolk, and albumen percentages were altered back toward the control's by ANSA176. The unexpected ADFI decline in the ANSA176 supplementation group could be due to the palatability change. Although little is known on the palatability effects of bacteria supplements in feed diet, it has been accepted that microbial fermentation could change palatability and consumption [37,38]. In addition, a pair-feeding model indicated that the adverse effects of mycotoxin on growth performance, antioxidative status, and inflammation reaction were more severe compared to the merely FI reduction [39].

The challenge of OTA has been shown to result in nephrotoxic, hepatotoxic, and immunotoxic issues [14]. Several biochemical indicators in the serum can be used to reflect the tissue damage and inflammatory status in animals. Tubular enzymes AAP and LAP were markers of nephron injury [40]. The high β2-MG level was considered a clinical marker for nephropathy, and its excretion was related to OTA in Balkan endemic nephropathy (BEN) [41]. Our results showed increases in serum AAP, LAP, and β2-MG concentrations in OTA-fed layers, indicating OTA-induced damage. In comparison, biodegradable ANSA176 significantly reduced AAP and LAP, revealing a mitigating effect. Antibody responses are initiated via immune cells and lead to the production of immunoglobulins. It has been reported that OTA-contaminated diet significantly increased IgA, IgG, and IgM, suggesting an immune response against mycotoxin [31]. Cytokines such as IL-2 and TNF-α were involved in systemic inflammation and immunity [42,43]. Al-Anati et al. demonstrated that OTA released TNF-α in a dose- and time-dependent fashion via the classical inflammatory signaling cascade [44]. In the current study, increases were observed in serum β2-MG, IgG, and TNF-α concentrations in layers exposed to dietary OTA, indicating the OTA induced immune response and inflammation. However, the supplementation of ANSA176 can reduce levels of IgG, LZM, IL-2, and TNF-α to present protective effects.

Oxidative stress and antioxidant potential are sensitive indicators reflecting the imbalance between systemic reactive oxygen species (ROS) and the biological ability to detoxify intermediates or repair damage [14,15]. Nowadays, the various detrimental effects of OTA have been associated with the toxin itself, as well as the influence of ROS and oxidative stress [14]. Both in vitro and in vivo studies have suggested that the OTA-induced oxidative stress and reduced antioxidant ability may be implicated in the renal toxicity and carcinogenicity [14,39,45]. Different antioxidant enzymes and total antioxidant status (TAS) reflected the disturbances in animals [15]. Considering that OTA enhances the production of free radicals, the activity of antioxidant enzymes such as SOD may be affected. A previous study showed that feeding OTA to broiler chicks resulted in a dose-dependent reduction of TAS and SOD in plasma and almost all the tissues [15]. Although limited information was available on the OTA-induced oxidative stress in poultry, studies on rats have been carried out [14]. The enzymatic antioxidant SOD was significantly decreased in the kidney of OTA-challenged rat, while the administration of alleviators restored it [46,47]. In addition, pretreatment of antioxidant inhibited OTA-induced ROS overproduction and SOD reduction [48]. Furthermore, the injection of antioxidants SOD and catalase provided protection against OTA in rats [49]. In the present study, levels of T-AOC and SOD were higher in the ANSA176-treated group, suggesting an ameliorative effect of oxidative stress inside layers. It also supported the hypothesis that the modulation of oxidative stress might be due to the change in antioxidant enzyme, which has been proposed before [15,50].

4. Conclusions

In summary, our findings showed that the OTA-biodegrading strain *A. faecalis* ANSA176 could be used as a potential bioproduct to alleviate OTA-induced kidney and liver damage, inflammation, and oxidative stress. The application of the ANSA176 in grain or feed needs to be further developed, especially in resolving the consumption decline. Studies on the identified OTA degradation enzymes from ANSA176 also have a promising future.

5. Materials and Methods

5.1. Isolation and Identification of the OTA-Degrading Strain

5.1.1. Culture Media and Reagents

The LB broth containing 10 g/L peptone, 5 g/L yeast extract, and 10 g/L NaCl was used for bacteria growth. All the reagents used in this study were purchased from Sigma-Aldrich (St Louis, MO, USA), unless otherwise indicated. The OTα standard was purchased from Biopure, Romer Labs (Tulln, Austria).

5.1.2. Isolation of Microorganisms with OTA Degradation Ability

Bacterial strains were isolated from the intestine of animals. About 0.05 g of intestinal content was added to 5 mL of phosphate saline buffer and mixed by vortex. The solutions were further diluted and cross-streaked in Petri dishes containing LB agar to generate individual colonies of bacteria. Then the colonies obtained were individually transferred to LB broth and incubated on a rotary shaker (n = 200 rpm) at 37 °C for 24 h. The cultures were obtained through two successive inoculations (inoculum size at 2%, v/v). An aliquot of 980 μL of the culture was mixed with 20 μL of 50 mg/mL OTA standard solution. Sterile LB medium was used as a control. The inoculated mixtures were incubated at 37 °C at 200 rpm for 24 h. An equal volume of methanol (1 mL) was added at the end of incubation.

5.1.3. Determination of OTA and OTα by Using HPLC

After incubation, cells of microbes were removed by centrifugation at 12,000 rpm for 4 min. The supernatants were filtered through filters with a 0.22 μm pore size (Millipore, Temecula, CA, USA) and analyzed by using high-performance liquid chromatography (HPLC). The HPLC was performed on an instrument (RF-20A, SHIMADZU Corporation, Kyoto, Japan) that contained a fluorescence detector set at 333 nm excitation and 477 nm emission wavelengths. The analysis was carried out with a reversed-phase Agilent Eclipse

Plus C18 column (4.6 × 150 mm, 5 μm) at an injection volume of 20 μL, and the mobile phase was water/acetonitrile/acetic acid (99:99:2, *v*/*v*/*v*) at a flow rate of 1.0 mL/min. During the HPLC analysis of OTA with fluorescence detection, the peak of OTα production was also clearly observed.

5.1.4. Identification of the OTA-Degrading Strain

The morphological observation of the strain named ANSA176 with high OTA-degrading ability was carried out. To identify the bacterial strain, the 16S rRNA gene sequence was amplified by using primer set 27f-1492r. The inspection and appraisal report were given by the Institute of Microbiology (Chinese Academy of Sciences, Beijing, China). The strain was preserved in 20% glycerol and stored at −20 °C. Bacteria were revived and grown in LB broth at 37 °C for 24 h, with shaking (n = 200 rpm), for experiment.

5.1.5. Characterization of *A. faecalis* ANSA176 Growth with Different Temperature and pH Levels

In order to determine the growth of ANSA176 under different conditions, a range of temperature (17, 22, 27, 32, 37, and 42 °C) and pH (2.5, 3, 4, 5, 6, 7, 8, 9, and 10) levels were investigated during a period of 24 h. Optical densities (ODs) were measured and recorded at 600 nm. Subsequently, the number of bacteria was calculated from the equation of plate counts against OD and expressed as $\log_{10}$ CFU/mL.

5.2. Animal Trial in Layer Hens

5.2.1. Dietary Treatments of Animal Trial

According to Qing et al. [31], an *Aspergillus ochraceus* (CGMCC 3.4412) strain was used to produce OTA by artificial infection of sterile maize for 21 days, at 25–28 °C, and then the maize was dried and smashed. The concentration of OTA in maize powder was measured by HPLC, which was later added into the basal diet at 18% to meet the predicted concentration and verified by HPLC (predicted = 250 μg/kg; measured = 247.8 μg/kg). This dose was intended to observe the adverse effect of OTA contamination and the alleviation effect of *A. faecalis* ANSA176 supplementation.

The broth of strain *A. faecalis* ANSA176 was transformed into a freeze-dried powder. The number of bacteria in the powder was later determined as 3.1×10^8 CFU/g. Then it was added to the contaminated basal diet up to an overdose of 2 kg/T feed to ensure the effectiveness of degradation.

5.2.2. Design of Animal Experiments

All procedures were reviewed and approved by the Laboratory Animal Welfare and Animal Experimental Ethical Committee of China Agricultural University (No. AW 13301202-1-7). The trial strictly complied with the standard operating procedures for experimental animals of the Ministry of Science and Technology (Beijing, China), and every effort was made to minimize suffering.

A total of 36 Jingfen No.1 layers (26 weeks of age) were randomly allocated to three feeding treatments and 12 replicate pens per treatment (separately feeding): (1) basal diet group (A) fed the basal diet without OTA contamination; (2) OTA-contaminated group (B) fed the contaminated diet containing exceeded limit dose of OTA (about 250 μg/kg); and (3) biodegradable agent group (C) fed the OTA contaminated diet added 2 kg *A. faecalis* ANSA176 freeze-dried culture per ton (6.2×10^8 CFU/kg diet of ANSA176). For the nutritive values and feeding procedures, we referred to the NY/T 33-2004 (China) and recommendations for Jingfen No.1 layers (Huadu yukou, Beijing, China). The composition and nutrient levels of the basal and contaminated diets are shown in Appendix A Table A1. The feeding trial lasted 70 days. On the final day of the feeding trial, blood samples were collected from each layer via the wing vein. After that, layers were euthanized for tissue collection, following the sodium pentobarbital injection (0.4 mL/kg BW; Sile Biological Technology Co., Ltd., Guangzhou, China).

5.2.3. Laying Performance and Egg Quality

The laying performance was recorded and calculated. At the end of the experiment, 30 eggs were randomly selected from each treatment group to assess the egg quality, as previously described [51]. Shell color was determined with a QCR color reflectometer (QCR SPA, TSS, York, England). Thickness and strength were tested by the eggshell tester (ESTG-01, Orka Technology Ltd., Ramat Hasharon, Israel). Haugh unit and yolk color were measured by a multifunctional egg quality tester (EA-01, Orka Technology Ltd., Ramat Hasharon, Israel). Then the yolk was separated with a separator and weighed. The relative shell, yolk, and albumen proportions were calculated.

5.2.4. Residues of OTA in Eggs

Residues of OTA and OTα in eggs were determined weekly during the animal trial, as previously conducted [31]. Briefly, homogenized sample was weighed and added into an acetonitrile/water solution (60:40, v/v). The mixture was shaken, followed by filtering to get the supernatant. The cleanup step was performed by passing the extracted sample through the immunoaffinity column (OchraTestWB, VICAM, Watertown, MA, USA) at a rate of 1 or 2 drops per second, under gentle vacuum pressure. Then the column was washed with 10 mL of water–methanol (90:10, v/v) and dried under nitrogen gas (N_2) for 5 min. Finally, OTA was eluted by 2 mL of methanol for HPLC analysis. The HPLC parameter settings were as described above. The determined limit of detection (LOD) for OTA and OTα was0.1 μg/kg (based on a signal–noise ratio of 3).

5.2.5. Blood Sampling and Serum Biochemical Analysis

Blood samples were centrifuged at 3000 rpm for 15 min at room temperature, and then serum samples were stored at −20 °C until analyzed. Serum contents of AST, ALT, ALP, PEPCK, Cr, TP, ALB, LZM, T-AOC, SOD, and GR were measured by using diagnostic kits (Nanjing Jiancheng Bioengineering Institute, Nanjing, China), according to the manufacturer's instructions. Serum activities of LAP, AAP, MDA, T-GSH, GSH-Px, globulins (β2-MG, IgA, IgG, and IgM), and cytokines (TNF-α, IL-2, and IL-10) were determined by ELISA kits (Nanjing Jiancheng Bioengineering Institute, Nanjing, China), following the manufacturer's instructions.

5.2.6. Liver and Kidney Histological Assessment

Part of the liver and kidney samples were fixed in 10% neutral-buffered formalin solution and embedded in paraffin. Then these sections were stained with hematoxylin and eosin for histopathological examination, as previously described [52].

5.3. *Statistical Analysis*

The experimental data were analyzed through one-way analysis of variance (ANOVA), using the Statistical Analysis System software (SAS) package version 9.4. Tukey multiple comparison analysis was performed to determine significance of differences among treatment means. Differences were considered significant if the p-value was < 0.05 and demonstrated a trend if the p-value was < 0.10. The GraphPad Prism 9 software was used to generate graphs.

Author Contributions: R.Z., methodology, formal analysis and data curation during screening and identification of *Alcaligenes faecalis* ANSA176, and writing—original draft preparation; H.Q., methodology, formal analysis and data curation during feeding trial evaluation, and writing—review and editing; Q.M., conceptualization, writing—review and editing, project administration, and funding acquisition; L.Z., C.J. and J.Z., validation and supervision; X.H. and S.H., writing—review and editing. All authors have read and agreed to the published version of the manuscript.

Funding: This research was funded by the Taishan Leading Industry Talents—Agricultural Science of Shandong Province (LJNY202022), Beijing Municipal Science Foundation (No. 6172017), Anhui Province Key Laboratory of Livestock and Poultry Product Safety Engineering (No. XM2004), and China Agricultural Research System program (CARS-40-K08).

Institutional Review Board Statement: The study was conducted according to the guidelines for experimental animals of the Ministry of Science and Technology (Beijing, China) and approved by the Laboratory Animal Welfare and Animal Experimental Ethical Committee of China Agricultural University (No. AW 13301202-1-7, 25 November 2019).

Data Availability Statement: The data presented in this study are available on request from the corresponding author.

Acknowledgments: We thank all the researchers at our laboratory for their help with sample collection.

Conflicts of Interest: The authors declare no conflict of interest.

Appendix A

Table A1. Composition and nutrient level of basal and contaminated diet.

Ingredient (%)	Basal Diet	Contaminated Diet
Normal corn	65.05	47.05
Contaminated corn [1]	0	18
Soybean meal (43% CP)	24.20	24.20
Limestone (38% Ca)	8.20	8.20
Calcium hydrophosphate (21% Ca, 16% P)	1.70	1.70
NaCl	0.30	0.30
DL-methionine	0.12	0.12
Mineral premix [2]	0.30	0.30
Vitamin premix [3]	0.03	0.03
Choline chloride	0.10	0.10
Total	100.00	100.00
Nutritional content (%) [4]		
Crude protein	16.04	16.04
ME (MJ/kg)	11.25	11.25
Calcium	3.60	3.60
Total phosphorus	0.65	0.65
Non-phytin phosphorus	0.39	0.39
Methionine	0.38	0.38
Methionine + cystine	0.65	0.65
Lysine	0.78	0.78
Tryptophan	0.16	0.16
Threonine	0.59	0.59

[1] When the contaminated corn's supplemental level was 18%, the total OTA content in the diet was 247.8 μg/kg. [2] Provided per kilogram of diet: 6.8 mg of Cu, 66 mg of Fe, 83 mg of Zn, 80 mg of Mn, 1 mg of I, and 0.3 mg of Se. [3] Provided per kilogram of diet: 11,700 IU of vitamin A, 3600 IU of vitamin D_3, 21 IU of vitamin E, 4.2 mg of vitamin K_3, 3 mg of vitamin B_1, 10.2 mg of vitamin B_2, 0.9 mg of folic acid, 15 mg of calcium pantothenate, 45 mg of niacin, 5.4 mg of vitamin B_6, 24 μg of vitamin B_{12}, and 150 μg of biotin. [4] All nutrient levels were calculated.

References

1. Iqbal, S.Z.; Asi, M.R.; Hanif, U.; Zuber, M.; Jinap, S. The Presence of Aflatoxins and Ochratoxin A in Rice and Rice Products; and Evaluation of Dietary Intake. *Food Chem.* **2016**, *210*, 135–140. [CrossRef] [PubMed]
2. Li, X.; Ma, W.; Ma, Z.; Zhang, Q.; Li, H. The Occurrence and Contamination Level of Ochratoxin A in Plant and Animal-Derived Food Commodities. *Mol. Basel Switz.* **2021**, *26*, 6928. [CrossRef] [PubMed]
3. Chen, W.; Li, C.; Zhang, B.; Zhou, Z.; Shen, Y.; Liao, X.; Yang, J.; Wang, Y.; Li, X.; Li, Y.; et al. Advances in Biodetoxification of Ochratoxin A-A Review of the Past Five Decades. *Front. Microbiol.* **2018**, *9*, 1386. [CrossRef] [PubMed]
4. Pozzo, L.; Cavallarin, L.; Antoniazzi, S.; Guerre, P.; Biasibetti, E.; Capucchio, M.T.; Schiavone, A. Feeding a Diet Contaminated with Ochratoxin A for Broiler Chickens at the Maximum Level Recommended by the EU for Poultry Feeds (0.1 Mg/Kg). 2. Effects on Meat Quality, Oxidative Stress, Residues and Histological Traits. *J. Anim. Physiol. Anim. Nutr.* **2013**, *97* (Suppl. 1), 23–31. [CrossRef]

5. Tolosa, J.; Rodríguez-Carrasco, Y.; Ruiz, M.J.; Vila-Donat, P. Multi-Mycotoxin Occurrence in Feed, Metabolism and Carry-over to Animal-Derived Food Products: A Review. *Food Chem. Toxicol.* **2021**, *158*, 112661. [CrossRef]
6. IARC Working Group on the Evaluation of Carcinogenic Risks to Humans. Some Traditional Herbal Medicines, Some Mycotoxins, Naphthalene and Styrene. *IARC Monogr. Eval. Carcinog. Risks Hum.* **2002**, *82*, 1–556.
7. Pleadin, J. Mycotoxins in Grains and Feed: Contamination and Toxic Effect in Animals. *Biotechnol. Anim. Husb.* **2015**, *31*, 441–456. [CrossRef]
8. Duarte, S.C.; Lino, C.M.; Pena, A. Food Safety Implications of Ochratoxin A in Animal-Derived Food Products. *Vet. J.* **2012**, *192*, 286–292. [CrossRef]
9. European Commission. Commission Recommendation of 17 August 2006 on the Presence of Deoxynivalenol, Zearalenone, Ochratoxin A, T-2 and HT-2 and Fumonisins in Products Intended for Animal Feeding. *Off. J. Eur. Union* **2006**, *229*, 7–9.
10. Hassan, Z.U.; Khan, M.Z.; Khan, A.; Javed, I.; Hussain, Z. Effects of Individual and Combined Administration of Ochratoxin A and Aflatoxin B1 in Tissues and Eggs of White Leghorn Breeder Hens. *J. Sci. Food Agric.* **2012**, *92*, 1540–1544. [CrossRef]
11. Denli, M.; Blandon, J.C.; Guynot, M.E.; Salado, S.; Perez, J.F. Efficacy of a New Ochratoxin-Binding Agent (OcraTox) to Counteract the Deleterious Effects of Ochratoxin A in Laying Hens. *Poult. Sci.* **2008**, *87*, 2266–2272. [CrossRef]
12. Denli, M.; Perez, J.F. Ochratoxins in Feed, a Risk for Animal and Human Health: Control Strategies. *Toxins* **2010**, *2*, 1065–1077. [CrossRef]
13. Vasiljević, M.; Marinković, D.; Milićević, D.; Pleadin, J.; Stefanović, S.; Trialović, S.; Raj, J.; Petrujkić, B.; Trialović, J.N. Efficacy of a Modified Clinoptilolite Based Adsorbent in Reducing Detrimental Effects of Ochratoxin A in Laying Hens. *Toxins* **2021**, *13*, 469. [CrossRef]
14. Tao, Y.; Xie, S.; Xu, F.; Liu, A.; Wang, Y.; Chen, D.; Pan, Y.; Huang, L.; Peng, D.; Wang, X.; et al. Ochratoxin A: Toxicity, Oxidative Stress and Metabolism. *Food Chem. Toxicol.* **2018**, *112*, 320–331. [CrossRef]
15. Hameed, M.R.; Khan, M.Z.; Saleemi, M.K.; Khan, A.; Akhtar, M.; Hassan, Z.; Hussain, Z. Study of Ochratoxin A (OTA)-Induced Oxidative Stress Markers in Broiler Chicks. *Toxin Rev.* **2017**, *36*, 270–274. [CrossRef]
16. Samuel, M.S.; Jeyaram, K.; Datta, S.; Chandrasekar, N.; Balaji, R.; Selvarajan, E. Detection, Contamination, Toxicity, and Prevention Methods of Ochratoxins: An Update Review. *J. Agric. Food Chem.* **2021**, *69*, 13974–13989. [CrossRef]
17. Var, I.; Erginkaya, Z.; Kabak, B. Reduction of Ochratoxin A Levels in White Wine by Yeast Treatments. *J. Inst. Brew.* **2009**, *115*, 30–34. [CrossRef]
18. Piotrowska, M.; Zakowska, Z. The Elimination of Ochratoxin A by Lactic Acid Bacteria Strains. *Pol. J. Microbiol.* **2005**, *54*, 279–286.
19. Bejaoui, H.; Mathieu, F.; Taillandier, P.; Lebrihi, A. Biodegradation of Ochratoxin A by Aspergillus Section Nigri Species Isolated from French Grapes: A Potential Means of Ochratoxin A Decontamination in Grape Juices and Musts. *FEMS Microbiol. Lett.* **2006**, *255*, 203–208. [CrossRef]
20. Varga, J.; Péteri, Z.; Tábori, K.; Téren, J.; Vágvölgyi, C. Degradation of Ochratoxin A and Other Mycotoxins by Rhizopus Isolates. *Int. J. Food Microbiol.* **2005**, *99*, 321–328. [CrossRef]
21. Abrunhosa, L.; Serra, R.; Venâncio, A. Biodegradation of Ochratoxin A by Fungi Isolated from Grapes. *J. Agric. Food Chem.* **2002**, *50*, 7493–7496. [CrossRef] [PubMed]
22. Luz, C.; Ferrer, J.; Mañes, J.; Meca, G. Toxicity Reduction of Ochratoxin A by Lactic Acid Bacteria. *Food Chem. Toxicol.* **2018**, *112*, 60–66. [CrossRef] [PubMed]
23. Zhang, H.H.; Wang, Y.; Zhao, C.; Wang, J.; Zhang, X.L. Biodegradation of Ochratoxin A by Alcaligenes Faecalis Isolated from Soil. *J. Appl. Microbiol.* **2017**, *123*, 661–668. [CrossRef] [PubMed]
24. Angioni, A.; Caboni, P.; Garau, A.; Farris, A.; Orro, D.; Budroni, M.; Cabras, P. In Vitro Interaction between Ochratoxin A and Different Strains of Saccharomyces Cerevisiae and Kloeckera Apiculata. *J. Agric. Food Chem.* **2007**, *55*, 2043–2048. [CrossRef]
25. Xiong, K.; Wang, X.; Zhi, H.-W.; Sun, B.-G.; Li, X.-T. Identification and Safety Evaluation of a Product from the Biodegradation of Ochratoxin A by an Aspergillus Strain. *J. Sci. Food Agric.* **2017**, *97*, 434–443. [CrossRef]
26. Bellis, P.; Tristezza, M.; Haidukowski, M.; Fanelli, F.; Sisto, A.; Mule, G.; Greco, F. Biodegradation of Ochratoxin A by Bacterial Strains Isolated from Vineyard Soils. *Toxins* **2015**, *7*, 5079–5093. [CrossRef]
27. Loi, M.; Fanelli, F.; Liuzzi, V.C.; Logrieco, A.F.; Mulè, G. Mycotoxin Biotransformation by Native and Commercial Enzymes: Present and Future Perspectives. *Toxins* **2017**, *9*, 111. [CrossRef]
28. Zhang, H.; Zhang, Y.; Yin, T.; Wang, J.; Zhang, X. Heterologous Expression and Characterization of A Novel Ochratoxin A Degrading Enzyme, N-Acyl-L-Amino Acid Amidohydrolase, from *Alcaligenes Faecalis*. *Toxins* **2019**, *11*, 518. [CrossRef]
29. Chang, X.; Wu, Z.; Wu, S.; Dai, Y.; Sun, C. Degradation of Ochratoxin A by Bacillus Amyloliquefaciens ASAG1. *Food Addit. Contam. Part. Chem. Anal. Control. Expo. Risk Assess.* **2015**, *32*, 564–571. [CrossRef]
30. Liuzzi, V.C.; Fanelli, F.; Tristezza, M.; Haidukowski, M.; Picardi, E.; Manzari, C.; Lionetti, C.; Grieco, F.; Logrieco, A.F.; Thon, M.R.; et al. Transcriptional Analysis of Acinetobacter Sp. Neg1 Capable of Degrading Ochratoxin A. *Front. Microbiol.* **2017**, *7*, 2162. [CrossRef]
31. Qing, H.; Huo, X.; Huang, S.; Zhao, L.; Zhang, J.; Ji, C.; Ma, Q. *Bacillus subtilis* ANSB168 Producing d-Alanyl-d-Alanine Carboxypeptidase Could Alleviate the Immune Injury and Inflammation Induced by Ochratoxin A. *Int. J. Mol. Sci.* **2021**, *22*, 12059. [CrossRef]
32. Abrunhosa, L.; Santos, L.; Venâncio, A. Degradation of Ochratoxin A by Proteases and by a Crude Enzyme of Aspergillus Niger. *Food Biotechnol.* **2006**, *20*, 231–242. [CrossRef]

33. Dobritzsch, D.; Wang, H.; Schneider, G.; Yu, S. Structural and Functional Characterization of Ochratoxinase, a Novel Mycotoxin-Degrading Enzyme. *Biochem. J.* **2014**, *462*, 441–452. [CrossRef]
34. Stander, M.A.; Bornscheuer, U.T.; Henke, E.; Steyn, P.S. Screening of Commercial Hydrolases for the Degradation of Ochratoxin A. *J. Agric. Food Chem.* **2000**, *48*, 5736–5739. [CrossRef]
35. Abrunhosa, L.; Venâncio, A.; Teixeira, J.A. Optimization of Process Parameters for the Production of an OTA-Hydrolyzing Enzyme from Aspergillus Niger under Solid-State Fermentation. *J. Biosci. Bioeng.* **2011**, *112*, 351–355. [CrossRef]
36. Johri, T.S.; Swain, B. Effect of Varying Levels of Aflatoxin, Ochratoxin and Their Combinations on the Performance and Egg Quality Characteristics in Laying Hens. *Asian-Australas. J. Anim. Sci.* **2003**, *16*, 1015–1019. [CrossRef]
37. Jovanović, M.; Zlatanović, S.; Micić, D.; Bacić, D.; Mitić-Ćulafić, D.; Đuriš, M.; Gorjanović, S. Functionality and Palatability of Yogurt Produced Using Beetroot Pomace Flour Granulated with Lactic Acid Bacteria. *Foods Basel Switz.* **2021**, *10*, 1696. [CrossRef]
38. Shiferaw Terefe, N.; Augustin, M.A. Fermentation for Tailoring the Technological and Health Related Functionality of Food Products. *Crit. Rev. Food Sci. Nutr.* **2020**, *60*, 2887–2913. [CrossRef]
39. Qing, H.; Huang, S.; Zhan, K.; Zhao, L.; Zhang, J.; Ji, C.; Ma, Q. Combined Toxicity Evaluation of Ochratoxin A and Aflatoxin B1 on Kidney and Liver Injury, Immune Inflammation, and Gut Microbiota Alteration Through Pair-Feeding Pullet Model. *Front. Immunol.* **2022**, *13*, 3539. [CrossRef]
40. Emeigh Hart, S.G. 7.11–In Vivo Methodologies Used to Assess Renal Function and Injury. In *Comprehensive Toxicology*, 2nd ed.; McQueen, C.A., Ed.; Elsevier: Oxford, UK, 2010; pp. 263–303. ISBN 978-0-08-046884-6.
41. Yordanova, P.; Wilfried, K.; Tsolova, S.; Dimitrov, P. Ochratoxin A and B2-Microglobulin in BEN Patients and Controls. *Toxins* **2010**, *2*, 780–792. [CrossRef]
42. Marin, D.; Taranu, I. Ochratoxin A and Its Effects on Immunity. *Toxin Rev.* **2015**, *34*, 11–20. [CrossRef]
43. Darif, Y.; Mountassif, D.; Belkebir, A.; Zaid, Y.; Basu, K.; Mourad, W.; Oudghiri, M. Ochratoxin A Mediates MAPK Activation, Modulates IL-2 and TNF-α MRNA Expression and Induces Apoptosis by Mitochondria-Dependent and Mitochondria-Independent Pathways in Human H9 T Cells. *J. Toxicol. Sci.* **2016**, *41*, 403–416. [CrossRef]
44. Al-Anati, L.; Essid, E.; Stenius, U.; Beuerlein, K.; Schuh, K.; Petzinger, E. Differential Cell Sensitivity between OTA and LPS upon Releasing TNF-α. *Toxins* **2010**, *2*, 1279–1299. [CrossRef] [PubMed]
45. Kőszegi, T.; Poór, M. Ochratoxin A: Molecular Interactions, Mechanisms of Toxicity and Prevention at the Molecular Level. *Toxins* **2016**, *8*, 111. [CrossRef] [PubMed]
46. Abdel-Wahhab, M.A.; Aljawish, A.; El-Nekeety, A.A.; Abdel-Aziem, S.H.; Hassan, N.S. Chitosan Nanoparticles plus Quercetin Suppress the Oxidative Stress, Modulate DNA Fragmentation and Gene Expression in the Kidney of Rats Fed Ochratoxin A-Contaminated Diet. *Food Chem. Toxicol.* **2017**, *99*, 209–221. [CrossRef] [PubMed]
47. El-Shafie, M.; Elmasry, H.G.; Elsadek, M.F.B.; AlMajwal, A.M. Curative Effect of Orally Consumed Aloe Vera Juice on Ochratoxin A-Induced Nephrotoxicity in Rats. *Prog. Nutr.* **2015**, *17*, 128–136.
48. Yang, Q.; Shi, L.; Huang, K.; Xu, W. Protective Effect of N-Acetylcysteine against DNA Damage and S-Phase Arrest Induced by Ochratoxin A in Human Embryonic Kidney Cells (HEK-293). *Food Chem. Toxicol. Int. J. Publ. Br. Ind. Biol. Res. Assoc.* **2014**, *70*, 40–47. [CrossRef]
49. Baudrimont, I.; Betbeder, A.M.; Gharbi, A.; Pfohl-Leszkowicz, A.; Dirheimer, G.; Creppy, E.E. Effect of Superoxide Dismutase and Catalase on the Nephrotoxicity Induced by Subchronical Administration of Ochratoxin A in Rats. *Toxicology* **1994**, *89*, 101–111. [CrossRef]
50. Marin-Kuan, M.; Nestler, S.; Verguet, C.; Bezençon, C.; Piguet, D.; Mansourian, R.; Holzwarth, J.; Grigorov, M.; Delatour, T.; Mantle, P.; et al. A Toxicogenomics Approach to Identify New Plausible Epigenetic Mechanisms of Ochratoxin a Carcinogenicity in Rat. *Toxicol. Sci. Off. J. Soc. Toxicol.* **2006**, *89*, 120–134. [CrossRef]
51. Geng, S.; Huang, S.; Ma, Q.; Li, F.; Gao, Y.; Zhao, L.; Zhang, J. Alterations and Correlations of the Gut Microbiome, Performance, Egg Quality, and Serum Biochemical Indexes in Laying Hens with Low-Protein Amino Acid-Deficient Diets. *ACS Omega* **2021**, *6*, 13094–13104. [CrossRef]
52. Fan, Y.; Zhao, L.; Ji, C.; Li, X.; Jia, R.; Xi, L.; Zhang, J.; Ma, Q. Protective Effects of *Bacillus Subtilis* ANSB060 on Serum Biochemistry, Histopathological Changes and Antioxidant Enzyme Activities of Broilers Fed Moldy Peanut Meal Naturally Contaminated with Aflatoxins. *Toxins* **2015**, *7*, 3330–3343. [CrossRef]

toxins MDPI

Article

Patulin Detoxification by Recombinant Manganese Peroxidase from *Moniliophthora roreri* Expressed by *Pichia pastoris*

Shuai Wang [1], Xiaolu Wang [2], Leena Penttinen [3], Huiying Luo [2], Yuhong Zhang [1], Bo Liu [1], Bin Yao [2], Nina Hakulinen [3], Wei Zhang [1,*] and Xiaoyun Su [2,*]

[1] Biotechnology Research Institute, Chinese Academy of Agricultural Sciences, Beijing 100081, China; wangshuai9873@163.com (S.W.); zhangyuhong@caas.cn (Y.Z.); liubo01@caas.cn (B.L.)
[2] State Key Laboratory of Animal Nutrition, Institute of Animal Sciences, Chinese Academy of Agricultural Sciences, Beijing 100193, China; wangxiaolu@caas.cn (X.W.); luohuiying@caas.cn (H.L.); binyao@caas.cn (B.Y.)
[3] Department of Chemistry, Joensuu Campus, University of Eastern Finland, FIN-80101 Joensuu, Finland; leena.penttinen@uef.fi (L.P.); nina.hakulinen@uef.fi (N.H.)
* Correspondence: zhangwei02@caas.cn (W.Z.); suxiaoyun@caas.cn (X.S.)

Abstract: The fungal secondary metabolite patulin is a mycotoxin widespread in foods and beverages which poses a serious threat to human health. However, no enzyme was known to be able to degrade this mycotoxin. For the first time, we discovered that a manganese peroxidase (*Mr*MnP) from *Moniliophthora roreri* can efficiently degrade patulin. The *Mr*MnP gene was cloned into pPICZα(A) and then the recombinant plasmid was transformed into *Pichia pastoris* X-33. The recombinant strain produced extracellular manganese peroxidase with an activity of up to 3659.5 U/L. The manganese peroxidase *Mr*MnP was able to rapidly degrade patulin, with hydroascladiol appearing as a main degradation product. Five mg/L of pure patulin were completely degraded within 5 h. Moreover, up to 95% of the toxin was eliminated in a simulated patulin-contaminated apple juice after 24 h. Using *Escherichia coli* as a model, it was demonstrated that the deconstruction of patulin led to detoxification. Collectively, these traits make *Mr*MnP an intriguing candidate useful in enzymatic detoxification of patulin in foods and beverages.

Keywords: patulin; mycotoxin; manganese peroxidase; apple juice; detoxification

Key Contribution: *Mr*MnP efficiently degraded patulin, particularly in the malonate/Mn^{2+} system, lead to detoxification. It removed 95% of patulin in a simulated patulin-contaminated apple juice after 24 h treatment.

Citation: Wang, S.; Wang, X.; Penttinen, L.; Luo, H.; Zhang, Y.; Liu, B.; Yao, B.; Hakulinen, N.; Zhang, W.; Su, X. Patulin Detoxification by Recombinant Manganese Peroxidase from *Moniliophthora roreri* Expressed by *Pichia pastoris*. *Toxins* **2022**, *14*, 440. https://doi.org/10.3390/toxins14070440

Received: 8 June 2022
Accepted: 27 June 2022
Published: 29 June 2022

1. Introduction

Patulin (PAT) is a secondary metabolite and a food-born mycotoxin produced by at least 60 different filamentous fungi including *Penicillium expansum*, *Penicillium shell*, *Penicillium clavum*, and *Aspergillus clavatus* [1]. This mycotoxin has been detected in many kinds of fruits (such as apples, pears, grapes, kiwifruit, blueberries, and peaches) and their products (such as juice, jam, and cider) [2–4]. Based on cellular and animal toxicological studies, it has been found that patulin can cause genotoxicity, embryonic toxicity, cytotoxicity, neurotoxicity, immunotoxicity, carcinogenicity, and teratogenicity [5]. At the molecular level, patulin induces DNA damage, leading to cell-cycle arrest, which inhibits the activity of cell survival proteins and induces apoptosis until cell death [5]. Because of its toxicity and high frequency of contamination, the World Health Organization (WHO), some European countries, the Food and Drug Administration of the United States (FDA), and the Ministry of Health of China have all set up their recommended maximum concentrations of patulin in foods and beverages. For example, 50 μg/kg, 25 μg/kg, and 10 μg/kg of patulin are allowed in apple juice, solid apple products, and fruit baby foods, respectively, in the European Union [6]. However, it was noted that, although legal provisions enforced in Serbia are in line with the EU regulations, from the 142 kinds of fruit juices (apple or multiple fruits) collected from the market in three consecutive years (2013–2015), patulin was detected in 51.4% of the fruit juices, and 0.7% of the samples exceeded the legal limit of 50 μg/kg [7]. In Spain, when analyzing PAT in 161 apple juice, 77 solid apple food, and 146 apple baby foods, PAT was discovered in 42% of the apple sauce samples, 32% in multiple fruit plates, and 25% in apple juice [8]. In Italy, 65% of 120 fruit plates and fruit paste samples tested positive for patulin [9]. The wide existence of patulin, in conjunction with its detrimental effects to the human health, points to the necessity to eliminate this mycotoxin in foods.

To remove the patulin contamination in foods, many trials have been carried out previously. These can be classified as physical, chemical, and biological treatments. Ultraviolet radiation is a physical means approved by both Canada and the United States for degradation of patulin in food [10], but the turbidity of apple juice and cider, plus the ascorbic acid present in large quantities, can significantly diminish the effect of this treatment. Treatment with chemicals such as ozone was reported to rapidly remove up to 98% of patulin within 1 min [11]. However, this manipulation can also generate new, undefined, chemicals and cause loss of important nutrients, thereby limiting its wide application in the food industry. These drawbacks prompt the scientific community to pursue other economically viable, efficient, safe, and environmentally friendly ways to remove the patulin and ensure food safety. Thus, biological decomposition of patulin in food by microorganisms or enzymes is emerging as an attractive alternative.

Up to now, the microorganisms that have been reported to be able to effectively degrade patulin include *Pichia caribbica* [12], the marine yeast *Kodameae ohmeri* [13], the biocontrol yeast *Rhodosporidium kratochvilovae* [14], and *Rhodotorula mucilaginosa* [15]. The rate of decomposition can reach as high as 97.66% [13]. Although the use of microbial cells has been proven to be an effective strategy, the presence of abundant residual cells and high concentrations of metabolites after treatment may change the final quality of the product. In essence, microbes transform patulin by their encoding enzymes. Unlike the microorganisms, the enzyme biocatalysts retain the ability to degrade patulin but do not introduce unwanted cells or their metabolites. Therefore, the use of detoxifying enzymes, which has the potential to ensure food safety and quality, should be a promising strategy in controlling patulin contamination in foods. However, even in the patulin-degrading microorganisms, the enzymes responsible for eliminating patulin have not been identified, thus impeding the use of an enzyme to detoxify patulin. Moreover, some microbes employ a strategy to import the mycotoxin and degrade it within the cell. This commonly involves the participation of expensive co-factors such as NAD^+/NADH or $NADP^+$/NADPH, which may not be economically viable in practical applications [16,17].

Manganese peroxidases (MnP) are a kind of heme-containing peroxidases, which are generally produced by the lignocellulose-degrading basidiomycete filamentous fungi and symbolized by their ability to oxidize Mn^{2+} to Mn^{3+} [18]. In nature, the oxidized magnesium is chelated by dicarboxylic acids (such as oxalate [19]) excreted by the fungi, which can stabilize Mn^{3+} and penetrate into lignocellulose to attack the recalcitrant parts. Previously, we have demonstrated that manganese peroxidases can detoxify four major feed mycotoxins including aflatoxin, zearalenone, deoxynivalenol, and fumonisin [20], highlighting their potential to be used in feed. Moreover, this enzyme can act by either directly interacting with the substrate or indirectly through an oxidized Mn^{3+}. Neither of the two ways involves the necessity of an expensive co-factor, saving the cost in mycotoxin degradation. However, the ability of MnP to degrade patulin remains unknown. Therefore, to understand whether a manganese peroxidase can degrade patulin and whether the degradation leads to detoxification, in this study, *Mr*MnP, a manganese peroxidase originating from *Moniliophthora roreri* [21] and recombinantly prepared in *Pichia pastoris*, was explored for its ability to degrade patulin (Figure 1). The nature of the degradation products was analyzed by mass spectrometry and the toxicity of the degradation products was determined.

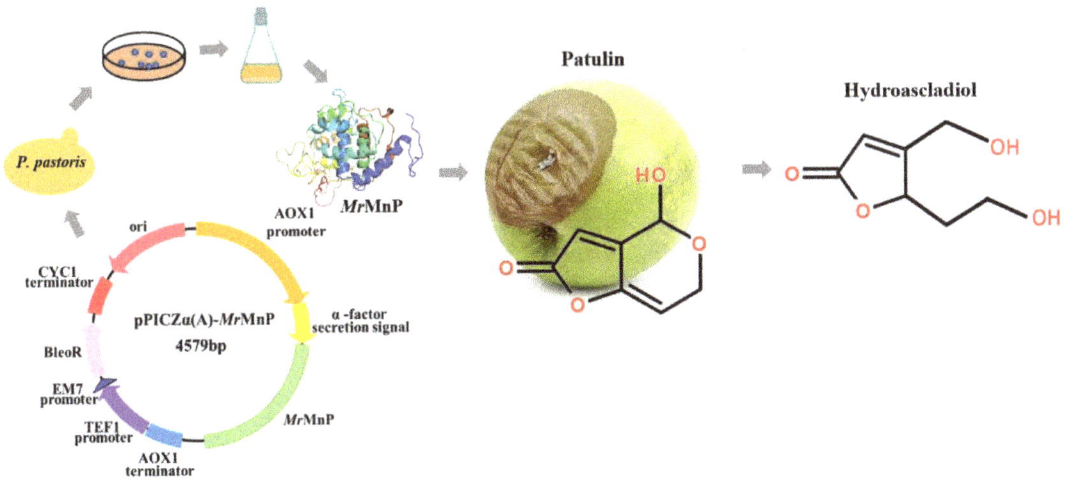

Figure 1. Schematic diagram showing the use of a recombinant *Mr*MnP expressed in *P. pastoris* to degrade and detoxify patulin, a mycotoxin commonly discovered in fruits.

2. Results and Discussion

2.1. Recombinant Production of the Manganese Peroxidase MrMnP

The most convenient way of characterizing the ability of a manganese peroxidase to degrade patulin would be to use a recombinant enzyme. However, the manganese peroxidases are notoriously difficult to be recombinantly produced, which might be, at least in part, ascribed to the presence of the heme prosthetic group in this kind of enzyme. Accordingly, many MnPs are expressed as insoluble, non-functional inclusion bodies in *Escherichia coli* and require tedious denaturation and refolding processes to obtain active enzymes [22–24]. However, in the commonly used eukaryotic microbial expression systems *P. pastoris* and *Aspergillus* spp., there were occasional reports indicating that a few MnPs could be functionally expressed, despite commonly being found in minor amounts [25–30]. The *M. roreri* MnP (*Mr*MnP) is such an enzyme that can be successfully produced in *P. pastoris* [21]. The gene encoding *Mr*MnP was thus artificially synthesized, cloned into the *Eco*RI and *Not*I restriction sites of pPICZα(A) to generate the recombinant plasmid pPICZα(A)-*Mr*MnP, and transformed into the *P. pastoris* X33 strain (Figure 2A). The transformant bearing the *Mr*MnP gene was first cultured in BMGY medium until the optical density of the culture at 600 nm (OD_{600}) reached 6.0. Then, the medium was changed to BMMY to induce *Mr*MnP and the culture was continued for 5 d. The recombinant strain produced 271.4 and 3659.5 U/L of MnP (using ABTS as the substrate) in flask and fed-batch fermentations (Figure 2B), respectively. This is comparable to that described by Agathe et al. [21], thus providing a sound basis for subsequent enzymatic characterization and degradation of patulin.

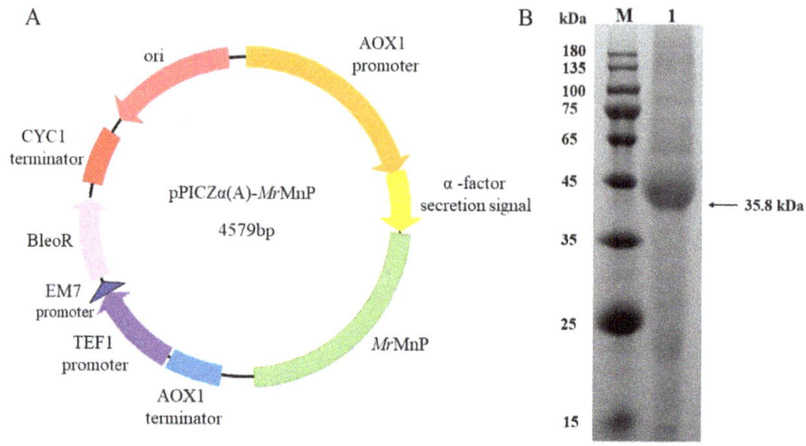

Figure 2. Expression of *Mr*MnP in *P. pastoris*. (**A**) The plasmid map of pPICZα(A)-*Mr*MnP. (**B**) SDS-PAGE analysis of the *Mr*MnP enzyme recombinantly produced in *P. pastoris*. The arrow indicates the recombinant *Mr*MnP protein. Lane M: protein molecular mass marker; 1: *Mr*MnP protein recombinantly produced in *P. pastoris*.

2.2. The Di-Carboxylic Acids Play a Critical Role in Degradation of Patulin by MrMnP

For an enzyme, the buffering system normally exerts a profound effect on its activity. Specifically, for manganese peroxidase, the buffer components may be directly involved in the reaction. For example, eight MnPs from different microbial sources have been demonstrated to be able to degrade four major feed mycotixins in presence of the di-carboxylic acid malonate [20], indicative of the potency of MnPs in detoxifying mycotoxins. However, in that study, the tested buffers were restricted to malonate, lactate, and acetate, while the effects of the buffers were only tested on aflatoxin B_1 and zearalenone. Herein, the effects of the buffer components on degradation of patulin were systematically examined by including the tri-carboxylic acid (citrate), di-carboxylic acids (malonic acid and oxalic acid), the α-hydroxyl carboxylic acid (lactate), the mono-carboxylic acid (acetate), the inorganic acid (phosphate), and the zwitterionic buffers (MES, standing for 2-morpholinoethanesulphonic acid; and HEPES, standing for 2-[4-(2-hydroxyethyl)-1-piperazinyl] ethanesulfonic acid). In the previous study, it was noted that zearalenone, but not aflatoxin B1, could be transformed at a similarly high rate in the acetate buffer to that in the malonate buffer. Similarly, *Mr*MnP was highly efficient in degrading patulin, eliminating all the mycotoxin after 5 h of incubation in the malonate buffer (Figure 3). Contrary to the situation for zearalenone, patulin was only slightly degraded in the acetate buffer (10.6 ± 0.1%). The degradation in oxalate was not as high as that in malonate, but still obvious (54.3 ± 0.3%). The degradation was reduced to 28.5 ± 0.1% in citrate buffer and decreased to be marginal in the lactic acid (15.2 ± 3.4%) and phosphate (11.5 ± 2.7%). The reason for ineffective degradation may best be explained by the knowledge that Mn^{2+} cannot form stable chelates with either of the two acids. Degradation of mycotoxins in oxalate and citrate is of physiological relevance since, in nature, the fungi expressing MnPs can also produce these two compounds in their cellular metabolism [22]. Additionally, it has been reported that gluconic acid, cellobionic acid, and other organic acids as well, can aid manganese peroxidases to exert their degradation on lignin molecules [31]. This allows the fungi to use manganese peroxidases, in assistance with these naturally occurring molecules, to break the lignin barrier apart and capture energy from the plant cell wall polysaccharides [32]. Unexpectedly, there were 42.6 ± 0.3% and 33.1 ± 0.5% degradation in the MES and HEPES buffers (Figure 3), higher than those in acetate, citrate, lactate, and phosphate buffers. In the univariate analysis, there was an extremely significant difference ($p < 0.001$), indicating that the degradation of patulin was affected by the buffer.

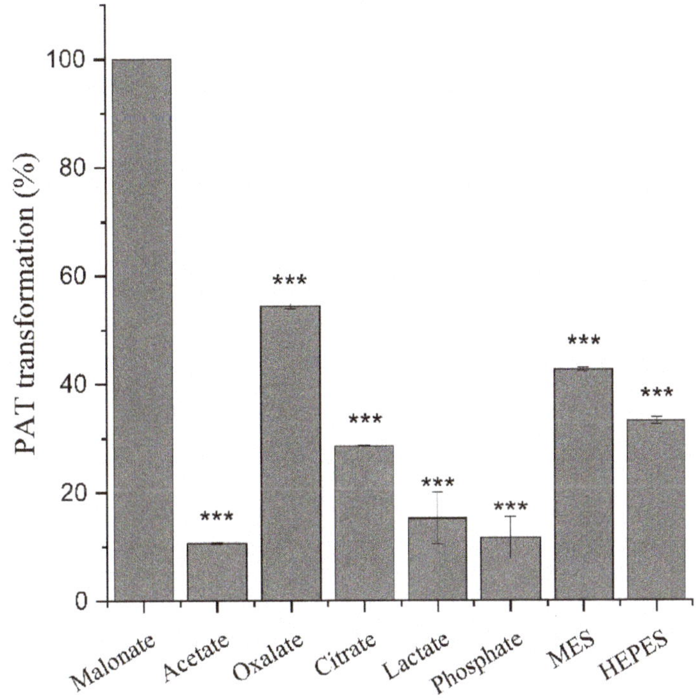

Figure 3. Effects of the buffer components on patulin transformation. The reactions were carried out by incubating 0.5 U/mL of *Mr*MnP with 5 mg/L of patulin in one of the buffers containing malonate, acetate, oxalate, citrate, lactate, phosphate, MES, and HEPES at 30 °C for 24 h. The data in the picture is the average ± standard deviation, *** $p < 0.001$, One Way anova test.

2.3. Mn^{2+} Is Another Key Determinant in Degradation of Patulin by MrMnP

Next, we sought to determine if the presence of Mn^{2+} would affect the activity of *Mn*MnP in patulin degradation. This is because the manganese peroxidases commonly have two substrate channels. One is the δ-heme edge, responsible for oxidation of hydrophobically-bound substrates including ABTS and phenolic compounds. The other one is the γ-heme edge, which is involved in catalysis of Mn^{2+} [33]. Patulin is a small chemical with ring structures, which might best fit for the first substrate channel, but could not enter the γ-heme edge. Incubation of *Mr*MnP with patulin in the acetate buffer, with or without Mn^{2+}, did not lead to obvious degradation of patulin. The degradation rate is 1.1 ± 0.2% in acetic acid without Mn^{2+} and 10.6 ± 0.1% in the system containing Mn^{2+}. The degradation was only observed in presence of both malonate and Mn^{2+} (Figure 4). The degradation rate reached 52 ± 0.3% at 2 h, and patulin was completely degraded at 5 h. These results collectively indicate that *Mr*MnP degraded patulin via a Mn^{3+}-mediated indirect way, and patulin could not enter the δ-heme edge for degradation. The enzyme first catalyzed oxidation of Mn^{2+} to Mn^{3+}, which then formed a chelate with malonate and subsequently oxidized patulin. Therefore, the enzyme's activity against patulin appeared to be related to the stability of Mn^{3+} in complex with the buffer components in the reaction systems. The stability and reactivity of the Mn^{3+} ion are strongly dependent on the nature and concentrations of the Mn^{3+}-complexing agents [18]. Both oxalate and malonate are dicarboxylic acids, but the degradation rate of patulin in malonate was higher, indicating that the chelate of Mn^{3+} and malonate is more stable and hence has higher activity for patulin. Based on this assumption, it was also suggested that, in the reactions with MES and HEPES serving the buffer systems, the newly generated Mn^{3+} was stabilized by these two buffers to a higher extent than that in acetate, citrate, lactate, and phosphate buffers. Although the reason for putatively increased Mn^{3+} stability remains unknown, it was noted that both chemicals have the sulfonate group, which could be involved in this stabilization.

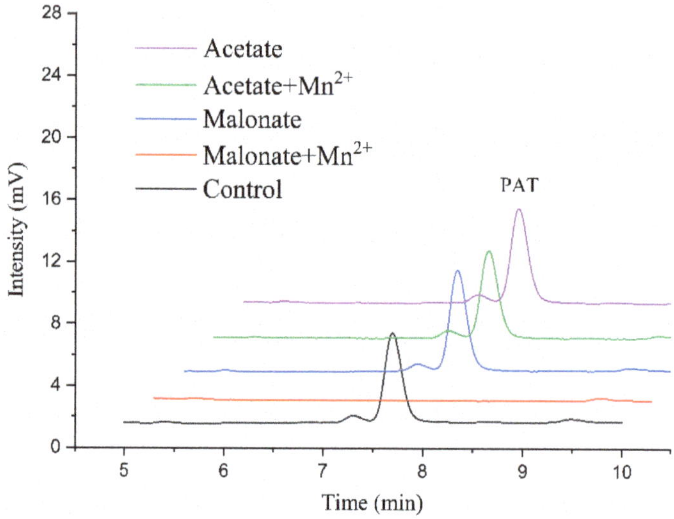

Figure 4. Mn^{2+} played an important role in degrading patulin. The reactions were carried out by incubating 0.5 U/mL of *Mr*MnP with 5 mg/L of patulin in the acetate or malonate buffer in absence or presence of Mn^{2+} at 30 °C for 24 h. Then the products were analyzed by HPLC.

2.4. *MrMnP-Catalyzed Degradation of Patulin Led to Detoxification*

The pre-requisite to apply *Mr*MnP in patulin detoxification is that the degradation products should have much less toxicity. *Escherichia coli* has been used as a microbial sensor system for successful monitoring of the toxicity of patulin [34]. In this study, increasing concentrations of untreated patulin were first incubated with *E. coli*. At 1 mg/L and 10 mg/L of patulin. There was no significant negative impact on the bacterial growth in a culturing period of 10 h. However, when 50 mg/L and 100 mg/L of patulin were added, the growth of *E. coli* was significantly retarded, as manifested by the drop of OD_{600} from 1.03 to 0.89 (for 50 mg/L) and 0.79 (for 100 mg/L) at 10 h after incubation (Figure 5A). These results indicated that *E. coli* could indeed be used to monitor the toxicity of patulin.

Patulin samples at final concentrations of 1, 10, 50, and 100 mg/L, respectively, were individually treated with 0.5 U/mL of *Mr*MnP. The HPLC analysis indicated that under all concentrations tested, the patulin was completely degraded (data not shown). It was observed that, at all these concentrations, the treated patulin no longer retarded the growth of *E. coli*, which suggested that the degradation of patulin by *Mr*MnP led to detoxification (Figure 5B).

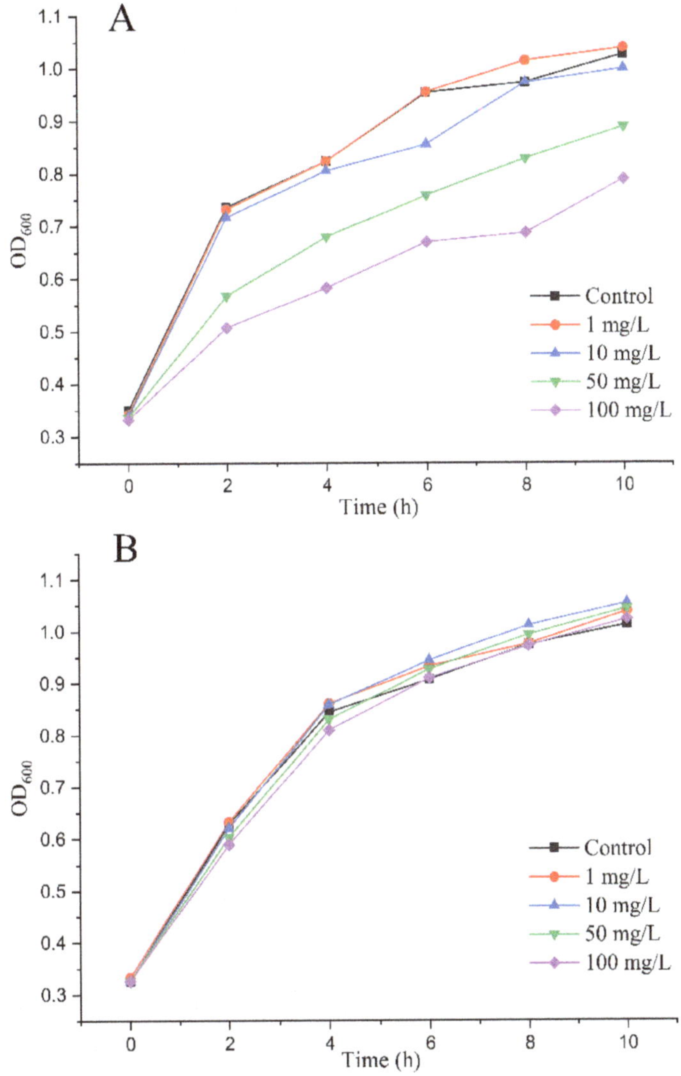

Figure 5. *Mr*MnP-catalyzed degradation of patulin led to detoxification. (**A**) Patulin was toxic to *E. coli* as demonstrated by retarded growth of the bacterium. A series of concentrations (1, 10, 50, and 100 mg/L) of patulin were added to equal amounts of *E. coli* and the culture was continued at 37 °C for 12 h. (**B**) *Mr*MnP-catalyzed degradation of patulin alleviated the retarding effect of patulin on *E. coli*. Patulin (1, 10, 50, and 100 mg/L) was first treated with 0.5 U/mL of *Mr*MnP at 30 °C for 24 h and then added to *E coli*.

2.5. Structural Analysis of the Degradation Products

As *Mr*MnP-catalyzed degradation of patulin led to detoxification, it would be interesting to know the chemical nature of the degradation products. Degradation and detoxification of mycotoxins are a process that, in essence, involves the transformation of mycotoxins into less-toxic or even non-toxic compounds [35]. Therefore, the degradation products of *Mr*MnP on PAT were further identified by UPLC-MS/MS. It was found that hydroascladiol (5-(2-hydroxyethyl)-4-(hydroxymethyl)furan-2(5*H*)-one) was one of the main intermediate degradation products of PAT. The parent ion appeared at *m*/*z* 157.1 [M−H]$^-$, producing daughter ions of 129.0 [M−H−CO]$^-$ and 113.1 [M−H−CO_2]$^-$ (Figure 6). Daughter ions were produced by continuous loss of carbon dioxide [36]. In a previous study using *Lactobacillus plantarum* to degrade PAT, hydroascladiol was also obtained as the degradation product [37]. However, other intermediates including (E)-ascladiol and (Z)-ascladiol identified in their study were not discovered in our study. The toxicity of patulin is related to the hemiacetal and lactone rings in its structure. The generation of hydroascladiol, and hence concurrent destruction of the hemiacetal ring, can significantly reduce the toxicity of patulin [38]. With the prolonged reaction, hydroascladiol was further diminished (data are not shown), leading to further destruction of the lactone acid in hydroascladiol and lower toxicity. This is consistent with the observed decreased toxicity of the degradation products on *E. coli*.

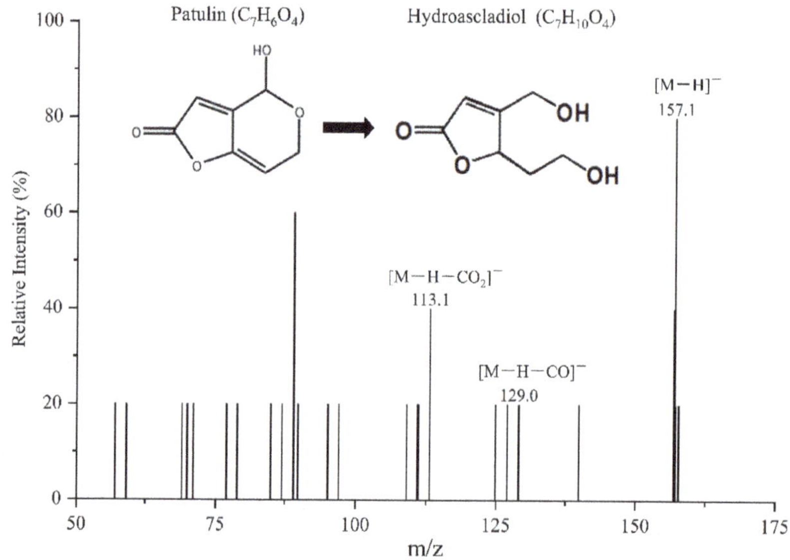

Figure 6. Identification of hydroascladiol as one major transformation product of patulin. PAT was incubated with 0.5 U/mL of *Mr*MnP in 50 mM malonate buffer (pH 5.0) supplemented with 1 mM $MnSO_4$ and 0.1 mM H_2O_2 and the reaction was carried out at 30 °C for 8 h. The degradation products were analyzed by HPLC-MS/MS.

2.6. Degradation in a Simulated Patulin-Contaminated Apple Juice

The ability of an enzyme to degrade a mycotoxin does not necessarily mean that the enzyme can efficiently degrade the mycotoxin in real foods/feeds. This is because the foods or feeds contain numerous components that could either adsorb the mycotoxins or act as competitors or inhibitors of the enzyme. For example, lignin phenolic compounds are naturally the substrate of MnPs [39–42]. In addition, many mycotoxins including aflatoxin B1, ochratoxin A, and zearalenone all have high affinity for lignocellulose, which also hinders the degradation process [43]. Therefore, to investigate whether *Mr*MnP can degrade patulin in the real environment, i.e., in foods with possibly interfering components, patulin was added to apple juice which then acted as a simulated patulin-contaminated beverage. In the control group, when no apple juice was added, rapid degradation of patulin was observed: the degradation rate was 76.8 ± 1.2% at 2 h of incubation and reached 100% after 5 h (Figure 7). When apple juice was present, the degradation rate of patulin was decreased to 19.8 ± 2.5% after incubation for 2 h. However, the degradation rate increased to 72.1 ± 4.3% after 5 h of incubation and then steadily increased to 95 ± 2.1% after 24 h. In the univariate analysis of 24 h data, there was a significant difference ($p < 0.05$).

Therefore, it was evident that some components negatively affected degradation of patulin. However, the degradation rate of patulin was still highly comparable to that in absence of apple juice at 12 h and 24 h of incubation. Therefore, *Mr*MnP-catalyzed degradation of patulin can serve as an effective means to control the pollution of patulin in fruit juice.

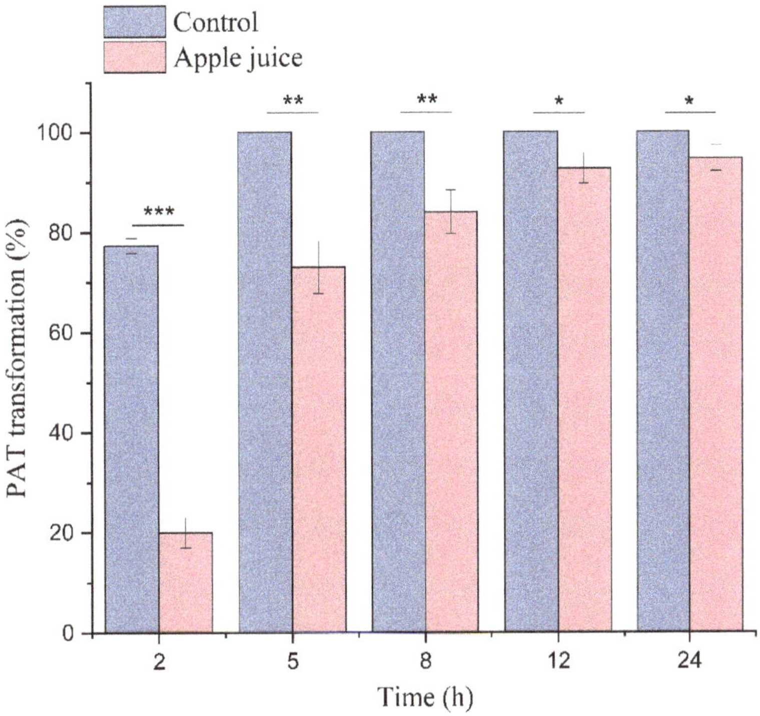

Figure 7. *Mr*MnP efficiently catalyzed degradation of patulin in a simulated patulin-contaminated apple juice. The reactions were carried out by incubating 0.5 U/mL of *Mr*MnP with 5 mg/L in absence (or presence) of apple juice. The samples were periodically taken out for HPLC analysis. The data in the picture is the average ± standard deviation, *** $p < 0.001$, ** $p < 0.01$, * $p < 0.05$, One Way anova test.

Manganese peroxidases are present in many different microorganisms, such as *Irpex lacteus* [44], *Rhizoctonia* sp. [45], *Stereum Ostrea* [46], and *Phanerochaete chrysosporium* [47]. Regardless of the source of manganese peroxidases, they can all catalyze the oxidation of Mn^{2+} to Mn^{3+}, which can be stabilized by forming complexes with a specific chelator in the reaction system. Therefore, it is expected that manganese peroxidase from other sources can also degrade patulin, and these enzymes will be a rich resource of candidate MnPs with physiochemical properties satisfying the demands of practical applications.

3. Conclusions

*Mr*MnP can degrade patulin most rapidly in the malonate/Mn^{2+} system, with 0.5 U/mL enzyme completely removing 5 mg/L of pure patulin within 5 h. One major degradation intermediate was identified by mass spectrometry to be hydroascladiol. In a simulated patulin-contaminated apple juice, 95% of patulin was eliminated after a 24 h treatment. Use of the enzyme to eliminate patulin contamination in juice appears to be advantageous to other detoxifying strategies such as microbial cells. Importantly, as the degradation led to detoxification, *Mr*MnP, and perhaps other manganese peroxidases, may serve as candidates for enzymatic detoxification of patulin in foods and beverages. As MnPs are commonly fragile, and the expression level of *Mr*MnP is still not comparable to those of other proteins such as glycoside hydrolases, in future efforts should be made to improve the stability and expression level of MnPs to reduce the cost in juice detoxification.

4. Materials and Methods

4.1. Strains and Plasmids

The *Escherichia coli* Trans1-T1 (TransGen, Beijing, China) was used for gene cloning and plasmid propagation. The *E. coli* DH5α (Vazyme, Nanjing, China) was used to detect the residual toxicity of patulin after *Mr*MnP treatment. The yeast used for recombinant *Mr*MnP expression was the *Pichia pastoris* X-33 strain (Invitrogen, Carlsbad, CA, USA). The plasmid used for construction of the expression plasmid was pPICZα(A) (Invitrogen, Carlsbad, CA, USA).

4.2. Cloning and Expression of MrMnP

The coding sequence of the *Mr*MnP manganese peroxidase gene (GenBank accession number: ESK95360.1) from *M. roreri* was codon-optimized according to the codon bias of *Pichia pastoris* and synthesized by the GenScript Biotech Corp. (Nanjing, China). Then, the *Mr*MnP gene was amplified from the synthesized gene by gene specific primers MrMnP-F and MrMnP-R (MrMnP-F: 5′-GCGGAATTCGCTGTTCCACAAAGAGTTGCTT-3′, where the underlined sequence indicates the *Eco*RI restriction site; MrMnP-R: 5′-GCGGCGGCCGCAGATGGTGGAACAGCTGGAAC-3′, where the underlined sequence indicates the *Not*I restriction site). The PCR product was treated with *Eco*RI and *Not*I and ligated into the expression vector pPICZα(A) pre-digested with the same two restriction enzymes to generate the recombinant plasmid pPICZα(A)-MrMnP, which was transformed into *E. coli* Trans1-T1 for cloning and sequencing. The integrity of the recombinant plasmid was confirmed by DNA sequencing. Then, the *Dra*I-linearized plasmid was transformed into the *P. pastoris* X33 competent cells by electroporation. Clones were selected on YPDS agar-plates including 100 μg/mL of zeocin.

A single colony of the transformant bearing the *Mr*MnP gene was inoculated into 10 mL of the YPD (yeast potato dextrose) medium, cultured at 30 °C overnight with an agitation of 200 rpm, and then transferred to 50 mL BMGY medium (containing 1% yeast extract, 2% peptone, 1% glycerol, 1.34% YNB, 100 mM sodium phosphate buffer pH 6, 0.4 mg/L biotin). The culture was continued at 30 °C and 200 rpm until the optical density of the culture at 600 nm (OD_{600}) reached 6.0. The cells were then collected by centrifugation and suspended in 50 mL BMMY medium (containing 1% yeast extract, 2% peptone, 1% methanol, 1.34% YNB, 100 mM sodium phosphate buffer pH 6.0, 0.4 mg/L biotin). Methanol was added every 24 h to a final concentration of 1% (v/w) and the induction of enzyme expression was continued at 30 °C for 5 d. The secreted crude *Mr*MnP enzyme was collected from the culture supernatant.

4.3. Fed-Batch Fermentation of MrMnP in a Bioreactor

The X33 transformant integrated with the *Mr*MnP gene was inoculated in 50 mL YPD and incubated at 30 °C for 48 h with shaking at 220 rpm. This "primary seed" culture was transferred into another three 200 mL fresh YPD medium (with a ratio of one tenth, v/v) and the culture was continued overnight, which served as the "secondary seed". This secondary seed culture was added into a bioreactor filled with 6 L of minimal salt medium, and the culture parameters were set with the pH to 4.0, temperature to 30 °C, and the rotation speed to 300 rpm. After the carbon source in the culture medium was exhausted, a mixture of glucose/methanol (glucose: 40%; glucose:methanol = 6:1) and 75 μM heme solution were added at a flow rate of 36 mL/(L h). The fermentation lasted for 120 h, during which the heme solution was replenished appropriately.

4.4. Degradation of Patulin by MrMnP

To determine the ability of *Mr*MnP to degrade patulin, the manganese peroxidase activity was first calibrated using 2,2′-azino-bis (3-ethylbenzothiazoline-6-sulfonic acid) (ABTS), which was carried out by monitoring the oxidation of ABTS ($\varepsilon 420$ = 36,000 M^{-1} cm^{-1}) at 420 nm in a buffer containing 50 mM malonate, 1 mM ABTS, 1 mM $MnSO_4$, and 0.1 mM H_2O_2 (pH 5.0 and 25 °C) as described in Qin et al. [48]. Then, *Mr*MnP with an activity of 0.5 U/mL against ABTS was incubated with 5 mg/L of patulin in 50 mM malonate buffer, 1 mM $MnSO_4$, and 0.1 mM H_2O_2. The reaction was carried out at 30 °C. At the end of the reaction, 3 volumes of methanol were added to the mixture for termination and the reaction products were analyzed by HPLC.

4.5. Effect of the Buffer Componesnts on Degradation of Patulin by MrMnP

To determine the effect of the buffer components on the degradation of patulin, the malonate (50 mM) was replaced with one of the other buffers, which included acetate, lactate, citrate, oxalate, phosphate, MES, and HEPES. The pH of the buffer was adjusted to 5.0 in all tested reaction systems. The reaction was carried out at 30 °C for 24 h. The reaction products were analyzed by HPLC and the degradation rate of patulin was then determined.

4.6. Effect of Mn^{2+} on Degradation of Patulin

In order to study the effect of Mn^{2+} on the degradation of patulin, the degradation rate of patulin in the malonate and acetate systems with or without Mn^{2+} was tested. All the reactions were carried out at 30 °C, and the samples were periodically (2 h, 5 h, 8 h, and 24 h) taken out for analysis of the degradation rate of patulin.

4.7. Toxicity Assay

The *E. coli* DH5α was used as a microbial sensor system to determine if the *Mr*MnP-catalyzed degradation products of patulin would still have toxicity [34]. *E. coli* was cultured in a Luria–Bertani broth (LB: 1.0% tryptone, 0.5% yeast extract, and 1.0% sodium chloride) to an OD_{600} of 0.3. Then, 0.5 U/mL of *Mr*MnP was added to varying concentrations (0, 1, 10, 50, and 100 mg/L) of patulin and the reaction was incubated at 30 °C for 24 h. Then, 5 mL of *E. coli* cells were mixed with 1 mL of patulin treated with or without *Mr*MnP. The growth of *E. coli* was carried out at 37 °C in a 96-well microplate. At 0, 2, 4, 6, 8, and 10 h, the OD_{600} was measured as an indicator of the growth of *E. coli* and toxicity of patulin to the bacterium.

4.8. HPLC and LC-MS/MS Analyses

Degradation of patulin was analyzed by High-Performance Liquid Chromatography (HPLC), which was carried out using a SHIMADZU 20A series instrument (Kyoto, Japan) with an Agilent ZORBAX SB-C18 column (5 μm, 4.6 mm × 250 mm) (Santa Clara, CA, USA). The elution condition included the use of 10% acetonitrile as the mobile phase, and the flow rate was set to 0.75 mL/min. Patulin was monitored under ultraviolet light of wavelength 276 nm.

The degradation products of patulin were further analyzed by using LC-MS/MS, which was carried out by coupling a SHIMADZU Nexera UHPLC system (Kyoto, Japan) to an AB-SCIEX 5600+ Triple TOF mass spectrometer (Waltham, MA, USA). The chromatographic column was XBrige BHE C18 (2.5 μm, 2.1 mm × 150 mm) and the column temperature was 40 °C. The mobile phase A was acetonitrile and the mobile phase B was 0.1% formic acid. One μL of the sample was injected and the flow rate was 0.3 mL/min. The elution procedure was as follows: initial 10% phase A; 0.5 min, 10% phase A; 1.5 min, 50% phase A; 5.0 min, 90% phase A; 6.0 min, 90% phase A; 6.2 min, 10% phase A; 8.0 min, 10% phase A. The detection conditions of mass spectrometry were as follows: negative ion; TOF-mass (Da) 50–500; ion source: Duo Spray Ion Source; ion source gas 1:50; ion source gas 2:50; curtain gas: 25; temperature: 450, IonSpray voltage floating (ISVF): −4500; declustering potential: −57.0; collision energy: −12.0; accumulation time: 0.1 s. ion scanning conditions: declustering potential: −80.0; collision energy: −30.0; collision energy spread: 0; ion release delay: 67; ion release width: 25.

4.9. MrMnP-Catalyzed Degradation of Patulin in a Simulated Patulin-Contaminated Apple Juice

The commercially available fresh apple juice (Huiyuan, Beijing, China) was obtained from a local retail market. No patulin was detected in this juice by HPLC analysis. Patulin was added to this product to a final concentration of 5 mg/L to simulate mycotoxin-contaminated juice. Then, the reaction was carried out by adding a final concentration of 0.5 U/mL of *Mr*MnP to the juice in the presence of 50 mM of malonic acid solution and 1 mM of $MnSO_4$. The reaction was initiated by addition of 0.1 mM of H_2O_2. At 0, 2, 5, 8, and 24 h, samples were taken out for HPLC analysis.

4.10. Statistical Analysis

The results of three repetitions were expressed as mean ± standard deviation (SD). The data were analyzed by one-way analysis of variance (ANOVA).

Author Contributions: S.W. conducted experiments and wrote the manuscript. X.W., L.P., H.L., Y.Z., B.L., B.Y. and N.H. conducted experiments. W.Z. and X.S. conceived and designed the research. All authors contributed to data analyses, results interpretation. All authors have read and agreed to the published version of the manuscript.

Funding: This study was supported by the National Key R&D Program of China (No. 2021YFA0910602, 2021YFC2103002 and 2021YFC2102400) and the National Chicken Industry Technology System of China (CARS-41).

Institutional Review Board Statement: Not applicable.

Informed Consent Statement: Not applicable.

Data Availability Statement: The data presented in this study are available on request from the corresponding author.

Conflicts of Interest: The authors declare no conflict of interest.

Abbreviations

PAT	Patulin
MnP	Manganese Peroxidases
ABTS	2,2′-Azino-bis (3-ethylbenzothiazoline-6-sulfonic acid)
H_2O_2	Hydrogen Peroxide
MES	2-Morpholinoethanesulphonic acid
HEPES	2-[4-(2-Hydroxyethyl)-1-piperazinyl] Ethanesulfonic acid

References

1. Iqbal, S.Z.; Malik, S.; Asi, M.R.; Selamat, J.; Malik, N. Natural occurrence of patulin in different fruits, juices and smoothies and evaluation of dietary intake in Punjab, Pakistan. *Food Control* **2018**, *84*, 370–374. [CrossRef]
2. Li, N.; Cui, R.; Zhang, F.; Meng, X.; Liu, B. A novel enzyme from *Rhodotorula mucilaginosa* Aldolase: Isolation, identification and degradation for patulin in apple juice. *Process Biochem.* **2022**, *116*, 148–156. [CrossRef]
3. Zhong, L.; Carere, J.; Lu, Z.; Lu, F.; Zhou, T. Patulin in apples and apple-based food products: The burdens and the mitigation strategies. *Toxins* **2018**, *10*, 475. [CrossRef] [PubMed]
4. Yang, Q.; Ma, J.; Solairaj, D.; Fu, Y.; Zhang, H. Efficacy of *Meyerozyma guilliermondii* in controlling patulin production by *Penicillium* expansum in shuijing pears. *Biol. Control* **2022**, *168*, 104856. [CrossRef]
5. Pal, S.; Singh, N.; Ansari, K.M. Toxicological effects of patulin mycotoxin on the mammalian system: An overview. *Toxicol. Res.* **2017**, *6*, 764–771. [CrossRef] [PubMed]
6. Van Egmond, H.P. Worldwide regulations for mycotoxins. *Adv. Exp. Med. Biol.* **2002**, *504*, 257–269.
7. Torovic, L.; Dimitrov, N.; Lopes, A.; Martins, C.; Alvito, P.; Assuncao, R. Patulin in fruit juices: Occurrence, bioaccessibility, and risk assessment for Serbian population. *Food Addit. Contam. Part A Chem. Anal. Control Expo. Risk Assess.* **2018**, *35*, 985–995. [CrossRef]

8. Cano-Sancho, G.; Marin, S.; Ramos, A.J.; Sanchis, V. Survey of patulin occurrence in apple juice and apple products in Catalonia, Spain, and an estimate of dietary intake. *Food Addit. Contam. Part B Surveill.* **2009**, *2*, 59–65. [CrossRef]
9. Barreira, M.J.; Alvito, P.C.; Almeida, C.M.M. Occurrence of patulin in apple-based-foods in Portugal. *Food Chem.* **2010**, *121*, 653–658. [CrossRef]
10. Ioi, J.D.; Zhou, T.; Tsao, R.; Marcone, F.M. Mitigation of patulin in fresh and processed foods and beverages. *Toxins* **2017**, *9*, 157. [CrossRef]
11. Karaca, H.; Sedat Velioglu, Y. Effects of some metals and chelating agents on patulin degradation by ozone. *Ozone Sci. Eng.* **2009**, *31*, 224–231. [CrossRef]
12. Cao, J.; Zhang, H.; Yang, Q.; Ren, R. Efficacy of *Pichia caribbica* in controlling blue mold rot and patulin degradation in apples. *Int. J. Food Microbiol.* **2013**, *162*, 167–173. [CrossRef] [PubMed]
13. Dong, X.; Jiang, W.; Li, C.; Ma, N.; Xu, Y.; Meng, X. Patulin biodegradation by marine yeast *Kodameae ohmeri*. *Food Addit Contam. Part A Chem. Anal. Control Expo. Risk Assess.* **2015**, *32*, 352–360. [PubMed]
14. Castoria, R.; Mannina, L.; Duran-Patron, R.; Maffei, F.; Sobolev, A.P.; De Felice, D.V.; Pinedo-Rivilla, C.; Ritieni, A.; Ferracane, R.; Wright, S.A. Conversion of the mycotoxin patulin to the less toxic desoxypatulinic acid by the biocontrol yeast *Rhodosporidium kratochvilovae* strain LS11. *J. Agric. Food Chem.* **2011**, *59*, 11571–11578. [CrossRef]
15. Li, X.; Tang, H.; Yang, C.; Meng, X.; Liu, B. Detoxification of mycotoxin patulin by the yeast *Rhodotorula mucilaginosa*. *Food Control* **2019**, *96*, 47–52. [CrossRef]
16. Ito, M.; Sato, I.; Ishizaka, M.; Yoshida, S.; Koitabashi, M.; Yoshida, S.; Tsushima, S. Bacterial cytochrome P450 system catabolizing the *Fusarium* toxin deoxynivalenol. *Appl. Environ. Microbiol.* **2013**, *79*, 1619–1628. [CrossRef]
17. Carere, J.; Hassan, Y.I.; Lepp, D.; Zhou, T. The Identification of DepB: An Enzyme Responsible for the Final Detoxification Step in the Deoxynivalenol Epimerization Pathway in *Devosia mutans* 17-2-E-8. *Front. Microbiol.* **2018**, *9*, 1573. [CrossRef]
18. Roy, B.P.; Paice, M.G.; Archibald, F.S.; Misra, S.K.; Misiak, L.E. Creation of metal-complexing agents, reduction of manganese dioxide, and promotion of manganese peroxidase-mediated Mn(III) production by cellobiose:quinone oxidoreductase from *Trametes versicolor*. *J. Biol. Chem.* **1994**, *269*, 19745–19750. [CrossRef]
19. Kondo, R.; Harazono, K.; Sakai, K. Bleaching of hardwood kraft pulp with manganese peroxidase secreted from *Phanerochaete sordida* YK-624. *Appl. Environ. Microbiol.* **1994**, *60*, 4359–4363. [CrossRef]
20. Wang, X.; Qin, X.; Hao, Z.; Luo, H.; Yao, B.; Su, X. Degradation of four major mycotoxins by eight manganese peroxidases in presence of a dicarboxylic acid. *Toxins* **2019**, *11*, 566. [CrossRef]
21. Bronikowski, A.; Koschorreck, K.; Urlacher, V.B. Redesign of a new manganese peroxidase highly expressed in *Pichia pastoris* towards a lignin-degrading versatile peroxidase. *ChemBioChem* **2018**, *19*, 2481–2489. [CrossRef] [PubMed]
22. Qin, X.; Su, X.; Luo, H.; Ma, R.; Yao, B.; Ma, F. Deciphering lignocellulose deconstruction by the white rot fungus *Irpex lacteus* based on genomic and transcriptomic analyses. *Biotechnol. Biofuels* **2018**, *11*, 58. [CrossRef] [PubMed]
23. Wang, N.; Ren, K.; Jia, R.; Chen, W.; Sun, R. Expression of a fungal manganese peroxidase in *Escherichia coli*: A comparison between the soluble and refolded enzymes. *BMC Biotechnol.* **2016**, *16*, 87. [CrossRef]
24. Whitwam, R.; Tien, M. Heterologous expression and reconstitution of fungal Mn peroxidase. *Arch. Biochem. Biophys.* **1996**, *333*, 439–446. [CrossRef] [PubMed]
25. Jiang, F.; Kongsaeree, P.; Charron, R.; Lajoie, C.; Xu, H.; Scott, G.; Kelly, C. Production and separation of manganese peroxidase from heme amended yeast cultures. *Biotechnol. Bioeng.* **2008**, *99*, 540–549. [CrossRef] [PubMed]
26. Jiang, F.; Kongsaeree, P.; Schilke, K.; Lajoie, C.; Kelly, C. Effects of pH and temperature on recombinant manganese peroxidase production and stability. *Appl. Biochem. Biotechnol.* **2008**, *146*, 15–27. [CrossRef]
27. Hilden, K.; Makela, M.R.; Lundell, T.; Kuuskeri, J.; Chernykh, A.; Golovleva, L.; Archer, D.B.; Hatakka, A. Heterologous expression and structural characterization of two low pH laccases from a biopulping white-rot fungus *Physisporinus rivulosus*. *Appl. Microbiol. Biotechnol.* **2013**, *97*, 1589–1599. [CrossRef]
28. Xu, H.; Guo, M.Y.; Gao, Y.H.; Bai, X.H.; Zhou, X.W. Expression and characteristics of manganese peroxidase from *Ganoderma lucidum* in *Pichia pastoris* and its application in the degradation of four dyes and phenol. *BMC Biotechnol.* **2017**, *17*, 19. [CrossRef]
29. Conesa, A.; Punt, P.J.; van den Hondel, C.A. Fungal peroxidases: Molecular aspects and applications. *J. Biotechnol.* **2002**, *93*, 143–158. [CrossRef]
30. Cortes-Espinosa, D.V.; Absalon, A.E.; Sanchez, N.; Loera, O.; Rodriguez-Vazquez, R.; Fernandez, F.J. Heterologous expression of manganese peroxidase in *Aspergillus niger* and its effect on phenanthrene removal from soil. *J. Mol. Microbiol. Biotechnol.* **2011**, *21*, 120–129. [CrossRef]
31. Ansari, Z.; Karimi, A.; Sedghi, S.; Razzaghi, M.; Ebrahimi, S. Glucose oxidase effect on treatment of textile effluent containing reactive azo dyes by *Phanerochaete chrysosporium*. *J. Chem. Technol. Biotechnol.* **2017**, *92*, 1721–1726. [CrossRef]
32. Zhang, J.; Presley, G.N.; Hammel, K.E.; Ryu, J.S.; Menke, J.R.; Figueroa, M.; Hu, D.; Orr, G.; Schilling, J.S. Localizing gene regulation reveals a staggered wood decay mechanism for the brown rot fungus *Postia placenta*. *Proc. Natl. Acad. Sci. USA* **2016**, *113*, 10968–10973. [CrossRef] [PubMed]
33. Gumiero, A.; Murphy, E.J.; Metcalfe, C.L.; Moody, P.C.; Raven, E.L. An analysis of substrate binding interactions in the heme peroxidase enzymes: A structural perspective. *Arch. Biochem. Biophys.* **2010**, *500*, 13–20. [CrossRef] [PubMed]
34. Zhu, R.; Feussner, K.; Wu, T.; Yan, F.; Karlovsky, P.; Zheng, X. Detoxification of mycotoxin patulin by the yeast *Rhodosporidium paludigenum*. *Food Chem.* **2015**, *179*, 1–5. [CrossRef]

35. Hathout, A.S.; Aly, S.E. Biological detoxification of mycotoxins: A review. *Ann. Microbiol.* **2014**, *64*, 905–919. [CrossRef]
36. Malysheva, S.V.; Diana Di Mavungu, J.; Boonen, J.; De Spiegeleer, B.; Goryacheva, I.Y.; Vanhaecke, L.; De Saeger, S. Improved positive electrospray ionization of patulin by adduct formation: Usefulness in liquid chromatography-tandem mass spectrometry multi-mycotoxin analysis. *J. Chromatogr. A* **2012**, *1270*, 334–339. [CrossRef]
37. Hawar, S.; Vevers, W.; Karieb, S.; Ali, B.K.; Billington, R.; Beal, J. Biotransformation of patulin to hydroascladiol by *Lactobacillus plantarum*. *Food Control* **2013**, *34*, 502–508. [CrossRef]
38. Diao, E.; Hou, H.; Hu, W.; Dong, H.; Li, X. Removing and detoxifying methods of patulin: A review. *Trends Food Sci. Technol.* **2018**, *81*, 139–145. [CrossRef]
39. Crestini, C.; D'Annibale, A.; Sermanni, G.G.; Saladino, R. The reactivity of phenolic and non-phenolic residual kraft lignin model compounds with Mn(II)-peroxidase from *Lentinula edodes*. *Bioorg. Med. Chem.* **2000**, *8*, 433–438. [CrossRef]
40. Kapich, A.N.; Korneichik, T.V.; Hatakka, A.; Hammel, K.E. Oxidizability of unsaturated fatty acids and of a non-phenolic lignin structure in the manganese peroxidase-dependent lipid peroxidation system. *Enzym. Microb. Technol.* **2010**, *46*, 136–140. [CrossRef]
41. Kong, W.; Chen, H.; Lyu, S.; Ma, F.; Yu, H.; Zhang, X. Characterization of a novel manganese peroxidase from white-rot fungus *Echinodontium taxodii* 2538, and its use for the degradation of lignin-related compounds. *Process Biochem.* **2016**, *51*, 1776–1783. [CrossRef]
42. Wang, X.; Yao, B.; Su, X. Linking enzymatic oxidative degradation of lignin to organics detoxification. *Int. J. Mol. Sci.* **2018**, *19*, 3373. [CrossRef] [PubMed]
43. Kostić, A.; Stojanović, M.; Lopičić, Z.; Milojković, J.; Petrović, M. Biomass waste material as potential adsorbent for sequestering pollutants. *Zaštita Materijala* **2012**, *53*, 231–237.
44. Qin, X.; Zhang, J.; Zhang, X.; Yang, Y. Induction, purification and characterization of a novel manganese peroxidase from *Irpex lacteus* CD2 and its application in the decolorization of different types of dye. *PLoS ONE* **2014**, *9*, e113282. [CrossRef] [PubMed]
45. Cai, Y.; Wu, H.; Liao, X.; Ding, Y.; Sun, J.; Zhang, D. Purification and characterization of novel manganese peroxidase from *Rhizoctonia* sp. SYBC-M3. *Biotechnol. Bioprocess Eng.* **2011**, *15*, 1016–1021. [CrossRef]
46. Praveen, K.; Usha, K.Y.; Viswanath, B.; Reddy, B.R. Kinetic properties of manganese peroxidase from the mushroom *Stereum ostrea* and its ability to decolorize dyes. *J. Microbiol. Biotechnol.* **2012**, *22*, 1540–1548. [CrossRef]
47. Ürek, R.Ö.; Pazarlioğlu, N.K. Purification and partial characterization of manganese peroxidase from immobilized *Phanerochaete chrysosporium*. *Process Biochem.* **2004**, *39*, 2061–2068. [CrossRef]
48. Qin, X.; Sun, X.; Huang, H.; Bai, Y.; Wang, Y.; Luo, H.; Yao, B.; Zhang, X.; Su, X. Oxidation of a non-phenolic lignin model compound by two *Irpex lacteus* manganese peroxidases: Evidence for implication of carboxylate and radicals. *Biotechnol. Biofuels* **2017**, *10*, 103. [CrossRef]

toxins

MDPI

Article

Sustainable Strategies to Counteract Mycotoxins Contamination and Cowpea Weevil in Chickpea Seeds during Post-Harvest

Claudia Pisuttu [1], Samuele Risoli [1,2], Lorenzo Moncini [3], Cristina Nali [1,4], Elisa Pellegrini [1,4,*] and Sabrina Sarrocco [1,4]

1 Department of Agriculture, Food and Environment, University of Pisa, Via del Borghetto 80, 56124 Pisa, Italy
2 University School for Advanced Studies IUSS, Piazza della Vittoria 15, 27100 Pavia, Italy
3 Biotechnical Instruments in Agriculture and Forestry Research Centre (CRISBA), ISIS "Leopoldo II di Lorena", Cittadella dello Studente, 58100 Grosseto, Italy
4 Nutrafood Research Center, University of Pisa, Via del Borghetto 50, 56124 Pisa, Italy
* Correspondence: elisa.pellegrini@unipi.it

Abstract: Mycotoxins contamination and pest infestation of foods and feeds represent a pivotal threat for food safety and security worldwide, with crucial implications for human and animal health. Controlled atmosphere could be a sustainable strategy to reduce mycotoxins content and counteract the vitality of deleterious organisms in foodstuff. Ozone treatment (O_3, 500 ppb for 30, 60 or 90 min) and high nitrogen concentration (N_2, 99% for 21 consecutive days) were tested in the post-harvest management of four batches of *Cicer arietinum* grains to control the presence of mycotoxigenic fungi and their secondary metabolites, as well as pest (i.e., *Callosobruchus maculatus*) infestation. At the end of the treatment, O_3 significantly decreased the incidence of *Penicillium* spp. (by an average of −50%, independently to the time of exposure) and reduced the patulin and aflatoxins content after 30 min (−85 and −100%, respectively). High N_2 concentrations remarkably reduced mycotoxins contamination (by an average of −94%) and induced pest mortality (at 100% after 5 days of exposure). These results confirm the promising potential of O_3 and N_2 in post-harvest conservation strategies, leading to further investigations to evaluate the effects on the qualitative characteristics of grains.

Keywords: *Cicer arietinum* L.; mycotoxigenic fungi; mycotoxin occurrence; pest attack; nitrogen-controlled atmosphere; ozone

Key Contribution: The findings of the present study are relevant in enhancing the shelf-life of chickpea seeds by controlling fungal growth, mycotoxin contamination and pest infestation with eco-friendly and low-cost storage practices such as ozone treatment and high nitrogen concentrations.

Citation: Pisuttu, C.; Risoli, S.; Moncini, L.; Nali, C.; Pellegrini, E.; Sarrocco, S. Sustainable Strategies to Counteract Mycotoxins Contamination and Cowpea Weevil in Chickpea Seeds during Post-Harvest. *Toxins* **2023**, *15*, 61. https://doi.org/10.3390/toxins15010061

Received: 21 December 2022
Revised: 5 January 2023
Accepted: 7 January 2023
Published: 11 January 2023

1. Introduction

The increasing occurrence of food/feed contaminants worldwide poses a huge threat to human and animal health. One of the major contaminants are mycotoxins, which annually cause enormous economic losses in the food industry and animal husbandry [1,2]. These low molecular weight metabolites produced by filamentous fungi (belonging to the phylum Ascomycota) contaminate various categories of foods and feeds [3]. Two groups of mycotoxigenic fungi exist: field fungi (such as *Fusarium* and *Aspergillus* spp.) that infect crops before harvest, and storage fungi (such as *Penicillium* spp.), which only occur after harvest [4,5]. According to a recent world survey based on around 97,000 analyses performed between January and December 2020 on more than 21,000 finished feed and raw commodity sources collected from 79 countries, the most prevalent mycotoxins were those produced by *Fusarium*, affecting more than the 60% of tested samples [Biomin, Inzersdorf-Getzersdorf, Austria, https://www.biomin.net/science-hub/world-mycotoxin-survey-impact-2021/, accessed on 4 January 2023], In 2020, Mesterhazy et al. highlighted how toxins are responsible for a loss of almost 700 mt during the harvest and storage of grains [6]. At any stage of the food production process (in the field, during harvest, during drying and transport, as well as during storage), the fungal production of mycotoxins can occur by exposing consumers to the risk of contamination, either directly through food consumption or indirectly through feed [7]. The most important mycotoxins are aflatoxins [mainly represented by aflatoxin B1 (AFB1), B2 (AFB2), G1 (AFG1), G2 (AFG2) and M1 (AFM1)], ochratoxins, fumonisins, trichothecenes, zearalenone, the emerging *Fusarium* mycotoxins, ergot alkaloids, *Alternaria* toxins, and patulin [8]. Of the approximately 400 compounds identified as mycotoxins, 30 have received significant consideration due to their harmful effects on both human and animal health (including genotoxicity and endocrine disruption [9]). Despite a huge number of published papers reporting the occurrence of mycotoxins on cereals and cereal-derived food products, in 2017, an analysis of 104 papers—from 2006 to 2016—was carried out, summarizing that mycotoxins are ubiquitously present in cereals and cereal-derived food products throughout the world [10]. If Africa and Asia showed the highest incidence (%) of cereals contaminated by aflatoxins and ochratoxins, respectively, South and North America registered the highest level of fumonisins and Europe the highest percentage of deoxynivalenol (trichothecenes)/zearalenone contamination [10].

Different physical, chemical and biological factors affect fungal colonization and mycotoxins production. Physical factors include environmental conditions such as temperature, relative humidity, pH, water activity, nutrients, insect infestation and other associated factors, which at specific rates are known to favor the growth of many types of fungi and the production of mycotoxins [2]. Biological factors are mainly related to the interactions between the colonizing toxic fungal species and the host, thus including features such as fungal species, strain specificity, levels of inoculation, the nature of the substrate, strain variation, the instability of fungal toxic properties, and insect damage [11].

Regulatory agencies have established strict legislative thresholds in order to keep the levels of mycotoxins in food/feed commodities under control. These limits range from below one to thousands of $\mu g\ kg^{-1}$, depending on the (i) mycotoxin, (ii) type of product and (iii) country considered [12,13]. Consequently, there is an urgent need to develop a feasible and highly sensitive analytical method for mycotoxins detection [14] and reduce the contamination of mycotoxins in food/feed, in order to protect/preserve their quality and safety. Overall, the control of mycotoxin contamination follows two strategies: the prevention of their production (i.e., microbial inactivation) and detoxification (e.g., mycotoxin degradation [3]). In pre-harvest, the control of mycotoxins is based on control of the contamination levels in crops to be used as food/feed components. Generally, these systems are based on preventive strategies (such as the use of resistant varieties, crop rotation, tillage and the management of irrigation) which aim to avoid the development of contamination, operating on the predisposing factors that facilitate the production of mycotoxins. Although pre-harvest approaches should be preferred, in the perspective of preventing mycotoxin contamination, the development of toxic fungi is inevitable under certain environmental conditions [15]. Therefore, appropriate storage practices and other post-harvest control systems (at the microbiological, physical and chemical level) are necessary to minimize the final mycotoxin content of foods/feeds [16]. These traditional methods for the elimination/inactivation of mycotoxins have some limitations regarding (potential) safety issues, loss in the nutritional value and the palatability of feeds, cost implication and limited efficacy [17]. In recent decades, various detoxification approaches have demonstrated to be (i) highly effective in degrading mycotoxins into less toxic products, (ii) economically favorable and (iii) not environmentally harmful [18]. Among these, cold plasma—containing reactive oxygen and nitrogen species and free radicals—has received attention in recent years for use on cereals during storage, due to its lethal effect on microorganisms and its potential to decontaminate surfaces and improve shelf-life [19]. Nevertheless, their practical application in food/feed matrices is limited, since the degradation process under conditions of large-scale production is much more complex, and the experiments at lab-scale might not always reflect practices in industrial processing [1]. Possible reasons for this are that the degradation process can be easily affected by multiple factors such as temperature, relative humidity, pH, water activity, nutrients, insect infestation and types of contamination [2]. The relevance of studying naturally contaminated samples is consistent with the actual distribution conditions of mycotoxins in the field and/or in grains.

Since gas composition is considered one of the most important abiotic conditions that impacts fungal and pest growth, ozonation (i.e., the application of gaseous ozone, O_3) is a simple technology for controlling insects and reducing mycotoxins in stored products, which does not leave harmful residues after application. Being unstable, O_3 quickly degrades into oxygen (and related cytotoxic radicals) in a short period, oxidizing the vital cellular components (such as unsaturated lipids and proteins) of pathogenic microbes and storage pests by causing lysis and rapid cell death [20]. Consequently, O_3 can inhibit fungal growth, sporulation and germination by offering a negligible loss of nutrients or sensory qualities in food/feed [21], making it a suitable candidate as a residue-free fumigant. For this reason, the application of O_3 in food chains has been considered safe and effective by the WHO and is now recognized as a "green technology" for the fumigation of grains, fruits and vegetables [22]. In fact, O_3-treated products are safe for consumption and their microbiological shelf life can be greatly enhanced. However, the efficacy of O_3 in fungi count reductions, mycotoxin degradation and insect control depend on the (i) method, concentration and timing of the O_3 application; (ii) microorganisms/contaminants to inactivate; (iii) the type and mass of food/feed processed; and (iv) other co-factors such as temperature, relative humidity and water activity [16,20]. Similarly, the use of a controlled or modified atmosphere by using a very high nitrogen (N_2) concentration is a valid alternative to chemical fumigation to control mycotoxigenic fungi contamination and pest challenge post-harvest [23]. Its effects on different stored products (such as wheat, maize, corn and rye) are well documented [2,24]. In particular, a N_2-controlled atmosphere can control fungal growth and proliferation by improving the quality of stored products [25]. The action of N_2 at high concentration is mainly due to the significant reduction in O_2 (1% or less [26]) and offers several advantages at the economic and environmental level [27]. A major advantage of N_2 is that all gas is free of pollutants, leaving no residue in food/feed. Consequently, N_2-treated products are safe for consumption and their microbiological shelf-life can be greatly enhanced [28]. Consequently, a N_2-controlled atmosphere might represent an eco-friendly tool that could be transferred to a large-scale system for grain storage as an alternative strategy to the use of conventional residue-producing chemical fumigants [29]. However, the efficacy of a N_2-controlled atmosphere in fungi count reductions, mycotoxin degradation and insect control depends on the (i) concentration of gas, (ii) the timing of the application, (iii) microorganisms/contaminants to inactive, (iv) the type and mass of food/feed processed, and (v) other co-factors such as temperature and water activity [30].

The chickpea (*Cicer arietinum* L.) is a legume of the family Fabaceae, subfamily Faboideae. It is one of the most cultivated pulses in terms of world production due to its low content of fat and sodium, absence of cholesterol and being an excellent source of both soluble and insoluble fiber, complex carbohydrates, vitamins, folate and minerals (such as calcium, phosphorus, iron and magnesium [31]). With a worldwide production of more than 12 million tons per year [FAOSTAT, https://www.atlasbig.com/en-in/countries-by-chickpea-production, accessed on 4 January 2023], chickpea represents one of the five leading pulses based on sales value. In 2021, India was the largest chickpea producer in the world, with around 11 million metric tons of production, followed by Turkey, with around 600,000 metric tons (https://www.statista.com/statistics/722203/chickpeas-production-volume-by-country-worldwide/, accessed on 4 January 2023). Chickpea is often attacked by fungi pre- and post-harvest, significantly affecting its productivity. Many fungal genera/species commonly isolated from chickpea seeds and chickpea by-products are potential mycotoxin-producers, especially of aflatoxins, ochratoxin A and patulin, so there would be a potential risk of contamination [32]. Another issue threatening chickpea quality is *Callosobruchus maculatus* (Fab.) (Coleoptera: Chrysomelidae: Bruchinae [33]), which is also known as the "cowpea weevil". The granivorous larvae of cowpea weevil are the considerable causative agent of severe losses in the grain germination, weight and nutritional values of chickpea (in some cases reaching 60% of the grain [34]). In addition, *C. maculatus* can favor the occurrence of infections due to mycotoxigenic fungi (*Aspergillus* and *Penicillium* [35]).

The aim of this work was to investigate the possibility of using a single pulse of O_3 or high N_2 concentrations as storage technologies (at lab- and large-scale) for the purpose of (i) containing the fungal population present on the chickpea seeds surface; (ii) reducing the mycotoxins content (such as aflatoxins and patulin; Figure 1); and (iii) limiting the *C. maculatus* infestations (only in the case of N_2 treatment). We postulated that O_3 and N_2 can be an alternative to traditional chemical-based fumigants for controlling spoilage pathogens and insects in stored chickpea seeds.

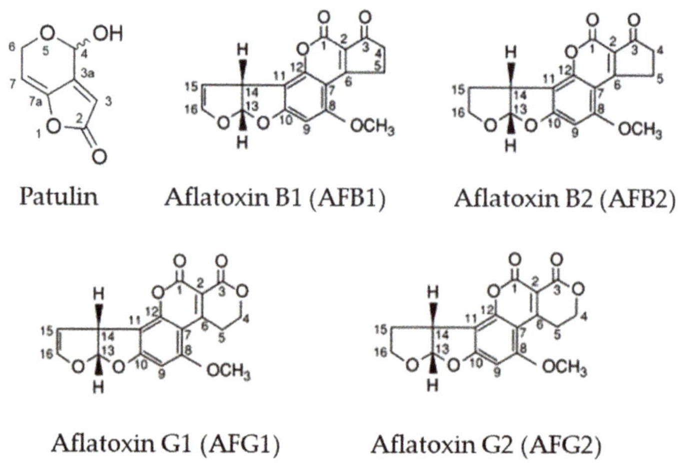

Figure 1. Molecular structures of measured mycotoxins (patulin, aflatoxin B1, aflatoxin B2, aflatoxin G1 and aflatoxin G2; modified by [3]), showing numbering of carbon atoms (3a and 7a represent small fragments).

2. Results

2.1. Effect of O_3 and N_2 Treatments on Fungal Infection

To evaluate the effect of O_3 treatment as well as of the conservation under a N_2-controlled atmosphere on a fungal population naturally occurring on chickpea seeds, with particular attention to mycotoxigenic fungi, a grain health test was performed on control stocks stored at 4 °C and on seeds after O_3 or N_2 treatments. After the morphological and molecular identification of single colonies isolated from seeds, all the batches were contaminated with *Penicillium* spp. isolates, but at a different extent depending on the quality of the batches. Molecular identification performed on the ITS region sequence confirmed the membership of all the isolates to this important fungal genus, with the species *P. pinophilum* and *P. polonicum* as the most represented.

Of particular interest was batch n. 2, where almost 80% of the analyzed seeds were infected by *Penicillium* isolates, thus, confirming the non-marketability of this batch for commercial purposes. The other three batches showed a percentage of natural occurring infection varying around 10–20% (Figure 2).

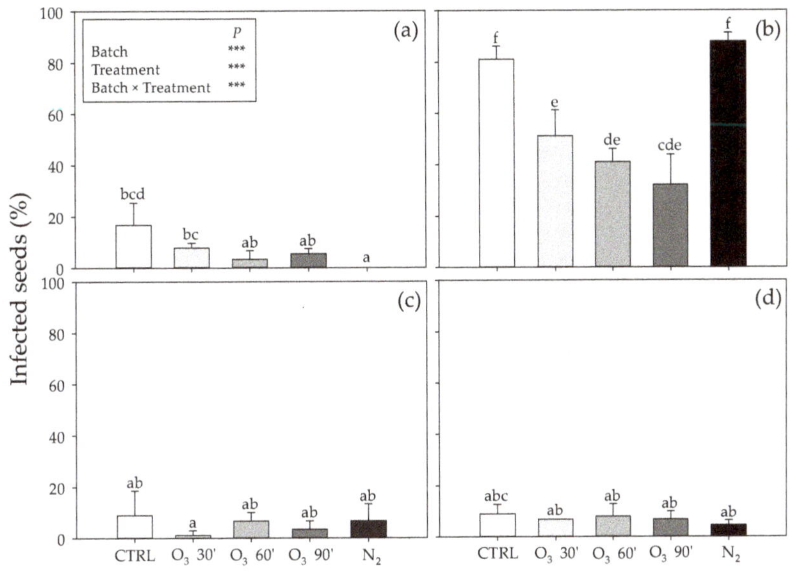

Figure 2. Contamination levels (% of infected seeds) by *Penicillium* spp. in four batches (n. 1 (**a**), n. 2 (**b**), n. 3 (**c**) and n. 4 (**d**)) of chickpea seeds (CTRL, white fill) exposed to ozone [500 ppb O_3 for 30 (O_3 30′, light grey fill), 60 (O_3 60′, grey fill) and 90 (O_3 90′,dark grey fill) minutes] or nitrogen treatment (99% N_2 for 21 consecutive days, dark fill). Data are shown as mean + standard deviation (n = 3). Results of two-way ANOVA are reported; asterisks show the significance of factors/interaction for: *** $p \leq 0.001$. According to Tukey's HSD post hoc test, different letters indicate significant differences ($p \leq 0.05$).

After O_3-exposure, as well as after incubation under the N_2-controlled atmosphere, the grain health test, in most cases, resulted in a reduction in the *Penicillium* contamination of the seeds. A significant effect of single factors and their interaction was observed. For batch n. 1, O_3 treatment did not result in any significant reduction if compared to the control, as well as among treatments. Conversely, a significant reduction was registered after N_2 incubation (−16% of infected grains compared with no seed developing *Penicillium* spp. colonies; Figure 2a). Much more evident was the effect of the O_3 treatment on batch n. 2, where, independently from the time of exposure, a significant reduction in the naturally occurring *Penicillium* population was observed (−55% as average), with the exposure lasting 90 min being the most efficient (−75%). On the other hand, any significant difference was observed after the incubation in the N_2-controlled atmosphere (Figure 2b). For batches n. 3 and 4, a similar behavior was observed, where any significant difference in the percentages of infected seeds was observed if compared with the controls, as well as among treatments (Figure 2c,d). However, in these last two cases, the initial infection was lower if compared with the other two batches, particularly with batch n. 2.

2.2. *Effects of O_3 and N_2 Treatments on Mycotoxin Levels*

In the four batches of chickpea grains maintained in filtered air or incubated in control silos, the mycotoxins patulin, AFG2, AFB2, AFG1 and AFB1 were found. Figure 3 shows an overview of the concentrations of these mycotoxins and their variability among the batches.

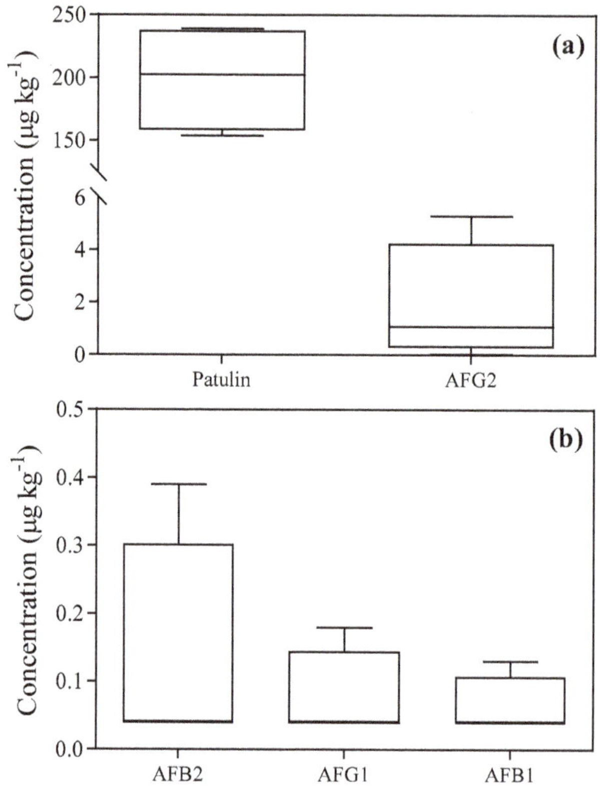

Figure 3. Box and whiskers representation of the average content of measured mycotoxins in four batches of chickpea seeds ((**a**) patulin and aflatoxin G2 (AFG2); (**b**) aflatoxin B2 (AFB2), aflatoxin G1 (AFG1) and aflatoxin B1 (AFB1)). For each mycotoxin, the top line represents the 90th percentile; the bottom line represents the 10th percentile; and the box represents the 75th percentile (upper side), the 25th percentile (lower side) and the median (50th percentile, central line), respectively.

The comparison between the content of selected mycotoxins and the batches of chickpea grains maintained in filtered air or incubated in control silos revealed that the concentrations of patulin were significantly higher in batch n. 4 than those in batch n. 1 (+55%; Table S1). No significant differences were observed among the remaining batches of chickpea grains. The levels of AFB1, AFB2 and AFG1 were significantly higher in batch n. 3 than those in the remaining batches of chickpea grains (about six-fold higher on average). Conversely, the levels of AFG2 were significantly higher in batch n. 2 than those in the remaining batches of chickpea grains (about 100-fold higher than batch n. 1, and five-fold higher than batches n. 2 and 3; Table S1).

The effects of the O_3 and N_2 treatments on the patulin levels in the four batches' grains are reported in Figure 4. The two-way ANOVA test revealed that the interaction "batch of chickpea grains × treatment" and the effects of each factor were significant. Ozone treatment induced a significant decrease in patulin in batch n. 2, independently of its duration (an average of −50% compared to chickpea grains maintained in filtered air; Figure 4b), and even more in the remaining batches (about 150-, 230- and 240-fold lower than CTRL in batches n. 1, 3 and 4, respectively). Similarly, a reduction in patulin was observed in batches n. 3 and 4 incubated in silos under high N_2 concentrations (−70 and −82% than CTRL, respectively; Figure 4c,d), and even more in the remaining batches (about 160-fold lower than those in batches n. 1 and 2 incubated in control silos, respectively).

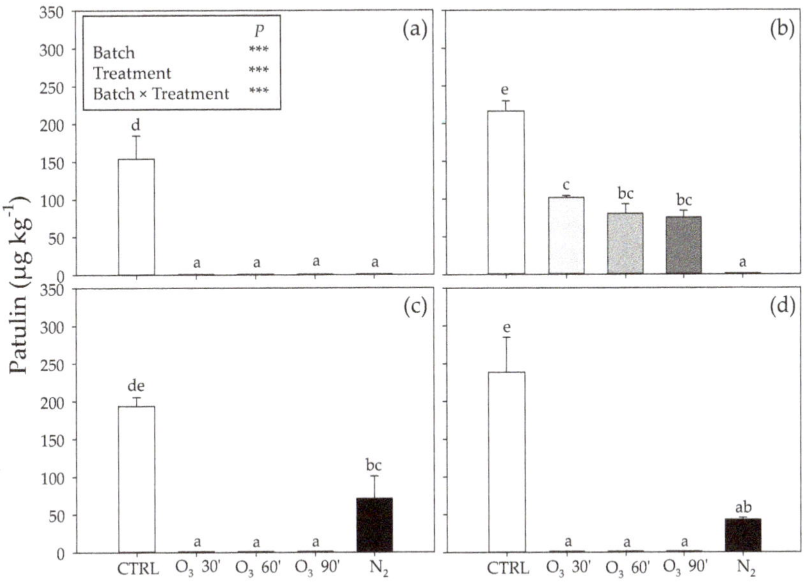

Figure 4. Patulin content in four batches (n. 1 (**a**), n. 2 (**b**), n. 3 (**c**) and n. 4 (**d**)) of chickpea seeds (CTRL, white fill) exposed to ozone (500 ppb O_3 for 30 [O_3 30′, light grey fill), 60 (O_3 60′, grey fill) and 90 (O_3 90′, dark grey fill) minutes] or nitrogen treatment (99% N_2 for 21 consecutive days, dark fill). Data are shown as mean + standard deviation (n = 3). Results of two-way ANOVA are reported; asterisks show the significance of factors/interaction for: *** $p \leq 0.001$. According to Tukey's HSD post hoc test, different letters indicate significant differences ($p \leq 0.05$).

The effects of the O_3 and N_2 treatments on the total aflatoxin levels in four batches of chickpea grains are reported in Figure S1. The two-way ANOVA test revealed that the interaction "batch of chickpea grains × treatment" and the effects of each factor were significant. Ozone treatment induced a complete reduction in total aflatoxins in batches n. 2, 3 and 4, independently of its duration (about 80-fold lower compared to chickpea grains maintained in filtered air). Similarly, a reduction in total aflatoxins was observed in batches n. 2, 3 and 4 incubated in silos under high N_2 concentrations (about 150-fold on average). No other significant differences were found in batch n. 1, independently of the kind of treatment.

2.3. Effect of N_2 Treatment on Pest Survival

The effect of 5-day exposure in the N_2-controlled atmosphere (in the 60 L lab-scale silos) on the number of emerged adults and on Abbott's index is reported in Figure 5.

A statistical analysis highlighted how not only the treatment and the time had a significant effect on the adult emergence, but also the interaction between these two sources of variability resulted as highly significant ($p \leq 0.001$). The exposure to the N_2-controlled atmosphere resulted in a reduction in the number of emerged adults from the first day of exposure. No significant differences were observed after 24 h of exposure. From the second until the fifth day, a continuous and significant reduction was observed, with the adult emergence close to zero after five days of exposure of the eggs within the silos under the N_2-controlled atmosphere.

With respect to Abbott's index, which is conventionally used to evaluate the effect of treatments on pest mortality compared to naturally occurring mortality, the incubation of eggs under high N_2 concentration resulted in an increasing mortality over time of exposure, with 100% reached at the end of the experiment, i.e., after 5 days of exposure (Figure 5, dashed line).

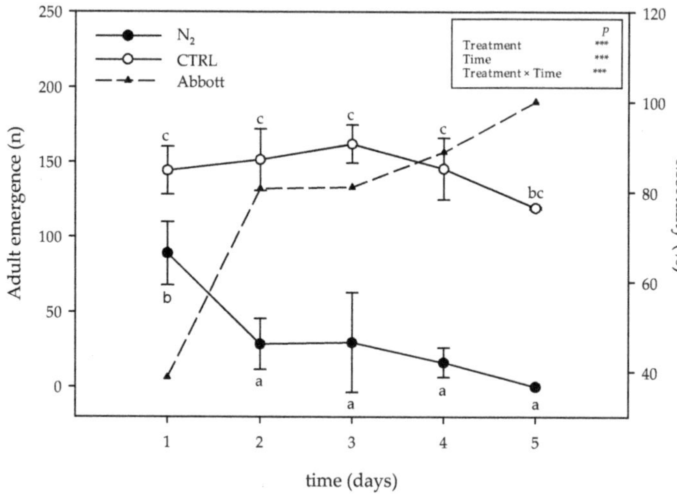

Figure 5. Number of emerged adults of *Callosobruchus maculatus* and mortality (measured using Abbott's index, dotted line and dark triangle) on batch n. 1 (CTRL, open circle) and exposed to nitrogen treatment (60 L lab-scale; dark circle). Data are shown as mean ± standard deviation (n = 3). Results of two-way ANOVA are reported; asterisks show the significance of factors/interaction for: *** $p \leq 0.001$. According to Tukey's HSD post hoc test, different letters indicate significant differences ($p \leq 0.05$).

When the effect of the high N_2 atmosphere was evaluated on *C. maculatus* adult number in real-scale silos after 5 days of exposure (Figure 6), a highly significant reduction in emergence was registered, compared to the control. This result was also confirmed by Abbott's index, which resulted in approximately 80% of mortality due to the N_2-controlled atmosphere.

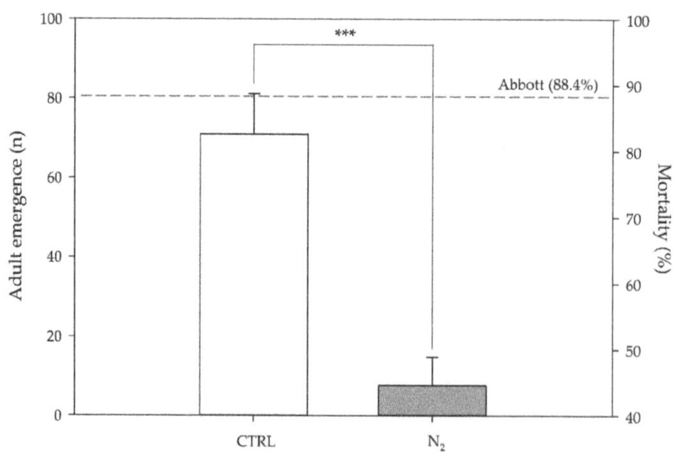

Figure 6. Number of emerged adults of *Callosobruchus maculatus* (Fab.) and mortality (measured using Abbott's index, dotted line) on batch n. 1 (CTRL, white fill) and exposed to nitrogen (N_2) treatment (real-scale; gray fill). Statistical differences were examined by paired Student's *t*-test; asterisks indicate statistical significance for: *** = $p \leq 0.001$.

3. Discussion

In the present study, the fungicidal efficacy of O_3 was only observed in batch n. 2 (the most infected seeds), as confirmed by the large reduction in *Penicillium* contamination. According to Mendez et al. [36], at the beginning of O_3 treatment, this gas reacts with a mass of grains and quickly decomposes. In its second phase, O_3 moves freely through the grains with little degradation. Moreover, O_3 reacts faster with the whole mass of grains when higher dosages are used. This was also confirmed by our results: a higher reduction in the naturally occurring fungal population was observed by using O_3 at 500 ppb for 90 min. To date, few studies have investigated the direct effect of O_3 on fungal growth and proliferation, preferring to focus on the reduction in mycotoxins in different products (such as wheat, corn, corn flour, peanuts and pistachio [37]). Most studies have been carried out on aflatoxins. Using O_3 in patulin, degradation has only been studied in apple juice, apples and pear with brown rot, flour, and malt feed [38]. In the present study, O_3 treatment induced a significant decrease in patulin in batch n. 2 (independently of its duration), and even more in the remaining batches. The mechanism of patulin degradation might be associated with the oxidation of a polyketide lactone on its structure, which made it highly susceptible to O_3 attack [39]. The experimental conditions used in this study (e.g., O_3 and/or the timing of treatment) were sufficient, by attacking two conjugated ethylenic double bonds on the chemical structure of patulin and inducing its partial or full degradation [40]. Similarly, all aflatoxins were easily attacked and degraded by O_3 (independently of its duration), confirming its efficacy as a detoxifying agent. The mechanism of AFB1 and AFG1 degradation might be associated with the oxidation of the C8-C9 double bond at the terminal furan, resulting in the production of primary ozonide [26]. This product may rapidly rearrange to a molozonide derivative, yielding a variety of carbonyl compounds or organic acids. Since AFB2 and AFG2 lack a susceptible double bond for oxidation, their degradation requires higher levels of O_3 and/or longer exposure until the lactone ring is opened [41]. The experimental conditions used in this study (e.g., O_3 concentration and/or the timing of treatment) were sufficient, by rapidly and effectively detoxifying aflatoxins without any difference in degradation rate between AFB1 and AFG1 with AFB2 and AFG2 (the most abundant aflatoxin in batches n. 2, 3 and 4). It is worth noting that the production of aflatoxins was not associated with the presence of the fungal itself, confirming that the absence of *Aspergillus* spp. from chickpea seeds does not guarantee the absence of aflatoxins because of their resistant chemical nature [42].

In the present study, high N_2 concentrations only induced a reduction in *Penicillium* contamination in batch n. 1 confirming that certain fungal species might continue to grow, albeit at a greatly reduced rate under low O_2 concentrations [43]. However, a significant decrease in patulin was observed in batches n. 3 and 4, and even more in the remaining ones, indicating that low O_2 concentrations may (partially or totally) depress patulin production by *Penicillium* spp. on chickpea seeds. In addition, the experimental conditions used in this study (e.g., N_2 concentration and/or the timing of treatment, temperature, relative humidity and water activity) were sufficient, by totally removing aflatoxins. Consequently, it is possible to speculate that high N_2 concentrations are more effective in inhibiting selected mycotoxins (aflatoxins > patulin) than in preventing the development of mycotoxigenic fungi [44]. Although a controlled atmosphere is used to control both mycotoxigenic fungi and insects in stored products, it has been documented that the experimental conditions sufficient for controlling fungal proliferation are not always effective against insect pests that can survive, due to the dependence on other environmental factors (e.g., temperature and humidity [45]). This is in line with the results of the present work, where the same experimental conditions (e.g., N_2 concentration and/or the timing of treatment, temperature, relative humidity and water activity) that were partially effective for reducing the growth of *Penicillium* spp. were effective for detoxifying mycotoxins (as previously reported) and limiting *C. maculatus* infestation (as confirmed by the reduction in adult emergence already observed starting from the first day of the experiment, and the concomitant pest mortality). This is of relevance, since it is generally recognized that pest attack can damage grains and favor moisture accumulation by creating suitable conditions for fungal development and mycotoxin production (as observed in batch n. 2 [46]). Consequently, it is possible to speculate that the experimental conditions used in this study were sufficient, by limiting the pest infestation and guaranteeing seed quality, as already reported in wheat [23,25]. In addition, the possibility of moving to a real-scale dimension, as here preliminary reported, with results comparable with those obtained in the 60 L lab-scale silos, marks a further step made in the direction of the scaling-up of the method. In fact, the set-up chosen for the present study is not only a suitable way to provide a proof-of-concept of its efficacy before scale-up, but also a valid choice for small and medium farms [23,25].

The findings of the present study are relevant in enhancing the shelf-life of chickpea seeds by controlling fungal growth, mycotoxin contamination and pest infestation with eco-friendly and low-cost storage practices (the mechanisms of action of O_3 and N_2 are summarized in Figure 7).

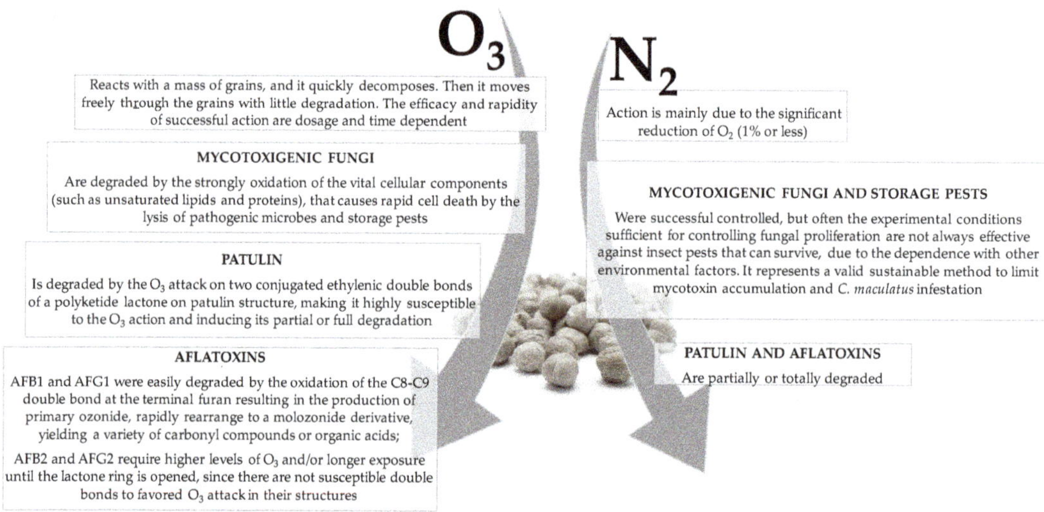

Figure 7. Gaseous ozone (O_3) and nitrogen (N_2) mechanisms of action during post-harvest of grains, and their effects on (i) mycotoxigenic fungi and storage pests (only N_2), (ii) patulin and (iii) aflatoxins.

In particular, O_3 can contain the fungal population present on the chickpea seeds surface and reduce the content of patulin and aflatoxins. This is a fundamental goal in the development of emerging new techniques, since these metabolites are recognized as a Group I carcinogen by the International Agency of Research on Cancer (Lyon, France), and their allowable levels in human foods and animal feedstuff are strictly regulated by governmental jurisdictions in about 100 countries [47]. It is worth noting that the antimicrobial activity of O_3 is highly dependent on vegetable/fungus species, growth stage, concentration and timing of exposure. Improvements and innovations in O_3 generation and application systems will be evaluated more effectively in the future by facilitating the enhanced control of both the quality and safety parameters of ozonized foods/feeds. Similarly, a N_2-controlled atmosphere represents a valid sustainable method to limit mycotoxin accumulation and *C. maculatus* infestation. This system requires low energetic costs, offers a negligible loss of nutrients or sensory qualities in food/feed, and demonstrates a reduced hazard to employees with no need for registration and no contamination of the environment [23,25]. Consequently, it can be considered as a promising alternative method that could be transferred to a large-scale grain storage system. For effective and safe use in processing, optimum O_3 and N_2 concentrations, contact time and other treatment conditions should be defined for foods and feeds. Here, a pilot test was conducted by offering scientific evidence to support the commercial application of these innovative strategies.

4. Conclusions

In conclusion, our pioneering study demonstrated that the experimental conditions used (e.g., O_3 and N_2 concentrations and/or the timing of treatment) were enough to rapidly and effectively (i) reduce the growth and proliferation of *Penicillium* spp. population present on the chickpea seeds' surface (in the case of O_3 treatment); (ii) detoxify patulin and aflatoxins; and (iii) limit *C. maculatus* infestation (only in the case of N_2 high concentrations). These are fundamental goals in the development of emerging new techniques and novel methods to control mycotoxin infections, intoxications and diseases, by offering a negligible loss of nutrients or sensory qualities in stored products and without leaving toxic chemical residues in the food and feed chain.

Therefore, our results suggest that a single pulse of O_3 and high N_2 concentrations are a promising alternative to traditional chemical-based fumigants for controlling spoilage pathogens and insects in stored chickpea seeds. An industrial facility of O_3 technology remains to be developed for the large-scale treatment of food/feed products, requiring input from different disciplines. For effective and safe use in processing, optimum O_3 and N_2 concentrations, contact time and other treatment conditions (e.g., vegetable/fungus species) should be defined for foods and feeds.

Additional research is obviously required to evaluate the responses of stored products to these effective and straightforward solutions, in order to control mycotoxin contamination and pest infestation at post-harvest. This would provide a clearer picture of the practical advantages of O_3 and N_2 treatment and allow further endorsement of our present results.

5. Materials and Methods

5.1. Reagents and Standards

Sodium hypochlorite, ethanol and streptomycin sulphate were supplied by Sigma-Aldrich (Milan, Italy). Potato Dextrose Agar was purchased from Biolife (Milan, Italy). Acetonitrile, methanol and water were HPLC-grade (Carlo Erba, Milan, Italy). Standards of patulin and aflatoxins were chromatographically pure and purchased from Sigma-Aldrich (Milan, Italy) and Romer Lab (Getzersdorf, Austria).

5.2. Raw Materials

Four *C. arietinum* batches, produced by a local farm located in Tuscany (Italy), were used in the present work. Before commercialization, all the batches (except in the case of n. 3) were submitted to a quality check by the producer, thus resulting in batches n. 1 and n. 4 being validated for sale, and batch n. 2 rendered non-compliant. All seeds were stored at 4 °C until submitted to a grain health test, mycotoxin determination and treatments under a controlled atmosphere.

5.3. Isolation and Identification of Fungal Contaminants Naturally Associated with Chickpea Seeds

To assess the presence of potential mycotoxigenic fungi naturally associated with chickpea seeds, all four batches were submitted to a grain health test. In detail, seeds from each batch were surface sterilized on a rotary shaker for 1 min in a solution containing NaClO (1% active chlorine) in 50% ethanol, then washed three times in sterile distilled water for 1 min each. After drying on filter paper, the seeds were transferred to 100 mm diameter Petri dishes containing Potato Dextrose Agar (PDA, 42 g L^{-1}) with the addition of 300 mg L^{-1} of streptomycin sulphate. Since batch n. 2 showed a profuse development of *Mucor* spp., the seeds from this sample were plated on PDA that contained the antibiotic, as previously described, and with hymexazol fungicide (at the final concentration of 300 mg L^{-1}; [48]). For each batch, four replicates (each consisting of twenty-five seeds) were made. The plates were incubated at room temperature (24 ± 2 °C). Then, from the second to the tenth day of incubation, colonies morphologically attributable to *Aspergillus* and *Penicillium* spp., developing from seeds, were transferred to new PDA + streptomycin sulphate plates and incubated under the same conditions previously reported. Then, when sporulated, they were used for single-spore cultivation. The single-spore *Penicillium* spp. colonies were used for molecular identification. Genomic DNA was extracted from each single-spore culture according to the Chelex 100 method [49]. For molecular identification, the complete internal transcribed spacers (ITS) 1 and 2 sequences—including the 5.8S gene—of the nuclear ribosomal DNA were amplified and sequenced as described in Sarrocco et al. [30]. All the sequences were then submitted to GeneBank (NCBI) to assign, where possible, the species for a preliminary evaluation of the risk of mycotoxin contamination that could occur on the seeds.

5.4. O_3 and N_2 Treatment Systems for Chickpea Seeds

5.4.1. O_3 Treatment System for Chickpea Seeds

Chickpea seeds (300 g from each batch) were placed in two Perspex chambers (60 × 60 × 110 cm) in a controlled environment fumigation facility. The system was adapted by including commercial colanders (34 × 23 cm, stainless steel), collocated in the middle of the chambers, in which seeds were placed to allow their complete exposure to O_3, maintained in the dark throughout the whole period of the experiment (temperature 25 ± 1 °C, relative humidity (RH) 50 ± 5%). The fumigation system was continuously ventilated (two complete air changes per min) with charcoal-filtered air. Fumigation was performed by generating O_3 from pure oxygen by electrical discharge, using a Fisher 500 air-cooled apparatus (Fisher America Inc., Houston, TX, USA). The O_3 concentration was monitored with a Serinus 10 analyzer (Ecotech Acoem Group, Milan, Italy) set at 500 ± 50 ppb of O_3 (for O_3, 1 ppb = 1.96 $\mu g\ m^{-3}$, at 20 °C and 101.325 kPa) for 30, 60 and 90 min, in which the chickpea seeds were mixed every 15 min. At the end of each treatment, the samples were collected and immediately used for the subsequent analyses (fungal infection and mycotoxin contamination). The entire methodology was performed according to Marchica et al. [50].

5.4.2. Nitrogen Treatment System for Chickpea Seeds

Nitrogen treatments were performed using a NitrosepAgri system (Eurosider sas, Grosseto, Italy) based on selective membrane (MNS, Membrane Nitrogen Separator) to separate N_2 from atmospheric air [25]. The N_2-enriched atmosphere was driven into silos where it was maintained under a slight overpressure. All the environmental parameters, such as temperature and RH, were constantly monitored, and the N_2 percentage could be set, automatically maintained and quickly reintegrated if needed. To perform the experiments here reported, two lab-scale (20 and 60 L silos, respectively) prototypes, already described in Moncini et al. [25], and one field-scale apparatus were used. For each chickpea batch, 300 g of seeds were transferred into a 1.5 L glass jar (9 cm diameter) that was closed with a micro-perforated nylon layer (350 μm pore size) to facilitate gas exchange and incubated under a 99% N_2 atmosphere for 21 consecutive days. For each treatment and for each batch, three replicates were made. The field-scale apparatus consisted of four 15 m^3 volume fiberglass silos connected to the MNS system. Each silos had a stainless steel (786 mm diameter) top hatch for grains charging, equipped with an overpressure valve and, on the bottom, with a stainless-steel ball (90 mm diameter) valve for discharging the product. The field-scale apparatus is located at the Azienda Agraria Macchiascandona (Castiglione della Pescaia, Grosseto, Italy).

5.5. Effect of O_3 and N_2 Treatments on Fungal Infection, Mycotoxin Contamination and Pest Survival

At the end of the O_3 and N_2 treatments, the seeds were collected and used to evaluate fungal contamination according to the grain health protocol. Mycotoxins determination was performed by using the clean-up aflatoxins and patulin (AFP) columns (OR SELL,

S.p.a., Modena, Italy), according to the manufacturing protocols, with a few modifications. The samples were extracted by adding 50 mL of acetonitrile:water (ACN:H_2O, 84:16 v/v) solution to 25 g of finely ground chickpea seeds, and vigorously vortexed for 6 min. The samples were centrifuged for 10 min at 12,000× *g* at room temperature, and the supernatants were filtered through Whatman® paper (Cytiva, Marlborough, MA, USA) and subsequently by using the clean-up AFP columns, which enable the contemporary purification of aflatoxins and patulin. The obtained solutions were equally separated, dried at 40 °C and finally resuspended in 400 µL of 45% methanol (v/v) or 75% ACN (v/v in HPLC-demineralized water) for aflatoxins and patulin separation, respectively. The separation was performed in a UHPLC Dionex UltiMate 3000 system (Thermo Scientific, Waltham, MA, USA) equipped with a ZORBAX Eclipse Plus C18 column (150 × 4.6 mm, 5 µm particle size, Agilent technologies, Santa Clara, CA, USA). Aflatoxins determination was carried out by using an UltiMate™ 3000 Fluorescence Detector (Thermo Scientific, Waltham, MA, USA) with excitation and emission at 362 and 420 nm, after post-column derivatization through a UVE™ Photochemical Reactor for Aflatoxin Analysis (254 nm lamp; 240 VAC, 50/60 Hz, LCTech, Obertaufkirchen, Germany). The run conditions were set at a flow rate of 0.8 mL min^{-1} of a mobile phase 45% methanol (v/v in HPLC-demineralized water), for 30 min at 30 °C. Patulin quantification was performed using a Dionex UV-Vis Detector (Dionex UVD 170 U UV-Vis detector, Thermo scientific, Waltham, MA, USA) at 276 nm, at a flow rate of 0.6 mL min^{-1}, for 30 min at 30 °C, and the same mobile phase reported above. Known amounts of pure (Patulin HPLC standard) or mixed (Aflatoxin Mix 4 solution) standards were injected into the UHPLC system (range 0.1−100 ng mL^{-1}), and the quantity of each mycotoxin was obtained by correlating the peak area to the related standard concentration by using the Chromeleon Chromatography Management System software, version 7.2.10-2019 (Thermo Scientific, Waltham, MA, USA). The sum of AFB1, AFB2, AFG1 and AFG2 was considered as a measure of the total aflatoxins content.

Stock cultures of *C. maculatus*, kindly provided by Graham J. Holloway (University of Reading, Reading Berkshire, UK), were maintained in a climatic chamber (25 ± 1 °C, 70 ± 5% RH) on chickpea seeds in a 150 mL PP jar (5.5 cm diameter) that was closed with a micro-perforated nylon layer (350 µm pore size) to facilitate air exchange. To obtain chickpea infested with eggs, pest sub-cultures were set up by placing 100 unsexed adults into glass Petri dishes (14 cm diameter) containing 100 g of chickpea seeds. Adults were allowed to oviposit for 24 h in a climatic chamber under the same conditions as described before. At the end of the oviposition period, the adults were removed. Seeds with the addition of age-synchronized eggs were immediately used for the N_2-controlled atmosphere test [51]. The lab-scale experiment was set up in six 60 L silos (three under a 99% N_2 atmosphere and three under an unmodified atmosphere used as control). Two bio-tests, each consisting of 15 g of chickpea seeds with the addition of 1-3 eggs (up to a total of 30 eggs) placed into a 30 mL PP jar (4.3 cm diameter) and covered with a micro-perforated nylon layer (350 µm pore size), were transferred into each of the 60 L silos, which were already full of chickpea seeds; one bio-test was placed at the center and the other 10 cm under the grains' surface. The infected seeds were incubated for 24, 48, 72, 96 and 120 days. To avoid the opening of the silos at each sampling, the experiment was repeated each time, with three replicates for each experiment.

With respect to the field-scale experiment, two silos filled with 9 tons of chickpea seeds (the same batch as the laboratory) were used for the tests: one of these was automatically kept at a 99% N_2-controlled atmosphere, and the other one, with the lid partially opened and disconnected from the N_2 supply, was used as a control. In order to simulate a storage condition inside the full grain mass, three bio-tests previously placed into a jute bag filled with 8 kg of chickpea seeds were transferred into the silos. The test was carried out for 5 days and replicated three times. At the end of each experiment (lab-scale and large-scale test), the bio-tests were removed from the silos and their contents transferred into larger PP jars (like those used for pest stock cultures), closed with a micro-perforated nylon layer and placed in a climate chamber at 25 ± 2 °C, 70 ± 5% RH. The adults' emergence was periodically recorded from 35 to 70 days after oviposition. The number of adults was registered, and mortality was corrected using Abbott's formula [52].

$$\text{Mortality (\%)} = [(N_c - N_t)/(N_c)] \times 100$$

where N_c = No. of emerged adults in control and N_t = No. of emerged adults in treatment.

5.6. Statistical Analysis

The robustness of data among the replicates was verified according to the results of the Shapiro–Wilk test for normality and Levene's tests for homogeneity of variance. Data were submitted to an analysis of variance (one-way or two-way ANOVA), and comparisons among the means were determined by Tukey's HSD post hoc test or Student *t*-test by using JMP Pro 14 software (SAS Institute Inc., Cary, NC, USA), in order to evaluate the effect of the treatments (control silos; O_3 exposure for 30, 60 and 90 min; and N_2 exposure for 21 consecutive days), batch (B1-4), and their interaction. For all the analyses, $p \leq 0.05$ was assumed as a significant level.

Supplementary Materials: The following supporting information can be downloaded at: https://www.mdpi.com/article/10.3390/toxins15010061/s1, Table S1: Analysis of variance of mycotoxins content in the four chickpea batches between the mean values determined on three different replicates; Figure S1: Total content of aflatoxins in four batches [n. 1 (a), n. 2 (b), n. 3 (c) and n. 4 (d)] of chickpea grains (CTRL, white fill), exposed to ozone [500 ppb O_3 for 30 (light grey fill), 60 (grey fill) and 90 (dark grey fill) minutes] and nitrogen treatment [99% N_2 for 21 consecutive days, dark fill]. Results of two-way ANOVA are reported, asterisks show the significance of factors/interaction for: *** $p \leq 0.001$. According to Tukey's HSD post hoc test, different letters indicate significant differences ($p \leq 0.05$).

Author Contributions: Conceptualization, L.M., C.N. and S.S.; methodology, software, validation, investigation and formal analysis, C.P., S.R., L.M. and S.S.; resources, C.N., S.S. and E.P.; data curation, C.P., S.R., L.M. and S.S.; writing—original draft preparation, C.P., S.R., S.S. and E.P.; writing—review and editing, L.M. and C.N. All authors have read and agreed to the published version of the manuscript.

Funding: The experiments were carried out at CRISBA–ISIS Leopoldo II di Lorena (GR) and the construction of the field scale-N_2 supplier device apparatus was funded by Fondazione CR Firenze.

Institutional Review Board Statement: Not applicable.

Informed Consent Statement: Not applicable.

Data Availability Statement: Not applicable.

Acknowledgments: We gratefully acknowledge Andrea Parrini and Grazia Puntoni for their technical support for the fumigation facilities and nitrogen treatment, respectively; Enrico Giussani for furnishing the raw material; and Luca Mussi for the experimental activities. We also acknowledge Gabriele Simone for the technical support in setting up the tests on pests.

Conflicts of Interest: Authors declare no conflict of interest.

References

1. Liu, L.; Xie, M.; Wei, D. Biological detoxification of mycotoxins: Current status and future advances. *Int. J. Mol. Sci.* **2022**, *23*, 1064. [CrossRef] [PubMed]
2. Hamad, G.M.; Mehany, T.; Simal-Gandara, J.; Abou-Alella, S.; Esua, O.K.; Abdel-Wahhab, M.A.; Hafez, E.E. A review of recent innovative strategies for controlling mycotoxins in foods. *Food Control* **2023**, *144*, 109350. [CrossRef]
3. Agriopoulou, S.; Stamatelopoulou, E.; Varzakas, T. Advances in occurrence, importance, and mycotoxin control strategies: Prevention and detoxification in foods. *Foods* **2020**, *9*, 137. [CrossRef] [PubMed]
4. Alshannaq, A.; Yu, J.H. Occurrence, toxicity, and analysis of major mycotoxins in food. *Int. J. Environ. Res. Public Health* **2017**, *13*, 632. [CrossRef] [PubMed]
5. Prencipe, P.; Siciliano, I.; Gatti, C.; Garibaldi, A.; Gullino, M.L.; Botta, R.; Spadaro, D. Several species of *Penicillium* isolated from chestnut flour processing are pathogenic on fresh chestnuts and produce mycotoxins. *Food Microbiol.* **2018**, *76*, 396–404. [CrossRef]

6. Mesterházy, Á.; Oláh, J.; Popp, J. Losses in the grain supply chain: Causes and solutions. *Sustainability* **2020**, *12*, 2342. [CrossRef]
7. Winter, G.; Pereg, L. A review on the relation between soil and mycotoxins: Effect of aflatoxin on field, food and finance. *Eur. J. Soil Sci.* **2018**, *70*, 882–897. [CrossRef]
8. Khaneghah, A.; Fakhri, Y.; Raeisi, S.; Armoon, B.; Sant'Ana, A.S. Prevalence and concentration of ochratoxin A, zearalenone, deoxynivalenol and total Aaflatoxin in cereal-based products: A systematic review and meta-analysis. *Food Chem. Toxicol.* **2018**, *118*, 830–848. [CrossRef]
9. Milicevic, D.; Nesic, K.; Jaksic, S. Mycotoxin contamination of the food supply chain—Implications for One Health programme. *Procedia Food Sci.* **2015**, *5*, 187–190. [CrossRef]
10. Jung, H.L.; Doing, R. Worldwide occurrence of mycotoxins in cereals and cereal-derived food products: Public health perspectives of their co-occurrence. *J. Agric. Food Chem.* **2017**, *65*, 7034–7051.
11. Walker, S.; Jaime, R.; Kagot, V.; Probst, C. Comparative effects of hermetic and traditional storage devices on maize grain: Mycotoxin development, insect infestation and grain quality. *J. Stored Prod. Res.* **2018**, *77*, 33–44. [CrossRef]
12. Mazumder, P.M.; Sasmal, D. Mycotoxins- Limits and regulations. *Anc. Sci. Life* **2001**, *20*, 1–19. [PubMed]
13. Iqbal, S.Z.; Asi, M.R.; Zuber, M.; Akhtar, J.; Saif, M.J. Natural occurrence of aflatoxins and ochratoxin A in commercial chilli and chilli sauce samples. *Food Control* **2013**, *30*, 621–625. [CrossRef]
14. Chen, B.H.; Inbaraj, B.S. Recent trends in analysis of mycotoxins in food using carbon-based nanomaterials. *J. Food Drug Anal.* **2022**, *30*. [CrossRef]
15. Adegoke, G.O.; Letuma, P. Strategies for the prevention and reduction of mycotoxins in developing countries. In *Mycotoxin and Food Safety in Developing Countries*; Makun, H.A., Ed.; InTech: Rijeka, Croatia, 2013; pp. 123–126.
16. Conte, G.; Fontanelli, M.; Galli, F.; Cotrozzi, L.; Pagni, L.; Pellegrini, E. Mycotoxins in feed and food and the role of ozone in their detoxification and degradation: An update. *Toxins* **2020**, *12*, 486. [CrossRef]
17. Ji, C.; Fan, Y.; Zhao, L. Review on biological degradation of mycotoxins. *Anim. Nutr.* **2016**, *2*, 127–133. [CrossRef]
18. Guo, Y.; Zhao, L.; Ma, Q.; Ji, C. Novel strategies for degradation of aflatoxins in food and feed: A review. *Food Res. Int.* **2021**, *140*, 109878. [CrossRef]
19. Kaur, M.; Hüberli, D.; Bayliss, K.L. Cold plasma: Exploring a new option for management of postharvest fungal pathogens, mycotoxins and insect pests in Australian stored cereal grain. *Crop Pasture Sci.* **2020**, *71*, 715–724. [CrossRef]
20. Guzel-Seydium, Z.B.; Greene, A.K.; Seydium, A.C. Use of ozone in the food industry. *LWT-Food Sci. Technol.* **2004**, *37*, 453–460. [CrossRef]
21. Wang, L.; Shao, H.; Luo, X.; Wang, R.; Li, Y.; Li, Y.; Luo, Y.; Chen, Z. Effect of ozone treatment on deoxynivalenol and wheat quality. *PLoS ONE* **2016**, *11*, e0147613. [CrossRef]
22. Nickhil, C.; Mohapatra, D.; Kar, A.; Giri, S.K.; Tripathi, M.K.; Sharma, Y. Gaseous ozone treatment of chickpea grains, part I: Effect on protein, amino acid, fatty acid, mineral content, and microstructure. *Food Chem.* **2021**, *345*, 128850. [CrossRef] [PubMed]
23. Moncini, L.; Sarrocco, S.; Pachetti, G.; Moretti, A.; Haidukowsky, M.; Vannacci, G. N_2 controlled atmosphere reduces postharvest mycotoxins risk and pests attack on cereal grains. *Phytoparasitica* **2020**, *48*, 555–565.
24. Cao, Y.; Xu, K.; Zhu, X.; Bai, Y.; Yang, W.; Li, C. Role of modified atmosphere in pest control and mechanisms of its effect on insects. *Front. Physiol.* **2019**, *10*, 206. [CrossRef] [PubMed]
25. Moncini, L.; Simone, G.; Romi, M.; Cai, G.; Guerriero, G.; Whittaker, A.; Benedettelli, S.; Berni, R. Controlled nitrogen atmosphere for the preservation of functional molecules during silos storage: A case study using old Italian wheat cultivars. *J. Stored Prod. Res.* **2020**, *88*, 101638. [CrossRef]
26. McKenzie, K.; Sarr, A.; Mayura, K.; Bailey, R.; Miller, D.; Rogers, T.; Norred, W.; Voss, K.; Plattner, R.; Kubena, L.; et al. Oxidative degradation and detoxification of mycotoxins using a novel source of ozone. *Food Chem. Toxicol.* **1997**, *35*, 807–820. [CrossRef]
27. Athanassiou, C.G.; Sakka, M.K. Using nitrogen for the control of stored product insects: One single application for multiple purposes. *Agrochemicals* **2022**, *1*, 22–28. [CrossRef]
28. Qu, C.; Li, W.; Yang, Q.; Xia, Y.; Lu, P.; Hu, M. Metabolic mechanisms of nitrogen modified atmosphere storage on delaying quality deterioration of rice grains. *Food Chem.* **2022**, *16*, 100519. [CrossRef]
29. Navarro, S. Modified atmospheres for the control of stored product insects and mites. In *Insect Management for Food Storage and Processing*; Heaps, J.W., Ed.; AACC International: St. Paul, MN, USA, 2006; pp. 105–145.
30. Sarrocco, S.; Diquattro, S.; Baroncelli, R.; Cimmino, A.; Evidente, A.; Vannacci, G.; Doveri, F. A polyphasic contribution to the knowledge of *Auxarthron* (Onygenaceae). *Mycol. Prog.* **2015**, *14*, 112. [CrossRef]
31. Ramirez, M.L.; Cendoya, E.; Nichea, M.J.; Zachetti, V.G.L.; Chulze, S.N. Impact of toxigenic fungi and mycotoxins in chickpea: A review. *Curr. Opin. Food Sci.* **2018**, *23*, 32–37. [CrossRef]
32. Urooj, R.; Achakzai, K.; Zaman, M.; Bibi, A.; Afreen, T.; Iram, S. Occurrences of mycotoxins contamination in crops from Pakistan: A review. *J. Agric. Environ. Sci.* **2015**, *15*, 1204–1212.
33. Zahra, T.; Ghramh, H.A.; Batool, M.; Abid, A.D.; Shahzad, S.; Shahbaz, M.; Faiz, A.H.; Saeed, Q.; Hashim, S. Temperature mediated influence of mycotoxigenic fungi on the life cycle attributes of *Callosobruchus maculatus* F. in stored chickpea. *J. King Saud. Univ. Sci.* **2023**, *35*, 102357. [CrossRef]
34. Shazali, M.E.H.; Imamura, T.; Miyanoshita, A. Mortality of eggs of the cowpea bruchid, *Callosobruchus maculatus* (F.) (Coleoptera: Bruchidae) in carbon dioxide under high pressure. *Appl. Entomol. Zool.* **2004**, *39*, 49–53. [CrossRef]

35. Kedia, A.; Prakash, B.; Mishra, P.K.; Singh, P.; Dubey, N.K. Botanicals as eco-friendly biorational alternatives of synthetic pesticides against *Callosobruchus* spp. (Coleoptera: Bruchidae)-A review. *J. Food Sci. Technol.* **2015**, *52*, 1239–1257. [CrossRef] [PubMed]
36. Mendez, F.; Maier, D.; Mason, L.J.; Woloshuk, C. Penetration of ozone into columns of stored grains and effects on chemical composition and processing performance. *J. Stored Prod. Res.* **2003**, *39*, 33–44. [CrossRef]
37. Trombete, F.M.; Freitas-Silva, O.; Venâncio, A.A.; Fraga, M.E. Ozone against mycotoxins and pesticide residues in food: Current applications and perspectives. *Int. Food Res. J.* **2016**, *23*, 2545–2556.
38. Cataldo, F. Ozone decomposition of patulin-A mycotoxin and food contaminant. *Ozone Sci. Eng.* **2008**, *30*, 197–201. [CrossRef]
39. Diao, E.; Hou, H.; Hu, W.; Dong, H.; Li, X. Removing and detoxifying methods of patulin: A review. *Trends Food Sci. Technol.* **2018**, *81*, 139–145. [CrossRef]
40. de Souza Sartori, J.A.; Angolini, C.F.F.; Eberlin, M.N.; De Aguiar, C.L. Criegee mechanisms as a safe pathway of color reduction in sugarcane juice by ozonation. *Food Chem.* **2017**, *225*, 181–187. [CrossRef]
41. Piemontese, L.; Messia, M.C.; Marconi, E.; Falasca, L.; Zivoli, R.; Gambacorta, L.; Perrone, G.; Solfrizzo, M. Effect of gaseous ozone treatments on DON, microbial contaminants and technological parameters of wheat and semolina. *Food Addit. Contam. Part A* **2018**, *35*, 761–772. [CrossRef]
42. Santos, A.R.; Carreiró, F.; Freitas, A.; Barros, S.; Brites, C.; Ramos, F.; Sanches Silva, A. Mycotoxins contamination in rice: Analytical methods, occurrence and detoxification strategies. *Toxins* **2022**, *14*, 647. [CrossRef]
43. Magan, N.; Lacey, J. Effects of gas composition and water activity on growth of field and storage fungi and their interactions. *Trans. Br. Mycol. Soc.* **1984**, *82*, 305–314. [CrossRef]
44. Paster, N.; Bullerman, L.B. Mould spoilage and mycotoxin formation in grains as controlled by physical means. *Int. J. Food Microb.* **1988**, *7*, 257–265. [CrossRef]
45. Chulze, S.N. Strategies to reduce mycotoxin levels in maize during storage: A review. *Food Addit. Contam. Part A. Chem. Anal. Control Expo. Risk Assess.* **2010**, *27*, 651–657. [CrossRef]
46. Drott, M.T.; Lazzaro, B.P.; Drown, D.L.; Carbone, I.; Milgroom, M.G. Balancing selection for aflatoxin in *Aspergillus flavus* is maintained through interference competition with, and fungivory by insects. *Proc. Biol. Sci.* **2017**, *284*, 20172408. [PubMed]
47. Waqas, M.; Iqbal, S.; Razis, A.A.; Pervaiz, W.; Ahmad, T.; Usman, S.; Ali, N.; Asi, M. Occurrence of aflatoxins in edible vegetable seeds and oil samples available in Pakistani retail markets and estimation of dietary intake in consumers. *Int. J. Environ. Res. Public Health* **2021**, *18*, 8015. [CrossRef] [PubMed]
48. Fiorini, L.; Guglielminetti, L.; Mariotti, L.; Curadi, M.; Picciarelli, P.; Scartazza, A.; Sarrocco, S.; Vannacci, G. *Trichoderma harzianum* T6776 modulates a complex metabolic network to stimulate tomato cv. Micro-Tom growth. *Plant Soil* **2016**, *400*, 351–366. [CrossRef]
49. Baroncelli, R.; Sarrocco, S.; Zapparata, A.; Tavarini, S.; Angelini, L.G.; Vannacci, G. Characterisation and epidemiology of *Colletotrichum acutatum* sensu lato (*C. chrysanthemi*) causing *Carthamus tinctorius* anthracnose. *Plant Pathol.* **2014**, *64*, 375–384. [CrossRef]
50. Marchica, A.; Ascrizzi, R.; Flamini, G.; Cotrozzi, L.; Tonelli, M.; Lorenzini, G.; Nali, C.; Pellegrini, E. Ozone as eustress for enhancing secondary metabolites and bioactive properties in *Salvia officinalis*. *Ind. Crops Prod.* **2021**, *170*, 113730. [CrossRef]
51. Pascua, G.F.S.; Bayogan, E.R.V.; Salaipeth, L.; Photchanachai, S. Pretreating *Callosobruchus maculatus* (F.) eggs in mung bean with modified atmosphere conditions influence its adult emergence and survival. *J. Stored Prod. Res.* **2021**, *91*, 101771. [CrossRef]
52. Abbott, W.S. A method of computing the effectiveness of an insecticide. *J. Econ. Ent.* **1925**, *18*, 265–267. [CrossRef]

MDPI
St. Alban-Anlage 66
4052 Basel
Switzerland
Tel. +41 61 683 77 34
Fax +41 61 302 89 18
www.mdpi.com

Toxins Editorial Office
E-mail: toxins@mdpi.com
www.mdpi.com/journal/toxins

www.ingramcontent.com/pod-product-compliance
Lightning Source LLC
Chambersburg PA
CBHW040741250926
51850CB00007BB/1375
* 9 7 8 3 0 3 6 5 8 4 1 4 0 *